Nonlinear Output Regulation

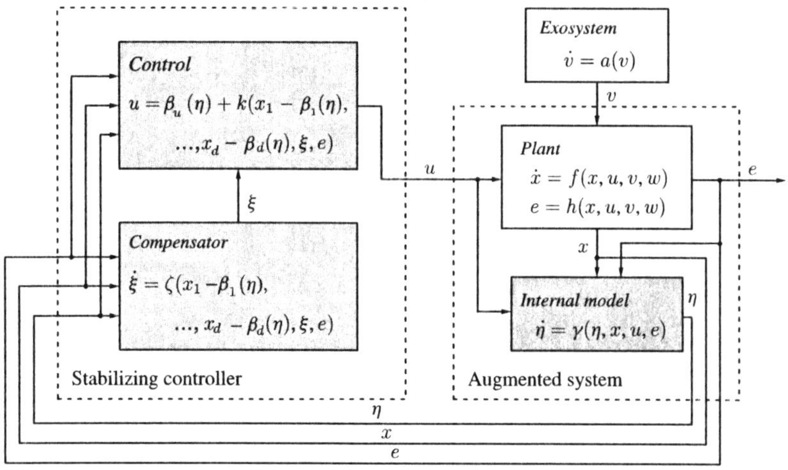

Advances in Design and Control

SIAM's Advances in Design and Control series consists of texts and monographs dealing with all areas of design and control and their applications. Topics of interest include shape optimization, multidisciplinary design, trajectory optimization, feedback, and optimal control. The series focuses on the mathematical and computational aspects of engineering design and control that are usable in a wide variety of scientific and engineering disciplines.

Series Volumes

Huang, J., *Nonlinear Output Regulation: Theory and Applications*
Haslinger, J. and Mäkinen, R. A. E., *Introduction to Shape Optimization: Theory, Approximation, and Computation*
Antoulas, A. C., *Lectures on the Approximation of Linear Dynamical Systems*
Gunzburger, Max D., *Perspectives in Flow Control and Optimization*
Delfour, M. C. and Zolésio, J.-P., *Shapes and Geometries: Analysis, Differential Calculus, and Optimization*
Betts, John T., *Practical Methods for Optimal Control Using Nonlinear Programming*
El Ghaoui, Laurent and Niculescu, Silviu-Iulian, eds., *Advances in Linear Matrix Inequality Methods in Control*
Helton, J. William and James, Matthew R., *Extending H^∞ Control to Nonlinear Systems: Control of Nonlinear Systems to Achieve Performance Objectives*

Nonlinear Output Regulation

Theory and Applications

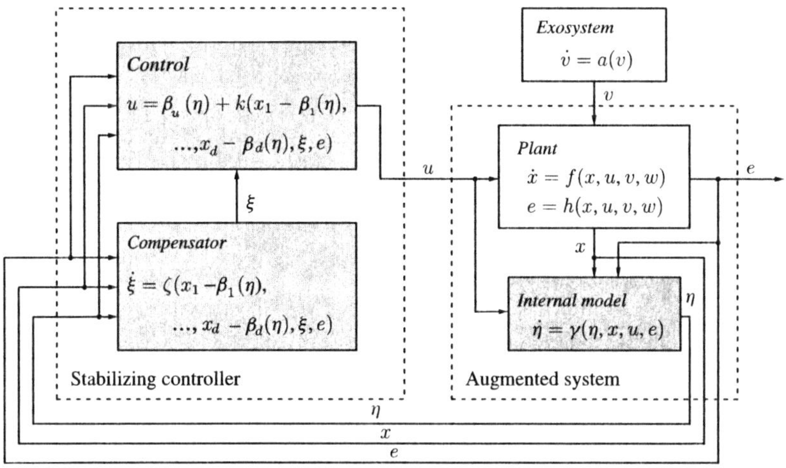

Jie Huang

The Chinese University of Hong Kong
Hong Kong

siam

Society for Industrial and Applied Mathematics
Philadelphia

10 9 8 7 6 5 4 3 2 1

Library of Congress Cataloging-in-Publication Data

Huang, Jie, 1955-
Nonlinear output regulation : theory and applications / Jie Huang.
p. cm. – (Advances in design and control)
Includes bibliographical references and index.
ISBN 0-89871-562-8
1. Servomechanisms–Design and construction. 2. Nonlinear functional analysis. I. Title. II. Series.

TJ214.H83 2004
629.8'323–dc22 2004052533

Contents

List of Figures

List of Tables

Notation

Symbol	*Usage*	*Meaning*
$\|\| \cdot \|\|$	$\|\|x\|\|$	the 2-norm of a vector x
$\|\| \cdot \|\|$	$\|\|A\|\|$	the induced 2-norm of a matrix A
$\mathcal{R}^n$	$x \in \mathcal{R}^n$	n-dimensional Euclidean space
$\mathcal{R}^{n\times m}$	$A \in \mathcal{R}^{n\times m}$	The set of all $n \times m$ matrix with elements in $\mathcal{R}^1$
I_n	λI_n	$n \times n$ identity matrix
$\sigma(\cdot)$	$\sigma(A)$	spectrum of matrix A
$\in$	$\lambda \in \sigma(A)$	λ is a member of $\sigma(A)$
$\notin$	$\lambda \notin \sigma(A)$	λ is not a member of $\sigma(A)$
$\otimes$	$A \otimes B$	Kronecker product
$\mid$	$\alpha(\lambda) \mid \beta(\lambda)$	$\alpha(\lambda)$ divides $\beta(\lambda)$
$\mathcal{C}_-$	$\sigma(A) \in \mathcal{C}_-$	open left half-complex plane
$\mathcal{C}_+$	$\sigma(A) \in \mathcal{C}_+$	open right half-complex plane
$\bar{\mathcal{C}}_-$	$\sigma(A) \in \bar{\mathcal{C}}_-$	closed left half-complex plane
$\bar{\mathcal{C}}_+$	$\sigma(A) \in \bar{\mathcal{C}}_+$	closed right half-complex plane
$\deg(\cdot)$	$\deg(\alpha(\lambda))$	degree of polynomial $\alpha(\lambda)$
$\dim(\cdot)$	$\dim(\mathcal{K})$	dimension of $\mathcal{K}$
rank	rank A	rank of matrix A

Preface

The output regulation problem, or alternatively, the servomechanism problem, addresses design of a feedback controller to achieve asymptotic tracking for a class of reference inputs and disturbance rejection for a class of disturbances in an uncertain system while maintaining closed-loop stability. This is a general mathematical formulation applicable to many control problems encountered in our daily life, for example, cruise control of automobiles, aircraft landing and taking-off, manipulation of robot arms, orbiting of satellites, motor speed regulation, and so forth. Study of the output regulation problem can be traced as far back as 1769, when James Watt devised a speed regulator for a steam engine. Yet rigorous formulation of this problem in a modern state-space framework was not available until the 1970s. In contrast to similar problems, such as trajectory tracking, where the trajectory to be tracked is assumed to be completely known, a distinctive feature of the output regulation problem is that the reference inputs and disturbances do not have to be known exactly so long as they are generated by a known, autonomous differential equation. In this book, the term "exogenous signals" will be used to refer to both reference inputs and disturbances when there is no need to distinguish them. The autonomous differential equation generating exogenous signals will be called the exosystem.

The output regulation problem was first studied for the class of linear systems under various names, such as the robust servomechanism problem (Davison) or the structurally stable output regulation problem (Francis and Wonham). It was completely solved by the collective efforts of several researchers, including Davison, Francis, and Wonham, to name just a few. Solvability conditions for the output regulation problem were worked out either in terms of the location of the transmission zeros of the system or in terms of the solvability of a set of Sylvester equations. A salient outcome of this research was the internal model principle, which includes classical PID (proportional-integral-derivative) control as a special case. From the control theoretic point of view, the significance of the internal model principle is that it enables the conversion of the output regulation problem into the well-known stabilization problem for an augmented linear system.

At almost the same time that research on the linear output regulation problem reached its peak, in the mid 1970s, Francis and Wonham considered the output regulation problem for a class of nonlinear systems for the special case when exogenous signals are constant. They showed that a linear regulator design based on the linearized plant can solve the robust output regulation problem for a weakly nonlinear plant while maintaining the local stability of the closed-loop system. In the late 1980s, Huang and Rugh further studied this problem for general nonlinear systems using a gain scheduling approach and related the solvability of this problem to solvability of a set of nonlinear algebraic equations.

To establish a general theory for the output regulation problem for uncertain nonlinear systems subject to time-varying exogenous signals, one must address three important issues: how to define and guarantee existence of the steady state of the system, and hence characterize the solvability of the problem; how to handle plant uncertainty when it is known that the linear internal model principle does not work for nonlinear systems in the general case; and how to achieve asymptotic tracking and disturbance rejection in a nonlinear system with arbitrarily large initial states of the plant, the exosystem, and the controller, in the presence of uncertain parameters that lie in an arbitrarily prescribed, bounded set.

None of these three issues can be dealt with by a simple extension of the existing linear output regulation theory. Because of these challenges, the output regulation problem for nonlinear systems has become one of the most exciting research areas since the 1990s. As a result of extensive work, these three issues have now been successfully addressed to a certain degree.

The difficulty associated with the first issue, existence of steady state, lies in the fact that the solution of a nonlinear system is not available. Isidori and Byrnes first addressed this issue for the case when the plant is assumed to be known exactly. By introducing center manifold theory, Isidori and Byrnes found that it is possible to use a set of mixed nonlinear partial differential and algebraic equations, called regulator equations in what follows, to characterize the steady state of the system. This discovery coupled with the zero dynamics theory of nonlinear systems leads to a solvability condition for the output regulation problem in terms of solvability of the regulator equations. It turns out that the regulator equations are a generalization of the Sylvester equations mentioned above. The solution of the regulator equations provided a feedforward control to cancel the steady-state tracking error. Based on the solution of the regulator equations, both state feedback and error feedback control laws can be readily synthesized to achieve asymptotic tracking and disturbance rejection for an exactly known plant while maintaining local stability of the closed-loop system.

The second issue is concerned with the plant uncertainty characterized by a set of unknown parameters. The feedforward control approach mentioned in the last paragraph cannot handle this case due to the presence of the unknown parameters. A design approach based on the linear internal model principle does not work either, as shown by a counterexample due to Isidori and Byrnes. Huang first revealed in 1991 that the linear internal model principle failed because, unlike the linear case, the steady-state tracking error in a nonlinear system is a nonlinear function of the exogenous signals. Based on this observation, Huang found that if the solution of the regulator equations is a polynomial in the exogenous signals, then it is possible to solve the output regulation problem for uncertain nonlinear systems by both state feedback and output feedback control. This approach effectively leads to a nonlinear version of the internal model principle. The robust output regulation problem was further pursued by Byrnes and Isidori, Delli Priscoli, and Khalil, generating various techniques and insights on this important issue.

While the first two issues have been intensively addressed since the 1990s, the investigation of the third issue, the output regulation problem with global stability, has just started and is rapidly unfolding. In the original formulation of the output regulation problem, as given by Isidori and Byrnes, only local stability is required for the closed-loop system. For this case, the stability issue can be easily handled by Lyapunov's linearization method. When a global stability requirement is imposed on the closed-loop system, the situation becomes much more complicated. Khalil studied the semiglobal robust output regulation problem for the class of feedback linearizable systems in 1994. His work was further extended to the class of lower triangular systems by Isidori in 1997. The output regulation problem with global stability was solved for the class of output strict feedback systems by Serrani and Isidori in 2000. Up to this point, the problem of output regulation with nonlocal stability was handled on a case-by-case basis, and only limited results were obtained. Recently, Huang and Chen have established a new framework that converts the robust output regulation problem for nonlinear systems into a robust stabilization problem. This new framework has offered greater flexibility to incorporate recent stabilization techniques, thus having set a stage for systematically tackling robust output regulation with global stability. This new framework has been successfully applied to solve the output regulation problem with global stability for several important classes of nonlinear systems.

The scope of research on the output regulation problem is constantly expanding, and the topic is made richer and more interesting with the injections of new ideas and techniques from other research areas such as stabilization, adaptive control, neural networks, and numerical mathematics. For example, the output regulation problem with uncertain exosystems was studied recently by Chen and Huang, Nikiforov, Serrani, Marconi and Isidori, and Ye and Huang, respectively. This scenario had not been studied previously, even for linear systems.

The output regulation problem arises from formulating daily engineering control problems. Therefore, in addition to the theoretical issues mentioned above, the application of this theory to practical design should be adequately addressed. A key issue critical to the applicability of the output regulation theory is the solvability of the regulator equations. Being a set of mixed nonlinear partial differential and algebraic equations, the solution of the regulator equations is usually unavailable. Thus it is necessary to develop approximation approaches to solving these equations. An approximation method based on Taylor series expansion was developed by Huang and Rugh in 1991 and was also considered by Krener in 1992. The effectiveness of these approximation methods has been demonstrated by many case studies, including benchmark nonlinear systems such as the ball and beam, the inverted pendulum on a cart, and the rotational/translational actuator.

This book will give a comprehensive and up-to-date treatment of the output regulation problem in a self-contained fashion. The book begins with an introduction to the linear output regulation theory in Chapter 1. Then a review of fundamental nonlinear control theory is given in Chapter 2. Chapters 3 and 4 are devoted to the output regulation problem and the approximate output regulation problem for continuous-time nonlinear systems, respectively. The robust output regulation problem for uncertain continuous-time nonlinear systems is presented in Chapters 5 and 6. In Chapter 7, the global robust output regulation is formulated and studied for uncertain continuous-time nonlinear systems. Chapter 8 presents both the output regulation problem and the robust output regulation problem for singular nonlinear systems. Finally, in Chapter 9, results on the output regulation problem and the robust output regulation problem are extended to discrete-time nonlinear systems. The author seeks to strike a balance between the theoretical foundations of the output regulation problem and practical applications of the theory. The treatment is accompanied by many examples, including practical case studies with numerical simulations based on the software platform MATLAB®.

This book can be used as a reference for graduate students, scientists, and engineers in the area of systems and control. Readers are assumed to have some fundamental knowledge of linear algebra, advanced calculus, and linear systems. Knowledge needed of nonlinear systems is summarized in Chapter 2. Some of the present chapters were used in the workshops of the 1999 IEEE Conference on Decision and Control, the 2004 World Congress on Intelligent Control and Automation, and graduate seminars at the Chinese University of Hong Kong.

The development of this book would not have been possible without the support and help of many people, including the author's master's thesis supervisor, Professor Xiangqiu Zeng; Ph.D. supervisor, Professor Wilson J. Rugh; and numerous colleagues and students. Professor Rugh not only guided the author into the area of nonlinear control, but also personally made substantial contributions to many results covered in Chapters 3 and 4. Some sections from Chapters 6–9 are adapted from joint publications of the author and some of his past and current students, including Zhiyong Chen, Guoqiang Hu, Weiyao Lan, Dan Wang, and Jin Wang. Three current students, Zhiyong Chen, Guoqiang Hu, and Weiyao Lan, have painstakingly proofread the manuscript several times and checked many examples with computer simulations. Professors Zhong-Ping Jiang, Zongli Lin, and Wilson J. Rugh have provided the author with valuable comments and suggestions. Professor Frank Lewis not only inspired and encouraged the author to embark on this project, but also introduced him to the SIAM acquisitions editor, Dr. Linda Thiel, who has been extremely helpful and enthusiastic. The SIAM Developmental Editor Simon Dickey and Production Editor Lisa Briggeman have done excellent work. The author is greatly indebted to Professor Alberto Isidori, whose seminal work on the output regulation problem with his coauthors has laid the foundation for this book.

The bulk of this research was supported by the Hong Kong Research Grants Council under grants CUHK 4316 /02E and CUHK 4168 /03E, and by National Natural Science Foundations of China under grant 60374038.

Jie Huang

Chapter 1

Linear Output Regulation

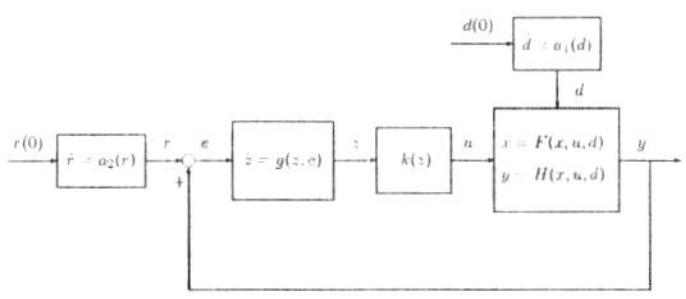

In this chapter, a concise but self-contained treatment of the subject of the output regulation problem for linear time-invariant systems is given. The output regulation problem was one of the central research topics in linear control theory in the 1970s. This research has generated a salient controller synthesis technique known as the *Internal Model Principle*. The purpose of this chapter is mainly to provide the background for understanding the nonlinear output regulation problem, and the chapter is organized as follows. In Section 1.1, a typical scenario that leads to the formulation of the problem is described. In Section 1.2, the precise definition of the output regulation problem is given and the solvability of the problem via both state feedback control and measurement output feedback control is presented. In Section 1.3, we further take into account model uncertainties, which leads to the formulation of the robust output regulation problem. We give the solution of this problem by both state feedback and error output feedback control. The robust output regulation problem is an enhanced version of the output regulation problem in the sense that it achieves the same objectives as the former even in the presence of model uncertainties. In Section 1.4, the solvability of the linear robust output regulation problem is further examined by introducing what is called the internal model principle. While the first four sections are devoted to continuous-time linear systems, results on the output regulation problem and on the robust output regulation problem for discrete-time linear systems are established in Sections 1.5 and 1.6.

1.1 Introduction

Many practical control problems such as trajectory planning of a robot manipulator, guidance of a tactic missile toward a moving target, attitude control of spacecraft subject to torque disturbance, weapon system pointing under firing disturbances, and so on, fall into the domain of the problem depicted in Figure 1.1. Here a plant is given that is subject to a disturbance $d(t)$, and a controller is to be designed so that the closed-loop system is *exponentially stable*, in the sense to be defined precisely later, and the output of the plant $y(t)$ asymptotically tracks a given reference input $r(t)$ in the following sense:

$$\lim_{t\to\infty} e(t) = \lim_{t\to\infty} (y(t) - r(t)) = 0. \tag{1.1}$$

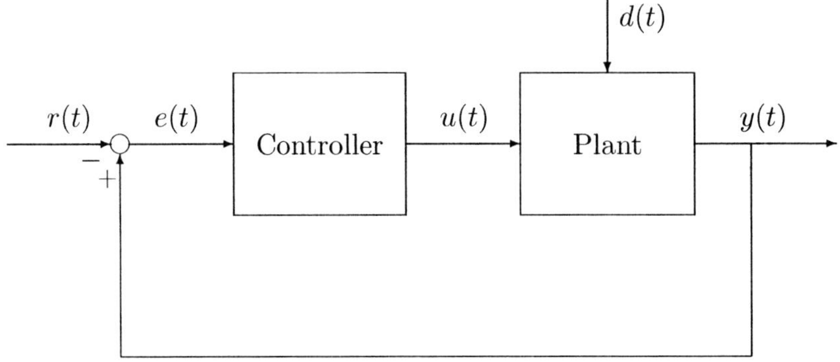

Figure 1.1. *Unity feedback control.*

This problem is conveniently called *asymptotic tracking and disturbance rejection* of the output. In the particular case where $r(t) = 0$, the problem is simply called *asymptotic regulation*.

A linear plant subject to a disturbance $d(t)$ can be modelled as follows:

$$\begin{aligned} \dot{x} &= Ax + Bu + E_d d, \\ y &= Cx + Du + F_d d. \end{aligned} \tag{1.2}$$

Thus, the tracking error is given by

$$e = Cx + Du + F_d d - r. \tag{1.3}$$

The controller can generally be modelled as follows:

$$\begin{aligned} u &= Kz, \\ \dot{z} &= \mathcal{G}_1 z + \mathcal{G}_2 e. \end{aligned} \tag{1.4}$$

This controller must guarantee the stability of the closed-loop system composed of (1.2) and (1.4) while assuring asymptotic tracking of $y(t)$ to $r(t)$ in the presence of the disturbance $d(t)$.

In practice, the reference input to be tracked and the disturbance to be rejected usually are not exactly known signals; for example, a disturbance in the form of a sinusoidal function can have any amplitudes and initial phases, or even any frequencies, and a reference input in the form of a step function can have arbitrary magnitudes. It is desirable that a single controller be able to handle a class of prescribed reference inputs and/or a class of prescribed disturbances. In this chapter, both the reference inputs and the disturbances are assumed to be generated by linear autonomous differential equations as follows:

$$\dot{r} = A_{1r} r, \quad r(0) = r_0, \quad \dot{d} = A_{1d} d, \quad d(0) = d_0,$$

where r_0 and d_0 are arbitrary initial states. The above autonomous equations can generate a large class of functions; for example, a combination of step functions of arbitrary magnitudes, ramp functions of arbitrary slopes, and sinusoidal functions of arbitrary amplitudes and initial phases.

Let

$$v = \begin{bmatrix} r \\ d \end{bmatrix}, \quad A_1 = \begin{bmatrix} A_{1r} & 0 \\ 0 & A_{1d} \end{bmatrix}.$$

Then the reference inputs and the disturbances can be lumped together as follows:

$$\dot{v} = A_1 v, \quad v(0) = \begin{bmatrix} r_0 \\ d_0 \end{bmatrix} \stackrel{def}{=} v_0. \tag{1.5}$$

Thus, the plant state and the tracking error can be put into the following form:

$$\begin{aligned} \dot{x} &= Ax + Bu + Ev, \\ e &= Cx + Du + Fv, \end{aligned} \tag{1.6}$$

where

$$\begin{bmatrix} E \\ F \end{bmatrix} = \begin{bmatrix} 0 & E_d \\ -I & F_d \end{bmatrix}.$$

Now the problem of asymptotic tracking of $y(t)$ to $r(t)$ can be treated as the problem of asymptotic regulation of $e(t)$ to the origin when $e(t)$ is viewed as the output of (1.6). Therefore, it suffices to study the regulation problem described by (1.6) while keeping in mind that the system (1.5), called the *exosystem* in what follows, can generate either the reference inputs or the disturbances or both. Thus, the problem of asymptotic tracking and disturbance rejection can be called simply the *output regulation* problem when the disturbances and the reference inputs are generated by (1.5). Alternatively, the output regulation problem is called a *servomechanism* problem.

In (1.6), the plant is defined by six finite-dimensional constant matrices A, B, E, C, D, and F. These matrices are usually obtained by linearizing a nonlinear system around an operating condition or by using a certain system identification approach. Due to the variations in the operating point or the limitations of system identification techniques, these matrices are invariably inaccurate. Typically, each entry of the matrices A, B, E, C, D, and F can take arbitrary values in an open neighborhood of its nominal value. Therefore, it is desirable to further require that the controller be able to maintain the property of asymptotic tracking and disturbance rejection in the closed-loop system regardless of small variations of the entries in the matrices A, B, E, C, D, and F. The problem of designing such controllers is called the *robust output regulation* problem or the *robust servomechanism* problem.

The discussion so far has exemplified a scenario of what is called the *output regulation* problem and its enhanced version the *robust output regulation* problem. The solvability of these two problems will be established in the remaining sections of this chapter.

1.2 Linear Output Regulation

Consider a class of linear time-invariant systems described by

$$\begin{aligned} \dot{x}(t) &= Ax(t) + Bu(t) + Ev(t), \quad x(0) = x_0, \quad t \geq 0, \\ e(t) &= Cx(t) + Du(t) + Fv(t), \end{aligned} \tag{1.7}$$

where $x(t)$ is the n-dimensional plant state, $u(t)$ the m-dimensional plant input, $e(t)$ the p-dimensional plant output representing the tracking error, and $v(t)$ the q-dimensional exogenous signal representing the reference inputs and/or the disturbances. The exogenous signal is generated by an exosystem of the form

$$\dot{v}(t) = A_1 v(t), \quad v(0) = v_0, \quad t \geq 0. \tag{1.8}$$

For convenience, we put the plant (1.7) and the exosystem (1.8) together into the following form:

$$\begin{aligned} \dot{x} &= Ax + Bu + Ev, \\ \dot{v} &= A_1 v, \\ e &= Cx + Du + Fv \end{aligned} \tag{1.9}$$

and call (1.9) a composite system with col(x, v) as the composite state.

Two classes of feedback control laws will be considered in this section, namely,

1. Static State Feedback:

$$u = K_x x + K_v v, \tag{1.10}$$

where $K_x \in \mathcal{R}^{m \times n}$ and $K_v \in \mathcal{R}^{m \times q}$ are constant matrices.

2. Dynamic Measurement Output Feedback:

$$u = Kz, \qquad \dot{z} = \mathcal{G}_1 z + \mathcal{G}_2 y_m, \tag{1.11}$$

where $z \in \mathcal{R}^{n_z}$ with n_z to be specified later, $y_m \in \mathcal{R}^{p_m}$ for some positive integer p_m is the measurement output, and $K \in \mathcal{R}^{m \times n_z}$, $\mathcal{G}_1 \in \mathcal{R}^{n_z \times n_z}$, $\mathcal{G}_2 \in \mathcal{R}^{n_z \times p_m}$ are constant matrices. It is assumed that y_m takes the following form:

$$y_m(t) = C_m x(t) + D_m u(t) + F_m v(t), \tag{1.12}$$

where $C_m \in \mathcal{R}^{p_m \times n}$, $D_m \in \mathcal{R}^{p_m \times m}$, and $F_m \in \mathcal{R}^{p_m \times q}$. A special case of the dynamic measurement output feedback control is the dynamic error output feedback control when $C_m = C$, $D_m = D$, $F_m = F$, that is, $y_m = e$. In many cases, the error output e is not the only measurable variable available for feedback control. Using the measurement output feedback control allows us to solve the output regulation problem for some systems that cannot be solved by the error output feedback control.

Denote the closed-loop system consisting of the plant (1.7), the exosystem (1.8), and the control law (1.10) or (1.11) as follows:

$$\begin{aligned} \dot{x}_c &= A_c x_c + B_c v, \quad x_c(0) = x_{c0}, \\ \dot{v} &= A_1 v, \\ e &= C_c x_c + D_c v, \end{aligned} \tag{1.13}$$

where, under the static state feedback, $x_c = x$ and

$$\begin{aligned} A_c &= A + BK_x, \quad B_c = E + BK_v, \\ C_c &= C + DK_x, \quad D_c = F + DK_v, \end{aligned} \tag{1.14}$$

and, under the dynamic measurement output feedback, $x_c = \text{col}(x, z)$ and

$$A_c = \begin{bmatrix} A & BK \\ \mathcal{G}_2 C_m & \mathcal{G}_1 + \mathcal{G}_2 D_m K \end{bmatrix}, \quad B_c = \begin{bmatrix} E \\ \mathcal{G}_2 F_m \end{bmatrix},$$
$$C_c = [C \quad DK], \quad D_c = F. \tag{1.15}$$

To describe the requirements on the closed-loop system (1.13), we first introduce the following definition.

Definition 1.1. *The closed-loop system (1.13) is said to be* exponentially stable *if we have the following.*

Property 1.1. The matrix A_c is Hurwitz, that is, all the eigenvalues of A_c have negative real parts.

The closed-loop system is said to have output regulation *property if the following holds.*

Property 1.2. For all x_{c0} and v_0, the trajectories of (1.13) satisfy

$$\lim_{t\to\infty} e(t) = \lim_{t\to\infty} (C_c x_c(t) + D_c v(t)) = 0.$$

Linear Output Regulation Problem (LORP): Design a control law of the form (1.10) or (1.11) such that the closed-loop system satisfies Properties 1.1 and 1.2.

Remark 1.2. In what follows, a control law that solves the linear output regulation problem will be called a *servoregulator*. In particular, if the control law is described by (1.10) or (1.11), then the controller will be called a *static state feedback servoregulator* or *dynamic measurement output feedback servoregulator*, respectively. ∎

At the outset, we list various assumptions needed for solving the linear output regulation problem.

Assumption 1.1. A_1 has no eigenvalues with negative real parts.

Assumption 1.2. The pair (A, B) is stabilizable.

Assumption 1.3. The pair

$$\left([C_m \quad F_m], \begin{bmatrix} A & E \\ 0 & A_1 \end{bmatrix} \right)$$

is detectable.

Remark 1.3. Assumption 1.1 is made only for convenience and loses no generality. In fact, if the linear output regulation problem is solvable by any controller under Assumption 1.1, then it is also solvable by the same controller even if Assumption 1.1 is violated. This is because Property 1.1 is simply a property of the plant data (A, B, C, D) and has nothing to do with the exosystem, and because Property 1.2 is only concerned with the asymptotic property of the closed-loop system. More specifically, the components of the exogenous signals corresponding to the modes associated with the eigenvalues of A_1 with negative real parts will exponentially decay to zero and will in no way affect the asymptotic behavior of the closed-loop system so long as the closed-loop system has Property 1.1. Assumption 1.2 is made so that Property 1.1, that is, the exponential stability of A_c, can be achieved by a state feedback. Assumption 1.3, together with Assumption 1.2, renders the exponential stability of A_c by the measurement output feedback. ∎

Lemma 1.4. *Under Assumption 1.1, consider the controller (1.10) or (1.11). Assume the closed-loop system (1.13) has Property 1.1. Then the following statements are equivalent:*

(i) *The closed-loop system has Property 1.2.*

(ii) *The controller solves the linear output regulation problem.*

(iii) *There exists a unique matrix X_c that satisfies the following matrix equations:*

$$\begin{aligned} X_c A_1 &= A_c X_c + B_c, \\ 0 &= C_c X_c + D_c. \end{aligned} \tag{1.16}$$

Proof. (i) $\leftrightarrow$ (ii). This is self-evident.

(ii) $\leftrightarrow$ (iii). The first equation of (1.16) is a Sylvester equation, which has a unique solution X_c if A_1 and A_c have no common eigenvalues (Appendix A). Since the closed-loop system satisfies Property 1.1, A_c is exponentially stable. Thus Assumption 1.1 and the exponential stability of A_c guarantee the existence of X_c, satisfying the first equation of (1.16). Let $\bar{x} = x_c - X_c v$. Then,

$$\begin{aligned} \dot{\bar{x}} &= A_c \bar{x}, \\ e &= C_c \bar{x} + (C_c X_c + D_c) v. \end{aligned}$$

Since A_c is exponentially stable, $\lim_{t\to\infty} \bar{x}(t) = 0$. To show (ii) $\leftarrow$ (iii), assume the matrix X_c also satisfies the second equation of (1.16); then

$$\lim_{t\to\infty} e(t) = \lim_{t\to\infty} C_c \bar{x}(t) = 0;$$

that is, the controller solves the linear output regulation problem. On the other hand, to show (ii) $\rightarrow$ (iii), assume the controller solves the linear output regulation problem; then,

$$\lim_{t\to\infty} (C_c X_c + D_c) v(t) = 0$$

for all $v(t) = e^{A_1 t} v(0)$ with any $v(0) \in \mathcal{R}^q$. Due to Assumption 1.1, $v(t)$ does not decay to zero for $v(0) \neq 0$. Therefore, necessarily, $C_c X_c + D_c = 0$. □

Remark 1.5.

(i) Lemma 1.4 gives a characterization of Property 1.2 in terms of the solvability of a set of linear matrix equations. This characterization allows the linear output regulation problem to be studied using the familiar mathematic tool of linear algebra. Further, it will be seen later that this lemma will render a natural translation of the requirements on the closed-loop system into the requirements on the controller, thus leading to the synthesis of the various controllers.

(ii) It is seen from the proof of Lemma 1.4 that if the output regulation problem is solvable, then there exists a subspace in $\mathcal{R}^{n+n_z+q}$ defined by the hyperplane $C_c x_c + D_c v = 0$ such that the trajectories $x_c(t)$ of the closed-loop system will approach this subspace asymptotically. ∎

Now let us first consider the static state feedback case where the controller is defined by two constant matrices K_x and K_v such that the closed-loop system is described by

$$\begin{aligned} \dot{x} &= (A + BK_x)x + (E + BK_v)v, \\ \dot{v} &= A_1 v, \\ e &= (C + DK_x)x + (F + DK_v)v. \end{aligned} \tag{1.17}$$

That is,

$$\begin{aligned} A_c &= A + BK_x, \quad B_c = E + BK_v, \\ C_c &= C + DK_x, \quad D_c = F + DK_v. \end{aligned}$$

The two matrices K_x and K_v will be called the *feedback gain* and the *feedforward gain*, respectively. The basic idea of designing the static state feedback controller is to use the feedback gain to make the closed-loop system satisfy Property 1.1 while using the feedforward gain to drive trajectories of the closed-loop system toward a subspace of $\mathcal{R}^{n+q}$ defined by the hyperplane $(C + DK_x)x + (F + DK_v)v = 0$. This idea is best illustrated by the following result.

Lemma 1.6. *Under Assumptions 1.1 and 1.2, let K_x render the exponential stability of $(A + BK_x)$. Then the linear output regulation problem is solvable by a static state feedback controller (1.10) if and only if there exist two constant matrices X_c and K_v that satisfy the following matrix equations:*

$$\begin{aligned} X_c A_1 &= (A + BK_x)X_c + BK_v + E, \\ 0 &= (C + DK_x)X_c + DK_v + F. \end{aligned} \tag{1.18}$$

Proof. Under Assumption 1.2, there exists K_x such that $(A + BK_x)$ is exponentially stable. Since equation (1.16) is exactly the same as equation (1.18) except that in (1.18) K_v is to be determined, if X_c and K_v satisfy (1.18), X_c also satisfies (1.16) for the two particular matrices K_x and K_v. On the other hand, if for some K_x and K_v, X_c satisfies (1.16), then X_c and K_v also satisfy (1.18). The proof thus follows from Lemma 1.4. □

Lemma 1.6 immediately suggests the following way of synthesizing the desired static state feedback controller.

Step 1. Find a feedback gain K_x such that $(A + BK_x)$ is stable.

Step 2. Solve for both X_c and K_v from the set of linear equations (1.18). Then the static state feedback controller is given by

$$u = K_x x + K_v v. \tag{1.19}$$

This approach, though straightforward to apply, has a drawback in that X_c and K_v depend on the feedback gain K_x. Thus, every time, a redesign of the feedback gain necessitates a recomputation of X_c and K_v. A better approach can be obtained by making the following linear transformation:

$$\begin{bmatrix} X \\ U \end{bmatrix} = \begin{bmatrix} I_n & 0_{n\times m} \\ K_x & I_m \end{bmatrix} \begin{bmatrix} X_c \\ K_v \end{bmatrix} \tag{1.20}$$

in equation (1.18), which leads to another set of linear matrix equations in unknown matrices X and U as follows:

$$\begin{aligned} XA_1 &= AX + BU + E, \\ 0 &= CX + DU + F. \end{aligned} \tag{1.21}$$

These equations are completely determined by the plant data A, B, E, C, D, F, and A_1. It is clear that there exist X and U satisfying (1.21) if and only if, for any $K_x \in \mathcal{R}^{m\times n}$, there exist X_c and K_v satisfying (1.18). Moreover, (X, U) and (X_c, K_v) are related to each other by equation (1.20).

Equations (1.21), known as the *regulator equations*, are instrumental to establishing the linear output regulation theory. In fact, in terms of the regulator equations, the above discussion can be summarized to yield the following result.

Theorem 1.7. *Under Assumptions 1.1 and 1.2, let the feedback gain K_x be such that $(A + BK_x)$ is exponentially stable. Then, the linear output regulation problem is solvable by a static state feedback control of the form*

$$u = K_x x + K_v v$$

if and only if there exist two matrices X and U that satisfy the linear matrix equations (1.21), with the feedforward gain K_v being given by

$$K_v = U - K_x X.$$

Remark 1.8. A systemic interpretation to the solution of (1.21) is given as follows. First consider the special case where the exogenous signal is constant. Since $A_1 = 0$, equations (1.18) and (1.21) become

$$\begin{aligned} 0 &= (A + BK_x)X_c + BK_v + E, \\ 0 &= (C + DK_x)X_c + DK_v + F, \end{aligned} \tag{1.22}$$

and, respectively,

$$\begin{aligned} 0 &= AX + BU + E, \\ 0 &= CX + DU + F. \end{aligned} \tag{1.23}$$

Equations (1.22) mean, for each constant v, that $X_c v$ is an equilibrium point of the closed-loop system at which the output is zero. Moreover,

$$\lim_{t\to\infty} x_c(t) = X_c v.$$

Thus, for each constant v, $X_c v$ is the *steady-state* state of the closed-loop system at which the output is zero. On the other hand, equations (1.23) mean, for each constant v, that Uv is the input under which the open-loop plant has an equilibrium state Xv at which the output is zero. Moreover, since $X_c = X$, and

$$\lim_{t\to\infty} u(t) = (K_x X + K_v)v = Uv,$$

for each constant v, whether or not the closed-loop system can be made to satisfy the output regulation property depends on the solvability of the regulator equations. The above interpretation can be extended to the general case. Under any controller that solves the linear output regulation problem, the trajectories of the closed-loop system from any initial state $x_c(0)$ and $v(0)$ satisfy

$$\lim_{t\to\infty} (x_c(t) - X_c v(t)) = \lim_{t\to\infty} (x_c(t) - Xv(t)) = 0.$$

Correspondingly, the control input satisfies

$$\lim_{t\to\infty} (u(t) - (K_x X + K_v)v(t)) = \lim_{t\to\infty} (u(t) - Uv(t)) = 0.$$

Thus, if the linear output regulation problem is solvable at all, necessarily, all trajectories of the closed-loop system approach $Xv(t)$, and the corresponding controls approach $Uv(t)$. Thus, the steady-state behavior of the closed-loop system is completely characterized by the solution of the regulator equations. For convenience, in what follows, $Xv(t)$ and $Uv(t)$ are called *zero-error constrained state* and *zero-error constrained control*, respectively. In particular, when v is constant, Xv is called *zero-error constrained equilibrium.* ■

An easily testable condition can be given with regard to the solvability of the regulator equations as shown below.

Theorem 1.9. *For any matrices E and F, the regulator equations (1.21) are solvable if and only if the following holds:*

Assumption 1.4. For all $\lambda \in \sigma(A_1)$, where $\sigma(A_1)$ denotes the spectrum of A_1,

$$\text{rank}\begin{bmatrix} A-\lambda I & B \\ C & D \end{bmatrix} = n + p. \tag{1.24}$$

Proof. The regulator equations (1.21) can be put into the following form:

$$\begin{bmatrix} I_n & 0_{n\times m} \\ 0_{p\times n} & 0_{p\times m} \end{bmatrix}\begin{bmatrix} X \\ U \end{bmatrix} A_1 - \begin{bmatrix} A & B \\ C & D \end{bmatrix}\begin{bmatrix} X \\ U \end{bmatrix} = \begin{bmatrix} E \\ F \end{bmatrix}. \tag{1.25}$$

Using the properties of the Kronecker product, which can be found in Appendix A, we can transform (1.25) into a standard linear algebraic equation of the form

$$Qx = b,$$

where

$$Q = A_1{}^T \otimes \begin{bmatrix} I_n & 0_{n\times m} \\ 0_{p\times n} & 0_{p\times m} \end{bmatrix} - I_q \otimes \begin{bmatrix} A & B \\ C & D \end{bmatrix},$$

$$x = \text{vec}\left(\begin{bmatrix} X \\ U \end{bmatrix} \right), \quad b = \text{vec}\left(\begin{bmatrix} E \\ F \end{bmatrix} \right).$$

Here the notation vec$(\cdot)$ denotes a vector-valued function of a matrix such that, for any $X \in \mathcal{R}^{n\times m}$,

$$\text{vec}\,(X) = \begin{bmatrix} X_1 \\ \vdots \\ X_m \end{bmatrix},$$

where for $i = 1, \ldots, m$, X_i is the ith column of X. Thus, equation (1.25) is solvable for any matrices E and F if and only if Q has full row rank. To obtain the condition under which Q has full row rank, we assume, without loss of generality, that A_1 is in the following Jordan form:

$$A_1 = \begin{bmatrix} J_1 & 0 & 0 & \cdots & 0 \\ 0 & J_2 & 0 & \cdots & 0 \\ \vdots & \vdots & \vdots & \vdots & \vdots \\ 0 & 0 & 0 & \cdots & J_k \end{bmatrix},$$

where J_i has dimension n_i such that $n_1 + n_2 + \cdots + n_k = q$ and is given by

$$J_i = \begin{bmatrix} \lambda_i & 1 & 0 & \cdots & 0 & 0 \\ 0 & \lambda_i & 1 & \cdots & 0 & 0 \\ \vdots & \vdots & \vdots & \vdots & \vdots & \vdots \\ 0 & 0 & 0 & \cdots & \lambda_i & 1 \\ 0 & 0 & 0 & \cdots & 0 & \lambda_i \end{bmatrix}.$$

A simple calculation shows that Q is a block lower triangular matrix of k blocks with its ith, $1 \le i \le k$, diagonal block having the form

$$\begin{bmatrix} \lambda_i\mathcal{E} - \mathcal{A} & 0 & 0 & \cdots & 0 & 0 \\ \mathcal{E} & \lambda_i\mathcal{E} - \mathcal{A} & 0 & \cdots & 0 & 0 \\ \vdots & \vdots & \vdots & \vdots & \vdots & \vdots \\ 0 & 0 & 0 & \cdots & \lambda_i\mathcal{E} - \mathcal{A} & 0 \\ 0 & 0 & 0 & \cdots & \mathcal{E} & \lambda_i\mathcal{E} - \mathcal{A} \end{bmatrix},$$

where

$$\mathcal{E} = \begin{bmatrix} I_n & 0_{n\times m} \\ 0_{p\times n} & 0_{p\times m} \end{bmatrix}, \quad \mathcal{A} = \begin{bmatrix} A & B \\ C & D \end{bmatrix}.$$

Clearly, Q has full row rank if and only if Assumption 1.4 holds. □

In conjunction with Theorem 1.7, Theorem 1.9 immediately leads to the following sufficient conditions for the solvability of the output regulation problem by the static state feedback control of the form (1.10).

Corollary 1.10. *Under Assumptions 1.1, 1.2, and 1.4, the linear output regulation problem is solvable by the static state feedback control (1.10).*

Remark 1.11. If the pair (A, B) is controllable and the pair (C, A) is observable, then those values of λ at which the matrix

$$\begin{bmatrix} A - \lambda I & B \\ C & D \end{bmatrix}$$

is not full rank are called the transmission zeros of the system. It is a generalization of the notion of zeros of the single-input, single-output systems to multi-input, multi-output systems. Thus Assumption 1.4 can be paraphrased by saying that the transmission zeros of the plant (1.7) do not coincide with the eigenvalues of the exosystem, and it is often simply called the *transmission zeros condition*. The plant (1.7) is called a minimum phase system if all of its transmission zeros are on the open left-half complex plane. Thus a minimum phase system always satisfies the transmission zeros condition. ∎

Remark 1.12. A systemic interpretation of Assumption 1.4 can also be given in the same spirit as Remark 1.8. First consider the special case where $A_1 = 0$. For this case, equation (1.24) actually takes the form

$$\operatorname{rank}\begin{bmatrix} A & B \\ C & D \end{bmatrix} = n + p \tag{1.26}$$

as $A_1 = 0$. Correspondingly, the regulator equations are given by (1.23). Thus, (1.26) is both necessary and sufficient for the plant to have a pair of zero-error constrained equilibrium and input for any E and F. A similar interpretation can be given to the case where $A_1 \neq 0$. For every $\lambda \in \sigma(A_1)$, let v_∞ be the eigenvector of A_1 associated with λ. Then the solution of the exosystem starting from $v(0) = v_\infty$ is $v(t) = v_\infty e^{\lambda t}$. Thus, if the closed-loop system has Properties 1.1 and 1.2, there exist $x_\infty \in \mathcal{R}^n$ and $u_\infty \in \mathcal{R}^m$ such that

$$\lim_{t\to\infty} (x_c(t) - x_\infty e^{\lambda t}) = 0,$$
$$\lim_{t\to\infty} (u(t) - u_\infty e^{\lambda t}) = 0.$$

Therefore, x_∞ and u_∞ must satisfy the following equations:

$$x_\infty \lambda e^{\lambda t} = A x_\infty e^{\lambda t} + B u_\infty e^{\lambda t} + E v_\infty e^{\lambda t},$$
$$0 = C x_\infty e^{\lambda t} + D u_\infty e^{\lambda t} + F v_\infty e^{\lambda t},$$

or, equivalently,

$$\begin{bmatrix} A - \lambda I & B \\ C & D \end{bmatrix} \begin{bmatrix} x_\infty \\ u_\infty \end{bmatrix} = -\begin{bmatrix} E \\ F \end{bmatrix} v_\infty. \tag{1.27}$$

Clearly, equation (1.27) has a solution x_∞ and u_∞ for any E and F if and only if Assumption 1.4 holds. It should be noted that, for a particular pair of (E, F), the regulator equations may still have a solution even if Assumption 1.4 fails. This happens when

$$\text{vec}\left(\begin{bmatrix} E \\ F \end{bmatrix}\right) \in \text{Im}(Q). \tag{1.28}$$

However, this case is not interesting since even arbitrarily small variations in (A, B, E, C, D, F) may fail (1.28). ∎

When the state x and the exogenous signal v are not available for feedback but Assumption 1.3 holds, the measurement output feedback control of the form (1.11) can be used to solve the linear output regulation problem. In this case,

$$A_c = \begin{bmatrix} A & BK \\ \mathcal{G}_2 C_m & \mathcal{G}_1 + \mathcal{G}_2 D_m K \end{bmatrix},$$
$$B_c = \begin{bmatrix} E \\ \mathcal{G}_2 F_m \end{bmatrix}, \quad C_c = [C \quad DK], \quad D_c = F. \tag{1.29}$$

Due to Lemma 1.4, we need to find a triple $(K, \mathcal{G}_1, \mathcal{G}_2)$ such that A_c is exponentially stable and (1.16) is solvable for X_c. To this end, we first translate the requirements on the closed-loop system as given by (1.16) into the requirements on the controller $(K, \mathcal{G}_1, \mathcal{G}_2)$ as given by the following result.

Lemma 1.13. *Under Assumption 1.1, suppose there exists a dynamic measurement output feedback controller $(K, \mathcal{G}_1, \mathcal{G}_2)$ such that the closed-loop system has Property 1.1. Then the following are equivalent:*

(i) *The linear output regulation problem is solvable by the measurement output feedback controller $(K, \mathcal{G}_1, \mathcal{G}_2)$.*

(ii) *There exists a matrix X_c that satisfies the following matrix equations:*

$$X_c A_1 = \begin{bmatrix} A & BK \\ \mathcal{G}_2 C_m & \mathcal{G}_1 + \mathcal{G}_2 D_m K \end{bmatrix} X_c + \begin{bmatrix} E \\ \mathcal{G}_2 F_m \end{bmatrix},$$
$$0 = [C \quad DK] X_c + F. \tag{1.30}$$

(iii) *There exist matrices (X, U, Z) such that X and U are the solution of the regulator equations*

$$XA_1 = AX + BU + E,$$
$$0 = CX + DU + F, \tag{1.31}$$

and Z is the solution of the Sylvester equation

$$ZA_1 = \mathcal{G}_1 Z + \mathcal{G}_2(C_m X + D_m U + F_m), \tag{1.32}$$

which satisfies

$$U = KZ. \tag{1.33}$$

Proof. (i) $\leftrightarrow$ (ii). This is actually Lemma 1.4 specialized to the measurement output feedback case.

(ii) $\leftrightarrow$ (iii). Assume (ii) holds. Partition X_c as

$$X_c = \begin{bmatrix} X \\ Z \end{bmatrix},$$

where $X \in \mathcal{R}^{n\times q}$ and $Z \in \mathcal{R}^{n_z\times q}$. Then (1.30) is the same as

$$\begin{aligned} XA_1 &= AX + BKZ + E, \\ ZA_1 &= \mathcal{G}_2 C_m X + (\mathcal{G}_1 + \mathcal{G}_2 D_m K)Z + \mathcal{G}_2 F_m, \\ 0 &= CX + DKZ + F, \end{aligned} \tag{1.34}$$

which is the same as

$$\begin{aligned} XA_1 &= AX + BKZ + E, \\ ZA_1 &= \mathcal{G}_1 Z + \mathcal{G}_2(C_m X + D_m KZ + F_m), \\ 0 &= CX + DKZ + F. \end{aligned} \tag{1.35}$$

Letting $U = KZ$ in (1.35) shows that X and U satisfy the regulator equations (1.31), and U and Z satisfy (1.32) and (1.33). This completes (ii) $\rightarrow$ (iii). On the other hand, assume (iii) holds. We will show that X and Z satisfy (1.34) or equivalently (1.35). Indeed, substituting $U = KZ$ into equation (1.31) shows that X and Z satisfy the first and third equations of (1.35), and substituting $U = KZ$ into (1.32) shows that Z satisfies the second equation of (1.35). $\square$

Now we turn to the construction of the triple $(K, \mathcal{G}_1, \mathcal{G}_2)$. Since we have already known how to synthesize a static state feedback controller which takes the plant state x and the exosystem state v as its inputs, we naturally seek to synthesize a measurement output feedback controller by estimating the state x and the exogenous signal v. To this end, lump the state x and exogenous signals v together to obtain the following system:

$$\begin{aligned} \begin{bmatrix} \dot{x} \\ \dot{v} \end{bmatrix} &= \begin{bmatrix} A & E \\ 0 & A_1 \end{bmatrix} \begin{bmatrix} x \\ v \end{bmatrix} + \begin{bmatrix} B \\ 0 \end{bmatrix} u, \\ y_m &= [C_m \ \ F_m] \begin{bmatrix} x \\ v \end{bmatrix} + D_m u. \end{aligned} \tag{1.36}$$

Employing the well-known Luenburger observer theory suggests the following observer:

$$\begin{aligned} u &= [K_x \ \ K_v] z, \\ \dot{z} &= \begin{bmatrix} A & E \\ 0 & A_1 \end{bmatrix} z + \begin{bmatrix} B \\ 0 \end{bmatrix} u + L(y_m - [C_m \ \ F_m] z - D_m u), \end{aligned} \tag{1.37}$$

where L is an observer gain matrix of dimension $(n + q)$ by p_m.

Clearly, (1.37) can be put into the form

$$u = Kz, \qquad \dot{z} = \mathcal{G}_1 z + \mathcal{G}_2 y_m,$$

with

$$K = [K_x \quad K_v],$$
$$\mathcal{G}_1 = \begin{bmatrix} A & E \\ 0 & A_1 \end{bmatrix} + \begin{bmatrix} B \\ 0 \end{bmatrix} K - L([C_m \; F_m] + D_m K), \quad \mathcal{G}_2 = L. \tag{1.38}$$

Theorem 1.14. *Under Assumptions 1.1, 1.2, and 1.3, the linear output regulation problem is solvable by a measurement output feedback controller* (K_x, K_v, L) *given by (1.37) (equivalently,* $(K, \mathcal{G}_1, \mathcal{G}_2)$ *given by (1.38)) if and only if there exists a pair of matrices* (X, U) *that satisfies the regulator equations*

$$\begin{aligned} XA_1 &= AX + BU + E, \\ 0 &= CX + DU + F. \end{aligned} \tag{1.39}$$

Proof. The "only if" part is a consequence of part (iii) of Lemma 1.13. To show the "if" part, first note that, by Assumption 1.2, there exists a state feedback gain K_x such that $(A + BK_x)$ is exponentially stable, and, by Assumption 1.3, there exist matrices L_1 and L_2 such that

$$A_L = \begin{bmatrix} A & E \\ 0 & A_1 \end{bmatrix} - \begin{bmatrix} L_1 \\ L_2 \end{bmatrix} [C_m \quad F_m] = \begin{bmatrix} A - L_1 C_m & E - L_1 F_m \\ -L_2 C_m & A_1 - L_2 F_m \end{bmatrix}$$

is exponentially stable. Now let (X, U) satisfy the regulator equations, and let $K_v = U - K_x X$, $K = [K_x, K_v]$, and

$$L = \begin{bmatrix} L_1 \\ L_2 \end{bmatrix}.$$

A simple calculation gives

$$\begin{aligned} A_c &= \begin{bmatrix} A & BK \\ \mathcal{G}_2 C_m & \mathcal{G}_1 + \mathcal{G}_2 D_m K \end{bmatrix} \\ &= \begin{bmatrix} A & BK_x & BK_v \\ 0 & A + BK_x & E + BK_v \\ 0 & 0 & A_1 \end{bmatrix} + \begin{bmatrix} 0 \\ L_1 \\ L_2 \end{bmatrix} \begin{bmatrix} C_m & -C_m & -F_m \end{bmatrix}. \end{aligned} \tag{1.40}$$

In (1.40), subtracting the first row from the second row and adding the second column to the first column shows that A_c is equivalent to the following matrix

$$\begin{aligned} &\begin{bmatrix} A + BK_x & BK_x & BK_v \\ 0 & A & E \\ 0 & 0 & A_1 \end{bmatrix} + \begin{bmatrix} 0 \\ L_1 \\ L_2 \end{bmatrix} \begin{bmatrix} 0 & -C_m & -F_m \end{bmatrix} \\ &= \begin{bmatrix} A + BK_x & BK_x & BK_v \\ 0 & A - L_1 C_m & E - L_1 F_m \\ 0 & -L_2 C_m & A_1 - L_2 F_m \end{bmatrix}. \end{aligned} \tag{1.41}$$

Thus $\sigma(A_c) = \sigma(A + BK_x) \cup \sigma(A_L)$; that is, we have shown that the triple (K_x, K_v, L) (equivalently, $(K, \mathcal{G}_1, \mathcal{G}_2)$) renders the closed-loop system Property 1.1. To show that the closed-loop system also satisfies Property 1.2, let

$$Z = \begin{bmatrix} X \\ I \end{bmatrix}. \tag{1.42}$$

We will show that the triple (X, U, Z) satisfies the conditions of part (iii) of Lemma 1.13. Since the pair (X, U) satisfies the regulator equations by assumption, it suffices to show that

$$ZA_1 = \mathcal{G}_1 Z + \mathcal{G}_2(C_m X + D_m U + F_m). \tag{1.43}$$

Indeed, using the definition of $\mathcal{G}_1$ given by (1.38) yields

$$\begin{aligned}\mathcal{G}_1 Z &= \begin{bmatrix} (A + BK_x)X + E + BK_v \\ A_1 \end{bmatrix} - L((C_m + D_m K_x)X + F_m + D_m K_v) \\ &= \begin{bmatrix} AX + B(K_x X + K_v) + E \\ A_1 \end{bmatrix} \\ &\quad - L(C_m X + D_m(K_x X + K_v) + F_m). \end{aligned} \tag{1.44}$$

Using $U = K_x X + K_v$ in (1.44) gives

$$\begin{aligned}\mathcal{G}_1 Z &= \begin{bmatrix} AX + BU + E \\ A_1 \end{bmatrix} - L(C_m X + D_m U + F_m) \\ &= \begin{bmatrix} X \\ I \end{bmatrix} A_1 - L(C_m X + D_m U + F_m) \\ &= ZA_1 - L(C_m X + D_m U + F_m)\end{aligned}$$

upon noting that X and U satisfy the regulator equations. The proof is completed by the equivalence of (i) and (iii) of Lemma 1.13. □

By Theorem 1.9, the solvability of the regulator equations is guaranteed by the satisfaction of the Assumption 1.4. Thus we have the following corollary.

Corollary 1.15. *Under Assumptions 1.1 to 1.4, the linear output regulation problem is solvable by a measurement output feedback controller* (K_x, K_v, L) *given by (1.37) (equivalently,* $(K, \mathcal{G}_1, \mathcal{G}_2)$ *given by (1.38)).*

1.3 Linear Robust Output Regulation

In this section, we will further consider the linear robust output regulation problem in which a controller has to be able to tolerate certain plant uncertainty. When the plant uncertainty is taken into consideration, the class of linear time-invariant systems is described by

$$\begin{aligned}\dot{x}(t) &= (A + \Delta A)x(t) + (B + \Delta B)u(t) + (E + \Delta E)v(t), \quad x(0) = x_0, \quad t \geq 0, \\ e(t) &= (C + \Delta C)x(t) + (D + \Delta D)u(t) + (F + \Delta F)v(t),\end{aligned} \tag{1.45}$$

where $x(t)$, $u(t)$, and $e(t)$ are the same as what are described in Section 1.2, and $v(t)$ is again generated by the same exosystem (1.8).

In (1.45), the matrices A, B, E, C, D and F represent the nominal part of the plant while ΔA, ΔB, and so forth represent the uncertain part. The entries of $(\Delta A, \Delta B, \Delta E, \Delta C, \Delta D, \Delta F)$ are allowed to take arbitrary values.

It is convenient to identify the system uncertainties with a vector w in the Euclidean space $\mathcal{R}^{n_w}$ with $w = \text{vec}\left(\begin{bmatrix} \Delta A & \Delta B & \Delta E \\ \Delta C & \Delta D & \Delta F \end{bmatrix}\right)$ and $n_w = (n+p) \times (n+m+q)$. Thus, we can adopt the following convenient notation:

$$\begin{aligned} A_w &= A + \Delta A, \quad B_w = B + \Delta B, \quad E_w = E + \Delta E, \\ C_w &= C + \Delta C, \quad D_w = D + \Delta D, \quad F_w = F + \Delta F \end{aligned}$$

with

$$\begin{aligned} A_0 &= A, \quad B_0 = B, \quad E_0 = E, \\ C_0 &= C, \quad D_0 = D, \quad F_0 = F. \end{aligned}$$

As a result, (1.45) can be written as follows:

$$\begin{aligned} \dot{x} &= A_w x + B_w u + E_w v, \\ e &= C_w x + D_w u + F_w v. \end{aligned} \tag{1.46}$$

For convenience of reference, the plant (1.46) and the exosystem (1.8) can be put together into the following:

$$\begin{aligned} \dot{x} &= A_w x + B_w u + E_w v, \\ \dot{v} &= A_1 v, \\ e &= C_w x + D_w u + F_w v, \end{aligned} \tag{1.47}$$

and (1.47) will be called the composite system.

We consider two classes of feedback control laws which are somehow different from those considered in the last section.

3. Dynamic State Feedback:

$$\begin{aligned} u &= K_1 x + K_2 z, \\ \dot{z} &= \mathcal{G}_1 z + \mathcal{G}_2 e, \end{aligned} \tag{1.48}$$

where $z \in \mathcal{R}^{n_z}$ with n_z to be specified later, and $(K_1, K_2, \mathcal{G}_1, \mathcal{G}_2)$ are constant matrices of appropriate dimensions.

4. Dynamic Output Feedback:

$$\begin{aligned} u &= K z, \\ \dot{z} &= \mathcal{G}_1 z + \mathcal{G}_2 e, \end{aligned} \tag{1.49}$$

where, again, $z \in \mathcal{R}^{n_z}$ with n_z to be specified later, and $(K, \mathcal{G}_1, \mathcal{G}_2)$ are constant matrices of appropriate dimensions.

Remark 1.16. Due to the presence of the uncertain parameter w, the robust output regulation problem that will be formulated shortly cannot be handled via the approach for solving the output regulation problem described in Section 1.2. It will be handled by a celebrated design methodology called the internal model principle. As a result, there exist no static state feedback control laws that can solve the robust output regulation problem, as will be shown in Lemma 1.21. On the other hand, as pointed out before, the measurement output feedback control is more general than the error output feedback case. However, in order to better illustrate the mechanism of the internal model principle, we will focus on the error feedback case when it comes to the robust output regulation problem. Remark 1.29 will give a clue on how to synthesize a measurement output feedback controller under some additional condition. To save the notation, we use the same notation z, $\mathcal{G}_1$, and $\mathcal{G}_2$ to describe the dynamic compensator in various controllers (1.11), (1.48), and (1.49). However, the dimension of z and the specific structure of the matrices $\mathcal{G}_1$ and $\mathcal{G}_2$ are totally different among these three different controllers. ∎

Denote the closed-loop system consisting of the plant (1.46), the exosystem (1.8), and the control law (1.48) or (1.49) as follows:

$$\begin{aligned} \dot{x}_c &= A_{cw}x_c + B_{cw}v, \\ \dot{v} &= A_1 v, \\ e &= C_{cw}x_c + D_{cw}v, \end{aligned} \tag{1.50}$$

where, under the dynamic state feedback, $x_c = \mathrm{col}(x, z)$ and

$$\begin{aligned} A_{cw} &= \begin{bmatrix} A_w + B_w K_1 & B_w K_2 \\ \mathcal{G}_2(C_w + D_w K_1) & \mathcal{G}_1 + \mathcal{G}_2 D_w K_2 \end{bmatrix}, & B_{cw} &= \begin{bmatrix} E_w \\ \mathcal{G}_2 F_w \end{bmatrix}, \\ C_{cw} &= [C_w + D_w K_1 \quad D_w K_2], & D_{cw} &= F_w, \end{aligned} \tag{1.51}$$

and under the dynamic output feedback, $x_c = \mathrm{col}(x, z)$ and

$$\begin{aligned} A_{cw} &= \begin{bmatrix} A_w & B_w K \\ \mathcal{G}_2 C_w & \mathcal{G}_1 + \mathcal{G}_2 D_w K \end{bmatrix}, & B_{cw} &= \begin{bmatrix} E_w \\ \mathcal{G}_2 F_w \end{bmatrix}, \\ C_{cw} &= [C_w \quad D_w K], & D_{cw} &= F_w. \end{aligned} \tag{1.52}$$

Correspondingly, we use $(A_{c0}, B_{c0}, C_{c0}, D_{c0})$ or simply (A_c, B_c, C_c, D_c) to denote the closed-loop system composed of the nominal plant and the control laws.

To describe the requirements on the closed-loop system (1.50), we first introduce the following definition.

Definition 1.17. *The closed-loop system (1.50) is said to be* exponentially stable *at* $w = 0$ *if the following property holds:*

Property 1.3. The matrix A_{c0} is Hurwitz, that is, all the eigenvalues of A_{c0} have negative real parts.

The closed-loop system is said to have robust output regulation *property at* $w = 0$ *if the following holds:*

Property 1.4. There exists an open neighborhood W of $w = 0$ such that, for all x_{c0} and v_0 and for all $w \in W$, the trajectories of (1.50) satisfy

$$\lim_{t \to \infty} e(t) = \lim_{t \to \infty} (C_{cw} x_c(t) + D_{cw} v(t)) = 0.$$

Remark 1.18. The set W does not have to be small in the statement of Property 1.4. It can be shown later in Lemma 1.4 that if the closed-loop system (1.50) satisfies Properties 1.3 and 1.4 for some open set W, then it also satisfies Property 1.4 for arbitrary set W in which A_{cw} is exponentially stable. In the following, we implicitly assume that W is an open set of w in which A_{cw} is exponentially stable. ∎

Now we are ready to state the problem precisely as follows.

Linear Robust Output Regulation Problem (LRORP): Design a control law of the form (1.48) or (1.49) such that the closed-loop system satisfies Properties 1.3 and 1.4.

Remark 1.19. Since Property 1.2 is clearly a particular case of Property 1.4, any controller that solves the linear robust output regulation problem also solves the linear output regulation problem. In what follows, a control law that solves the linear robust output regulation problem will be called a *robust servoregulator*. In particular, if the control law is described by (1.48) or (1.49), then the controller is called a *dynamic state feedback servoregulator*, or *dynamic output feedback servoregulator*. It is noted that the dynamic output feedback control law (1.49) is a special case of the dynamic measurement output feedback control law (1.11). ∎

In addition to Assumptions 1.1, 1.2, and 1.4 introduced in the last section, we need one more assumption in this section.

Assumption 1.5. The pair (C, A) is detectable.

This assumption is made so that Property 1.2 can be achieved by a dynamic output feedback control.

A result similar to Lemma 1.4 is given as follows.

Lemma 1.20. *Under Assumption 1.1, consider the controller (1.48) or (1.49). Assume the closed-loop system (1.50) has Property 1.3. Then the following statements are equivalent:*

(i) *The closed-loop system has Property 1.4.*

(ii) *The controller solves the linear robust output regulation problem .*

(iii) *For each* $w \in W$*, where* W *is an open neighborhood of* $w = 0$ *such that* A_{cw} *is exponentially stable, there exists a unique matrix* X_{cw} *that satisfies the following matrix equations:*

$$\begin{aligned} X_{cw} A_1 &= A_{cw} X_{cw} + B_{cw}, \\ 0 &= C_{cw} X_{cw} + D_{cw}. \end{aligned} \tag{1.53}$$

Proof. (i) $\leftrightarrow$ (ii). This is self-evident.

(ii) $\leftrightarrow$ (iii). Since the closed-loop system satisfies Property 1.3, there exists an open neighborhood W of $w = 0$ such that, for each $w \in W$, A_{cw} is exponentially stable. Note that, for each $w \in W$, the first equation of (1.53) is a Sylvester equation, which has a unique solution X_{cw} if and only if the spectra of A_1 and A_{cw} do not coincide. Thus Assumption 1.1 and the fact that A_{cw} is exponentially stable for $w \in W$ guarantee the existence of X_{cw} satisfying the first equation of (1.53) for $w \in W$. Let $\bar{x} = x_c - X_{cw}v$. Then,

$$\begin{aligned} \dot{\bar{x}} &= A_{cw}\bar{x}, \\ e &= C_{cw}\bar{x} + (C_{cw}X_{cw} + D_{cw})v. \end{aligned}$$

Since A_{cw} is exponentially stable for each $w \in W$, $\lim_{t\to\infty} \bar{x}(t) = 0$. Now if the matrix X_{cw} also satisfies the second equation of (1.53) for $w \in W$, then

$$\lim_{t\to\infty} e(t) = \lim_{t\to\infty} C_{cw}\bar{x}(t) = 0;$$

that is, the controller solves the linear robust output regulation problem. On the other hand, assume the controller solves the linear robust output regulation problem; then, for each $w \in W$, such that A_{cw} is exponentially stable,

$$\lim_{t\to\infty} (C_{cw}X_{cw} + D_{cw})v(t) = 0$$

for all $v(t) = e^{A_1 t}v(0)$ with any $v(0) \in \mathcal{R}^q$. Due to Assumption 1.1, $v(t)$ does not decay to zero for $v(0) \neq 0$. Therefore, necessarily, $C_{cw}X_{cw} + D_{cw} = 0$. □

Similar to Lemma 1.4, Lemma 1.20 gives a characterization of Property 1.4 in terms of the solvability of a set of linear matrix equations that depend on the uncertain parameter w. This characterization also allows a natural translation of the requirements on the closed-loop system into the requirements on the controller, thus leading to the synthesis of the various controllers. Nevertheless, the presence of the uncertain parameter w makes the solvability of the robust output regulation problem more difficult than the output regulation problem. In fact, let us first point out that the approach used in the last section cannot be carried over to the current case. As manifested by Lemma 1.6, under the static state feedback controller, the output regulation is achieved by appropriately designing a feedforward gain K_v that is able to annihilate the steady-state tracking error. However, the feedforward gain, as a solution of equations (1.18), is dependent on the plant parameters. As the plant parameters $(A_w, B_w, E_w, C_w, D_w, F_w)$ vary, the desired feedforward gain has to vary as a function of w, too. As a result, there exists no fixed-gain static feedback controller that solves the linear robust output regulation problem. The above argument can be formally stated in the following lemma.

Lemma 1.21. *There exists no static state feedback robust servoregulator for the linear robust output regulation problem.*

Proof. Assume there exists a static state feedback controller $u = K_x x + K_v v$ that solves the linear robust output regulation problem. We will lead to a contradiction by using

Lemma 1.20. To this end, note that since Lemma 1.20(iii) applies to an open neighborhood W of $w = 0$, it also applies to any subset of W. Now fix W, and define a subset of W, denoted by W_s, as follows:

$$W_s = \{w \in W \mid \Delta A = 0, \Delta B = 0, \Delta C = 0, \Delta D = 0\}. \tag{1.54}$$

By part (iii) of Lemma 1.20, for each $w \in W_s$ (hence, for each F_w and E_w), there must exist a matrix X_w such that

$$\begin{aligned} X_w A_1 &= (A + BK_x)X_w + BK_v + E_w, \\ 0 &= (C + DK_x)X_w + DK_v + F_w. \end{aligned} \tag{1.55}$$

Therefore, equations (1.55) define a surjective linear mapping $\mathcal{F} : \mathcal{R}^{n\times q} \to \mathcal{R}^{(n+p)\times q}$. But this is impossible since $n < (n + p)$. □

As a result, we have to employ other techniques to synthesize controllers that do not rely on the solution of the regulator equations. Again, our starting point is Lemma 1.20. In particular, part (iii) of Lemma 1.20 lends itself to the following idea of constructing a controller for the linear robust output regulation problem. Find a compensator $(\mathcal{G}_1, \mathcal{G}_2)$ such that the following augmented plant:

$$\begin{bmatrix} \dot{x} \\ \dot{z} \end{bmatrix} = \begin{bmatrix} A_w & 0 \\ \mathcal{G}_2 C_w & \mathcal{G}_1 \end{bmatrix} \begin{bmatrix} x \\ z \end{bmatrix} + \begin{bmatrix} B_w \\ \mathcal{G}_2 D_w \end{bmatrix} u + \begin{bmatrix} E_w \\ \mathcal{G}_2 F_w \end{bmatrix} v \tag{1.56}$$

has two properties:

(i) (1.56) can be stabilized by a state feedback control $u = K_1 x + K_2 z$ or by a partial state feedback control $u = Kz$.

(ii) For any state feedback control $u = K_1 x + K_2 z$ or any partial state feedback control $u = Kz$ that makes A_c exponentially stable, the unique solution of the first equation of (1.53) also satisfies the second equation of (1.53) so long as A_{cw} is exponentially stable.

In this section, we will show that, under Assumptions 1.1 to 1.3, such a compensator indeed exists. Further insights into the solvability of the linear robust output regulation problem will be provided in the next section.

Definition 1.22. *Given any square matrix $\hat{A}_1$, a pair of matrices $(\mathcal{G}_1, \mathcal{G}_2)$ is said to incorporate a p-copy internal model of the matrix $\hat{A}_1$ if the pair $(\mathcal{G}_1, \mathcal{G}_2)$ admits the following form:*

$$\mathcal{G}_1 = T \begin{bmatrix} S_1 & S_2 \\ 0 & G_1 \end{bmatrix} T^{-1}, \quad \mathcal{G}_2 = T \begin{bmatrix} S_3 \\ G_2 \end{bmatrix}, \tag{1.57}$$

where (S_1, S_2, S_3) are arbitrary constant matrices of any dimensions so long as their dimensions are compatible, T is any nonsingular matrix with the same dimension as $\mathcal{G}_1$, and (G_1, G_2) is described as follows:

$$G_1 = \text{block diag } \underbrace{[\beta_1, \ldots, \beta_p]}_{p\text{-tuple}}, \quad G_2 = \text{block diag } \underbrace{[\sigma_1, \ldots, \sigma_p]}_{p\text{-tuple}}, \tag{1.58}$$

where for $i = 1, \ldots, p$, β_i *is a constant square matrix of dimension* d_i *for some integer* d_i, *and* σ_i *is a constant column vector of dimension* d_i *such that*

(i) β_i *and* σ_i *are controllable.*

(ii) *The minimal polynomial of* $\hat{A}_1$ *divides the characteristic polynomial of* β_i.

Remark 1.23. Given any matrix $\hat{A}_1$ and any integer $p > 0$, it is always possible to find a p-copy internal model for the matrix $\hat{A}_1$. In fact, let

$$\alpha_m(\lambda) = \lambda^{n_m} + \alpha_1 \lambda^{(n_m - 1)} + \cdots + \alpha_{(n_m - 1)}\lambda + \alpha_{n_m} \tag{1.59}$$

be the minimal polynomial of $\hat{A}_1$,

$$\beta_i = \begin{bmatrix} 0 & 1 & \cdots & 0 \\ 0 & 0 & \cdots & 0 \\ \vdots & \vdots & \vdots & \vdots \\ 0 & 0 & \cdots & 1 \\ -\alpha_{n_m} & -\alpha_{(n_m-1)} & \cdots & -\alpha_1 \end{bmatrix}, \quad \sigma_i = \begin{bmatrix} 0 \\ 0 \\ \vdots \\ 0 \\ 1 \end{bmatrix}, \quad i = 1, \ldots, p. \tag{1.60}$$

Then, clearly, the pair (G_1, G_2) satisfies the conditions (i) and (ii) of Definition 1.22. Throughout this chapter, we will always assume $\hat{A}_1 = A_1$. It is clear that, under Assumptions 1.1 and 1.4, the matrix G_1 with β_i being described by (1.60) has the following property.

Property 1.5. For all $\lambda \in \sigma(G_1)$,

$$\operatorname{rank}\begin{bmatrix} A - \lambda I & B \\ C & D \end{bmatrix} = n + p. \quad \blacksquare \tag{1.61}$$

Remark 1.24. We allow the dimensions of the matrices S_1, S_2, S_3 to be zero and T be an identity matrix. Therefore, the pair (G_1, G_2) itself incorporates a p-copy internal model of the matrix A_1. In the following, we will call the pair (G_1, G_2) a minimal p-copy internal model of A_1 if the minimal polynomial of β_i, the characteristic polynomial of β_i, and the minimal polynomial of A_1 are the same for all $i = 1, \ldots, p$. $\blacksquare$

Definition 1.25. *A dynamic compensator of the form*

$$\dot{z} = \mathcal{G}_1 z + \mathcal{G}_2 e \tag{1.62}$$

is said to incorporate a p-copy internal model of the composite system (1.47) if the pair $(\mathcal{G}_1, \mathcal{G}_2)$ *incorporates a p-copy internal model of the matrix* A_1. *In particular, the dynamic compensator*

$$\dot{z} = G_1 z + G_2 e \tag{1.63}$$

is called a p-copy internal model of the composite system (1.47).

Lemma 1.26. *Under Assumptions 1.1 and 1.2, if the pair* (G_1, G_2) *incorporates a p-copy internal model of the matrix* A_1 *with* G_1 *satisfying Property 1.5, then the pair*

$$\left(\begin{bmatrix} A & 0 \\ G_2C & G_1 \end{bmatrix}, \begin{bmatrix} B \\ G_2D \end{bmatrix} \right) \tag{1.64}$$

is stabilizable.

Proof. Let

$$M(\lambda) = \begin{bmatrix} A-\lambda I & 0 & B \\ G_2C & G_1-\lambda I & G_2D \end{bmatrix}. \tag{1.65}$$

By the well-known PBH test, the pair (1.64) is stabilizable if and only if

$$\text{rank } M(\lambda) = n + n_z \quad \text{for all } \lambda \in \bar{\mathcal{C}}_+.$$

Since (A, B) is stabilizable, rank $[A - \lambda I \;\; B] = n$ for all $\lambda \in \bar{\mathcal{C}}_+$. Also, det $(G_1 - \lambda I) \neq 0$ for all $\lambda \notin \sigma(G_1)$. Thus

$$\text{rank } M(\lambda) = n + n_z \;\; \forall \lambda \notin \sigma(G_1) \text{ and } \forall \lambda \in \bar{\mathcal{C}}_+. \tag{1.66}$$

Write $M(\lambda) = M_1(\lambda)M_2(\lambda)$, where

$$M_1(\lambda) = \begin{bmatrix} I_n & 0 & 0 \\ 0 & G_2 & G_1-\lambda I \end{bmatrix}, \quad M_2(\lambda) = \begin{bmatrix} A-\lambda I & 0 & B \\ C & 0 & D \\ 0 & I_{n_z} & 0 \end{bmatrix}. \tag{1.67}$$

Since (G_1, G_2) is controllable, for all $\lambda \in \mathcal{C}$, $M_1(\lambda)$ has rank $n + n_z$. Since G_1 satisfies Property 1.5, $M_2(\lambda)$ has rank $n+n_z+p$ for all $\lambda \in \sigma(G_1)$. Hence, by Sylvester's inequality,[1]

$$\begin{aligned} n + n_z \geq \text{rank } M(\lambda) &\geq (n+n_z) + (n+n_z+p) - (n+n_z+p) \\ &= n + n_z \;\; \forall \lambda \in \sigma(G_1). \end{aligned}$$

Combining (1.66) and (1.68) gives

$$\text{rank } M(\lambda) = n + n_z \;\; \forall \lambda \in \bar{\mathcal{C}}_+. \tag{1.68}$$

Thus the pair (1.64) is stabilizable. □

Lemma 1.27. *Under Assumption 1.1, assume* $(\mathcal{G}_1, \mathcal{G}_2)$ *incorporates a p-copy internal model of* A_1. *Let*

$$A_c = \begin{bmatrix} \hat{A} & \hat{B} \\ \mathcal{G}_2\hat{C} & \mathcal{G}_1 + \mathcal{G}_2\hat{D} \end{bmatrix} \tag{1.69}$$

[1] $\text{rank}A + \text{rank}B - n \leq \text{rank}AB \leq \min\{\text{rank}A, \text{rank}B\}$ for any matrices $A \in \mathcal{R}^{m\times n}$ and $B \in \mathcal{R}^{n\times p}$.

be exponentially stable, where $\hat{A}$, $\hat{B}$, $\hat{C}$, $\hat{D}$ are any matrices with appropriate dimensions. Then, for any matrices $\hat{E}$ and $\hat{F}$ of appropriate dimensions, the following matrix equations:

$$\begin{aligned} XA_1 &= \hat{A}X + \hat{B}Z + \hat{E}, \\ ZA_1 &= \mathcal{G}_1 Z + \mathcal{G}_2(\hat{C}X + \hat{D}Z + \hat{F}) \end{aligned} \tag{1.70}$$

have a unique solution X and Z. Moreover, X and Z satisfy

$$0 = \hat{C}X + \hat{D}Z + \hat{F}. \tag{1.71}$$

Proof. Since A_c is exponentially stable, by Assumption 1.1, $\sigma(A_1) \cap \sigma(A_c) = \emptyset$. Therefore, there exist unique matrices X and Z that satisfy equation (1.70). We need to show that they also satisfy (1.71). To this end, let

$$\gamma = \hat{C}X + \hat{D}Z + \hat{F} \tag{1.72}$$

and

$$T^{-1}Z = \begin{bmatrix} \hat{\theta} \\ \bar{\theta} \end{bmatrix}, \tag{1.73}$$

where $\bar{\theta}$ has as many rows as those of G_1. Then (1.70) implies

$$\bar{\theta}A_1 - G_1\bar{\theta} = G_2\gamma. \tag{1.74}$$

Due to the block diagonal structure of G_1 and G_2, we can assume $p = 1$ without loss of generality. In this case, $G_1 = \beta_1$ and $G_2 = \sigma_1$. Since (G_1, G_2) is controllable, it can always be put into the following form:

$$G_1 = \begin{bmatrix} 0 & 1 & \cdots & 0 & 0 \\ 0 & 0 & \cdots & 0 & 0 \\ \vdots & \vdots & \vdots & \vdots & \vdots \\ 0 & 0 & \cdots & 0 & 1 \\ -a_{n_k} & -a_{(n_k-1)} & \cdots & -a_2 & -a_1 \end{bmatrix}, \quad G_2 = \begin{bmatrix} 0 \\ 0 \\ \vdots \\ 0 \\ 1 \end{bmatrix}, \tag{1.75}$$

where

$$\det(\lambda I - G_1) = \lambda^{n_k} + a_1\lambda^{(n_k-1)} + \cdots + a_{(n_k-1)}\lambda + a_{n_k}.$$

Let θ_j, $j = 1, \ldots, n_k$, denote the jth row of $\bar{\theta}$. Then expanding (1.74) gives

$$\begin{bmatrix} \theta_1 A_1 - \theta_2 \\ \theta_2 A_1 - \theta_3 \\ \vdots \\ \theta_{n_k-1}A_1 - \theta_{n_k} \\ \theta_{n_k}A_1 + a_{n_k}\theta_1 + \cdots + a_1\theta_{n_k} \end{bmatrix} = \begin{bmatrix} 0 \\ 0 \\ \vdots \\ 0 \\ \gamma \end{bmatrix}. \tag{1.76}$$

Equating the first $(n_k - 1)$ rows of (1.76) gives

$$\theta_j = \theta_1 A_1^{j-1}, \quad j = 2, \ldots, n_k. \tag{1.77}$$

Substituting (1.77) into the last row of (1.76) gives

$$\gamma = \theta_1 (A_1^{n_k} + a_1 A_1^{n_k - 1} + \cdots + a_{n_k} I). \tag{1.78}$$

Thus we have $\gamma = 0$ since the characteristic polynomial of G_1 is divisible by the minimal polynomial of A_1. As a result, X and Z must satisfy (1.71). □

Remark 1.28. Assume the compensator $\dot{z} = \mathcal{G}_1 z + \mathcal{G}_2 e$ incorporates a p-copy internal model of (1.47). Define an augmented system as follows:

$$\begin{aligned} \dot{x} &= Ax + Bu + Ev, \\ \dot{z} &= \mathcal{G}_1 z + \mathcal{G}_2 e, \\ e &= Cx + Du + Fv. \end{aligned} \tag{1.79}$$

Suppose a state feedback controller of the form $u = K_1 x + K_2 z$ stabilizes the augmented system (1.79). Then the closed-loop system matrix A_c takes the form (1.69) with $\hat{A} = A + BK_1$, $\hat{B} = BK_2$, $\hat{C} = C + DK_1$, $\hat{D} = DK_2$, $\hat{E} = E$, and $\hat{F} = F$. Since A_c is exponentially stable, by Lemma 1.27, the matrix equations (1.70) and (1.71) have a unique solution for any $\hat{E}$ and $\hat{F}$. But equations (1.70) and (1.71) can be put into the form

$$\begin{aligned} X_c A_1 &= A_c X_c + B_c \\ 0 &= C_c X_c + D_c, \end{aligned}$$

with

$$X_c = \begin{bmatrix} X \\ Z \end{bmatrix}, \quad B_c = \begin{bmatrix} E \\ \mathcal{G}_2 F \end{bmatrix}, \quad C_c = \begin{bmatrix} \hat{C} & \hat{D} \end{bmatrix}, \quad D_c = \hat{F}.$$

The solvability of the above equations means the solvability of equation (1.53) for any w in an open neighborhood of $w = 0$. By Lemma 1.20, the dynamic state feedback controller (1.48) solves the robust output regulation problem of the given system. Similarly, if an output feedback control of the form $u = Kz$ can stabilize the augmented system (1.79), then the output feedback control law (1.49) also solves the robust output regulation problem of the given system. The role of the internal model is to define the augmented system (1.79) whose stabilization solution leads to the solution of the robust output regulation problem of the original plant. ∎

Remark 1.29. Assume, instead of the error output feedback, that we consider the measurement output feedback. Then the augmented system would become

$$\begin{aligned} \dot{x} &= Ax + Bu + Ev, \\ \dot{z} &= \mathcal{G}_1 z + \mathcal{G}_2 y_m, \\ e &= Cx + Du + Fv. \end{aligned} \tag{1.80}$$

From the proof of Lemma 1.27, it is not difficult to see that, if $C_m X + D_m Z + F_m = 0$ implies $CX + DZ + F = 0$ (or, what is the same, that there exists a matrix T such that $C = TC_m$, $D = TD_m$, $F = TF_m$), then the stabilization solution of the augmented system (1.80) would still lead to the solution of the robust output regulation problem of the original plant. ∎

Combining Lemmas 1.20, 1.26, and 1.27 leads to the solvability conditions for the linear robust output regulation problem by a dynamic state feedback control as follows.

Theorem 1.30. *Under Assumptions 1.1 and 1.2, the following are equivalent:*

(i) *The transmission zeros condition (1.24) holds.*

(ii) *The linear robust output regulation problem is solvable by a dynamic state feedback controller* $(K_1, K_2, \mathcal{G}_1, \mathcal{G}_2)$*.*

(iii) *There exists an open neighborhood* W *of* $w = 0$ *such that for each* $w \in W$*, the following regulator equations:*

$$\begin{aligned} X_w A_1 &= A_w X_w + B_w U_w + E_w, \\ 0 &= C_w X_w + D_w U_w + F_w, \end{aligned} \tag{1.81}$$

have a solution (X_w, U_w)*.*

Proof. (i) $\rightarrow$ (ii). Due to Assumption 1.1 and the satisfaction of condition (1.24), there exists a pair (G_1, G_2) that is the minimal p-copy internal model of the composite system, for example, the pair described in Remark 1.23. Let $(\mathcal{G}_1, \mathcal{G}_2) = (G_1, G_2)$. Since the pair (G_1, G_2) is the minimal p-copy internal model of A_1 and G_1 satisfies Property 1.5, by Lemma 1.26, (1.64) is stabilizable. Thus, there exists (K_1, K_2) such that

$$A_c = \begin{bmatrix} A + BK_1 & BK_2 \\ G_2(C + DK_1) & G_1 + G_2 DK_2 \end{bmatrix} \tag{1.82}$$

is exponentially stable. It follows from Lemma 1.27 that there exists X_c that satisfies equations (1.70) and (1.71) with $\hat{A} = A + BK_1$, $\hat{B} = BK_2$, $\hat{C} = C + DK_1$, $\hat{D} = DK_2$, $\hat{E} = E$, and $\hat{F} = F$. By Remark 1.28, the dynamic state feedback controller (K_1, K_2, G_1, G_2) solves the linear robust output regulation problem.

(ii) $\rightarrow$ (iii). Assume that (K_1, K_2, G_1, G_2) solves the linear robust output regulation problem; then by the equivalence of (i) and (iii) of Lemma 1.20, there exists an open neighborhood W of $w = 0$ such that for each $w \in W$, equation (1.53) has a solution $X_{cw} = [X_w, Z_w]$ with $X_w \in \mathcal{R}^n$. Let $U_w = K_1 X_w + K_2 Z_w$; then, clearly, X_w and U_w satisfy (1.81).

(iii) $\rightarrow$ (i). Since (iii) holds for $w \in W$, it also holds for $w \in W_s$. This is the same as saying that the regulator equations have a solution for any (E, F). Thus, by Theorem 1.9 , (i) must hold. $\square$

When the state is not available for feedback, it is possible to construct an output feedback servoregulator on the basis of the state feedback regulator, as shown below.

Theorem 1.31. *Under Assumptions 1.1, 1.2, and 1.5, the following are equivalent:*

(i) *The transmission zeros condition (1.24) holds.*

(ii) *The linear robust output regulation problem is solvable by a dynamic output feedback controller* $(K, \mathcal{G}_1, \mathcal{G}_2)$*.*

(iii) *There exists an open neighborhood W of $w = 0$ such that for each $w \in W$, the following regulator equations:*

$$\begin{aligned} X_w A_1 &= A_w X_w + B_w U_w + E_w, \\ 0 &= C_w X_w + D_w U_w + F_w, \end{aligned} \tag{1.83}$$

have a solution (X_w, U_w).

Proof. We only need to show (i) $\rightarrow$ (ii) since the rest follows straightforwardly from the proof of Theorem 1.30. Due to Assumptions 1.1 and 1.2 and the satisfaction of condition (i), Theorem 1.30 guarantees the existence of a dynamic state feedback controller (K_1, K_2, G_1, G_2) that solves the linear robust output regulation problem. Thus (1.82) is exponentially stable. Also, by Assumption 1.5, there exists a constant matrix $L \in \mathcal{R}^{n \times p}$ such that $A - LC$ is exponentially stable. Let $K = (K_1, K_2)$, and let

$$\begin{aligned} u &= Kz, \\ \dot{z} &= \begin{bmatrix} A + BK_1 - L(C + DK_1) & (B - LD)K_2 \\ 0 & G_1 \end{bmatrix} z + \begin{bmatrix} L \\ G_2 \end{bmatrix} e \\ &\stackrel{def}{=} \mathcal{G}_1 z + \mathcal{G}_2 e. \end{aligned} \tag{1.84}$$

Then, clearly, the pair $(\mathcal{G}_1, \mathcal{G}_2)$ incorporates a p-copy internal model of the composite system. By Lemma 1.27, it suffices to show that

$$A_c = \begin{bmatrix} A & BK \\ \mathcal{G}_2 C & \mathcal{G}_1 + \mathcal{G}_2 DK \end{bmatrix} \tag{1.85}$$

is exponentially stable. Indeed, a simple calculation gives

$$A_c = \begin{bmatrix} A & BK_1 & BK_2 \\ LC & A + BK_1 - LC & BK_2 \\ G_2 C & G_2 DK_1 & G_1 + G_2 DK_2 \end{bmatrix}. \tag{1.86}$$

In (1.86), subtracting the first row from the second row and adding the second column to the first column gives

$$\begin{bmatrix} A + BK_1 & BK_1 & BK_2 \\ 0 & A - LC & 0 \\ G_2(C + DK_1) & G_2 DK_1 & G_1 + G_2 DK_2 \end{bmatrix}. \tag{1.87}$$

Thus the spectrum of (1.87) is given by those of (1.82) and $A - LC$. That is, A_c as defined by (1.86) is exponentially stable. Thus, by Lemma 1.27 and Remark 1.28, $(K, \mathcal{G}_1, \mathcal{G}_2)$ solves the linear robust output regulation problem. □

1.4 The Internal Model Principle

In the previous section, we first showed that there exists no static state feedback controller that solves the linear robust output regulation problem. Then we constructed both dynamic state feedback and output feedback controllers to solve the linear robust output regulation problem. One may wonder what the underlying idea is for suggesting the controllers of the form given by Theorem 1.30, and what the minimal order of the controller is. This section is aimed to respond to these questions. In fact, we will show that the controllers given in Theorem 1.30 are of the minimal order.

Lemma 1.32. *Under Assumption 1.1, assume that* $(K_1, K_2, \mathcal{G}_1, \mathcal{G}_2)$ *is any dynamic state feedback controller that solves the linear robust output regulation problem. Let* $\mathcal{S} : \mathcal{R}^{n_z \times q} \to \mathcal{R}^{n_z \times q}$ *be a Sylvester mapping such that*

$$\mathcal{S}(Z) = ZA_1 - \mathcal{G}_1 Z. \tag{1.88}$$

Let $\mathcal{K}$ *be the kernel of* $\mathcal{S}$, *that is,*

$$\mathcal{K} = \{Z \in \mathcal{R}^{n_z \times q} \mid \mathcal{S}(Z) = 0\}. \tag{1.89}$$

Then

$$\dim(\mathcal{K}) \geq pq. \tag{1.90}$$

Proof. Assume that the dynamic state feedback control $(K_1, K_2, \mathcal{G}_1, \mathcal{G}_2)$ solves the linear robust output regulation problem. Then, by part (iii) of Lemma 1.20, (1.53) holds in an open neighborhood W of $w = 0$, and hence holds in the subset W_s of W as defined in (1.54). Now partition X_{cw} as follows:

$$X_{cw} = \begin{bmatrix} X_w \\ Z_w \end{bmatrix}.$$

Then we can expand (1.53) for $w \in W_s$ as follows:

$$\begin{aligned} X_w A_1 &= (A + BK_1)X_w + BK_2 Z_w + E_w, \\ Z_w A_1 &= \mathcal{G}_1 Z_w, \\ 0 &= (C + DK_1)X_w + DK_2 Z_w + F_w. \end{aligned} \tag{1.91}$$

Equations (1.91) can be viewed as a linear mapping $\mathcal{F} : \mathcal{R}^{n \times q} \times \mathcal{K} \to \mathcal{R}^{(n+p) \times q}$ such that

$$\mathcal{F}\left(\begin{bmatrix} X_w \\ Z_w \end{bmatrix}\right) \overset{def}{=} \begin{bmatrix} X_w A_1 - (A + BK_1)X_w - BK_2 Z_w \\ -(C + DK_1)X_w - DK_2 Z_w \end{bmatrix} = \begin{bmatrix} E_w \\ F_w \end{bmatrix}. \tag{1.92}$$

Clearly (1.92) has a solution X_w and Z_w for any E_w and F_w only if

$$\dim(\mathcal{R}^{n \times q}) + \dim(\mathcal{K}) \geq \dim(\mathcal{R}^{(n+p) \times q}). \tag{1.93}$$

That is,

$$nq + \dim(\mathcal{K}) \geq (n + p)q.$$

Thus, necessarily,

$$\dim(\mathcal{K}) \geq pq. \quad \square$$

Theorem 1.33 (Internal Model Principle). *Under Assumption 1.1, assume that a dynamic state feedback control* $(K_1, K_2, \mathcal{G}_1, \mathcal{G}_2)$ *solves the linear robust output regulation problem, and the pair* $(\mathcal{G}_1, \mathcal{G}_2)$ *is controllable. Then* $\mathcal{G}_1$ *must have exactly* p *invariant factors, each of which is divisible by the minimal polynomial of* A_1.

Proof. Let $\{\delta_i, i = 1, \ldots, n_1\}$ and $\{\epsilon_j, j = 1, \ldots, n_2\}$ be the lists of invariant factors of $\mathcal{G}_1$ and A_1, respectively, such that

$$\begin{aligned} &\delta_{i+1} \mid \delta_i, \quad i = 1, \ldots, (n_1 - 1), \\ &\epsilon_{j+1} \mid \epsilon_j, \quad j = 1, \ldots, (n_2 - 1), \end{aligned}$$

where $\delta_{i+1} \mid \delta_i$ means δ_{i+1} divides δ_i. Let γ_{ij}, $i = 1, \dots, n_1$, $j = 1, \dots, n_2$, be the greatest common divisor of δ_i and ϵ_j. By the result on the kernel of the Sylvester map (Appendix A),

$$\dim(\mathcal{K}) = \sum_{i=1}^{n_1} \sum_{j=1}^{n_2} \deg(\gamma_{ij}). \tag{1.94}$$

Thus, using Lemma 1.32 gives

$$pq \leq \sum_{i=1}^{n_1} \sum_{j=1}^{n_2} \deg\,(\gamma_{ij}). \tag{1.95}$$

Since $\deg(\gamma_{ij}) \leq \deg(\epsilon_j)$, we have

$$\sum_{i=1}^{n_1} \sum_{j=1}^{n_2} \deg(\gamma_{ij}) \leq \sum_{i=1}^{n_1} \sum_{j=1}^{n_2} \deg(\epsilon_j) = n_1 q. \tag{1.96}$$

Combining (1.95) and (1.96) gives $p \leq n_1$. On the other hand, controllability of $(\mathcal{G}_1, \mathcal{G}_2)$ implies $n_1 \leq p$. Thus we have $n_1 = p$; that is, the matrix $\mathcal{G}_1$ has exactly p invariant factors. As a result, we can write (1.96) as

$$\sum_{i=1}^{p} \sum_{j=1}^{n_2} \deg(\gamma_{ij}) \leq pq. \tag{1.97}$$

Combining (1.95) and (1.97) gives

$$\sum_{i=1}^{p} \sum_{j=1}^{n_2} \deg(\gamma_{ij}) = pq. \tag{1.98}$$

Since

$$\sum_{j=1}^{n_2} \deg(\gamma_{ij}) \leq \sum_{j=1}^{n_2} \deg\,(\epsilon_j) = q \quad \forall i = 1, \dots, p, \tag{1.99}$$

equation (1.98) is possible only if

$$\deg(\gamma_{i1}) = \deg(\epsilon_1) \;\; \forall i = 1, \dots, p. \tag{1.100}$$

Since ϵ_1 is the minimal polynomial of A_1, equation (1.100) means that the minimal polynomial of A_1 divides each of p invariant factors of the matrix $\mathcal{G}_1$. □

Remark 1.34. Since $\mathcal{G}_1$ must have exactly p invariant factors, each of which is divisible by the minimal polynomial of A_1, and since $(\mathcal{G}_1, \mathcal{G}_2)$ is controllable, the pair $(\mathcal{G}_1, \mathcal{G}_2)$ necessarily takes the form given by (1.58) modulo coordinate transformations. Moreover, by Theorem 1.33, the minimal dimension of the matrix $\mathcal{G}_1$ is greater than or equal to pn_k, where n_k is the degree of the minimal polynomial of A_1. On the other hand, Theorem 1.30 has given a pair $(\mathcal{G}_1, \mathcal{G}_2)$ that defines a pn_k dimensional compensator. Thus, it is concluded that the minimal order dynamic state feedback control law is equal to pn_k, which is the degree of the minimal polynomial of A_1 multiplied by the dimension of the output e. ■

1.5 Output Regulation for Discrete-Time Linear Systems

The discrete-time counterpart of system (1.7) is described by

$$\begin{aligned} x(t+1) &= Ax(t) + Bu(t) + Ev(t), \quad x(0) = x_0, \quad t = 0, 1, \ldots, \\ e(t) &= Cx(t) + Du(t) + Fv(t), \end{aligned} \tag{1.101}$$

where $x(t)$ is the n-dimensional plant state, $u(t)$ the m-dimensional plant input, $e(t)$ the p-dimensional plant output representing the tracking error, and $v(t)$ the q-dimensional exogenous signal representing the reference inputs and/or disturbance, and is generated by an exosystem of the form

$$v(t+1) = A_1 v(t), \quad v(0) = v_0, \quad t = 0, 1, \ldots. \tag{1.102}$$

For convenience of reference, we can put the plant (1.101) and the exosystem (1.102) together as follows:

$$\begin{aligned} x(t+1) &= Ax(t) + Bu(t) + Ev(t), \\ v(t+1) &= A_1 v(t), \\ e(t+1) &= Cx(t) + Du(t) + Fv(t), \end{aligned} \tag{1.103}$$

and call (1.103) the composite system.

In this section, we will formulate the output regulation problem for discrete-time linear systems of the form (1.103) and present the solvability conditions for the problem. For this purpose, let us first describe two classes of feedback control laws as follows.

5. Static State Feedback:

$$u(t) = K_x x(t) + K_v v(t), \tag{1.104}$$

where $K_x \in \mathcal{R}^{m \times n}$ and $K_v \in \mathcal{R}^{m \times q}$ are constant matrices.

6. Dynamic Measurement Output Feedback:

$$\begin{aligned} u(t) &= Kz(t), \\ z(t+1) &= \mathcal{G}_1 z(t) + \mathcal{G}_2 y_m(t), \end{aligned} \tag{1.105}$$

where $z \in \mathcal{R}^{n_z}$ with n_z to be specified later, $(K, \mathcal{G}_1, \mathcal{G}_2)$ are constant matrices with appropriate dimensions, and $y_m(t) \in \mathcal{R}^{p_m}$ for some positive integer p_m is the measurable output. It is assumed that

$$y_m(t) = C_m x(t) + D_m u(t) + F_m v(t),$$

where $C_m \in \mathcal{R}^{p_m \times n}$, $D_m \in \mathcal{R}^{p_m \times m}$, and $F_m \in \mathcal{R}^{p_m \times q}$ are constant matrices.

Clearly, controllers (1.104) and (1.105) are discrete counterparts of (1.10) and (1.11), respectively.

Denote the closed-loop system consisting of the plant (1.101), exosystem (1.102), and control law (1.104) or (1.105) as follows:

$$
\begin{aligned}
x_c(t+1) &= A_c x_c(t) + B_c v(t), \\
v(t+1) &= A_1 v(t), \\
e(t) &= C_c x_c(t) + D_c v(t),
\end{aligned}
\tag{1.106}
$$

where the four matrices A_c, B_c, C_c, and D_c corresponding to various control laws are defined by exactly the same equations given in (1.14) and (1.15).

We can define the output regulation problem for discrete-time linear systems as follows.

Discrete-Time Linear Output Regulation Problem (DLORP): Design a control law of the form (1.104) or (1.105) such that the closed-loop system (1.106) satisfies the following two properties.

Property 1.6. The matrix A_c is Schur; that is, all the eigenvalues of A_c have modulus smaller than 1, and

Property 1.7. For all $x_c(0)$ and $v(0)$, the trajectories of (1.106) satisfy

$$\lim_{t\to\infty} e(t) = \lim_{t\to\infty} (C_c x_c(t) + D_c v(t)) = 0.$$

At the outset, we list the various assumptions needed for solving the above two problems.

Assumption 1.6. A_1 has no eigenvalues with modulus smaller than 1.

Assumption 1.7. The pair (A, B) is stabilizable.

Assumption 1.8. The pair

$$\left([C_m \quad F_m], \begin{bmatrix} A & E \\ 0 & A_1 \end{bmatrix} \right)$$

is detectable.

The solvability conditions for the discrete-time output regulation problem can be obtained in the same way as those for the continuous-time output regulation problem, and are thereby stated below without proof.

Theorem 1.35.

(i) *Under Assumptions 1.6 and 1.7, the discrete-time linear output regulation problem is solvable by a static state feedback controller of the form*

$$u = K_x x + K_v v \tag{1.107}$$

if and only if there exist two matrices X and U that satisfy the following linear matrix equations:

$$\begin{aligned} XA_1 &= AX + BU + E, \\ 0 &= CX + DU + F. \end{aligned} \tag{1.108}$$

(ii) *Under Assumptions 1.6, 1.7, and 1.8, the discrete-time linear output regulation problem is solvable by a measurement output feedback controller of the form (1.105) with*

$$\begin{aligned} K &= [K_x \quad K_v], \\ \mathcal{G}_1 &= \begin{bmatrix} A & E \\ 0 & A_1 \end{bmatrix} + \begin{bmatrix} B \\ 0 \end{bmatrix} K - L([C_m \ F_m] + D_m K), \quad \mathcal{G}_2 = L, \end{aligned} \tag{1.109}$$

if and only if there exist two matrices X and U that satisfy (1.108).

Remark 1.36. Equations (1.108) take exactly the same form as the regulator equations (1.21) for continuous-time linear systems, and they also play the same role in studying the discrete-time output regulation problem as equations (1.21) do in studying the continuous-time output regulation problem. Thus we will call (1.108) *discrete-time regulator equations.* Clearly, under Assumption 1.4, the discrete-time regulator equations are also solvable. In (1.107), the feedback gain K_x is such that $(A + BK_x)$ is Schur, and the feedforward gain K_v is given by $K_v = U - K_x X$. In (1.109), L is such that the matrix

$$\begin{bmatrix} A & E \\ 0 & A_1 \end{bmatrix} - L\,[C_m \quad F_m]$$

is Schur. ∎

1.6 Robust Output Regulation for Discrete-Time Linear Systems

The discrete-time counterpart of the uncertain linear system (1.45) is described by

$$\begin{aligned} x(t+1) &= (A + \Delta A)x(t) + (B + \Delta B)u(t) + (E + \Delta E)v(t), \quad t = 0, 1, \dots, \\ e(t) &= (C + \Delta C)x(t) + (D + \Delta D)u(t) + (F + \Delta F)v(t), \quad x(0) = x_0, \end{aligned} \tag{1.110}$$

where $x(t)$, $u(t)$, $e(t)$ are described as in equation (1.101) and $v(t)$ is also generated by the same exosystem (1.102).

As in (1.45), the matrices A, B, E, C, D, and F in (1.110) represent the nominal part of the plant, while ΔA, ΔB, and so forth the uncertain part of the plant. The entries of (ΔA, ΔB, ΔE, ΔC, ΔD, ΔF) are allowed to take arbitrary values. Let $w = \text{vec}\left(\begin{bmatrix} \Delta A & \Delta B & \Delta E \\ \Delta C & \Delta D & \Delta F \end{bmatrix}\right)$. Then $w \in \mathcal{R}^{n_w}$ with $n_w = (n+p) \times (n+m+q)$. We will also use the following convenient notation:

$$\begin{aligned} A_w &= A + \Delta A, \quad B_w = B + \Delta B, \quad E_w = E + \Delta E, \\ C_w &= C + \Delta C, \quad D_w = D + \Delta D, \quad F_w = F + \Delta F, \end{aligned}$$

with

$$\begin{aligned} A_0 &= A, \quad B_0 = B, \quad E_0 = E, \\ C_0 &= C, \quad D_0 = D, \quad F_0 = F. \end{aligned}$$

As a result, (1.110) can be written as follows:

$$\begin{aligned} x(t+1) &= A_w x(t) + B_w u(t) + E_w v(t), \\ e(t) &= C_w x(t) + D_w u(t) + F_w v(t). \end{aligned} \tag{1.111}$$

For convenience of reference, we can put the plant (1.111) and the exosystem (1.102) together as follows:

$$\begin{aligned} x(t+1) &= A_w x(t) + B_w u(t) + E_w v(t), \\ v(t+1) &= A_1 v(t), \\ e(t+1) &= C_w x(t) + D_w u(t) + F_w v(t), \end{aligned} \tag{1.112}$$

and call (1.112) the composite system.

As in the continuous-time case, we consider two classes of feedback control laws as follows.

7. Dynamic State Feedback:

$$\begin{aligned} u(t) &= K_1 x(t) + K_2 z(t), \\ z(t+1) &= \mathcal{G}_1 z(t) + \mathcal{G}_2 e(t), \end{aligned} \tag{1.113}$$

where $z \in \mathcal{R}^{n_z}$ with n_z to be specified later, and $(K_1, K_2, \mathcal{G}_1, \mathcal{G}_2)$ are constant matrices of appropriate dimensions.

8. Dynamic Output Feedback:

$$\begin{aligned} u(t) &= K z(t), \\ z(t+1) &= \mathcal{G}_1 z(t) + \mathcal{G}_2 e(t), \end{aligned} \tag{1.114}$$

where, again, $z \in \mathcal{R}^{n_z}$ with n_z to be specified later, and $(K, \mathcal{G}_1, \mathcal{G}_2)$ are constant matrices with appropriate dimensions.

Denote the closed-loop system consisting of the plant (1.111), exosystem (1.102), and control law (1.113) or (1.114) as follows:

$$\begin{aligned} x_c(t+1) &= A_{cw} x_c(t) + B_{cw} v(t), \\ v(t+1) &= A_1 v(t), \\ e(t) &= C_{cw} x_c(t) + D_{cw} v(t), \end{aligned} \tag{1.115}$$

where the four matrices A_{cw}, B_{cw}, C_{cw}, and D_{cw} corresponding to various control laws are defined by exactly the same equations given in (1.51) and (1.52), respectively. Also, we use $(A_{c0}, B_{c0}, C_{c0}, D_{c0})$, or simply (A_c, B_c, C_c, D_c), to denote the closed-loop system composed of the nominal plant and the control laws.

We can define the robust output regulation problem for discrete-time linear systems as follows.

Discrete-Time Linear Robust Output Regulation Problem (DLRORP): Design a control law of the form (1.113) or (1.114) such that the closed-loop system (1.115) satisfies the following two properties.

Property 1.8. The matrix A_{c0} is Schur.

Property 1.9. There exists an open neighborhood W of $w = 0$ such that, for all x_{c0} and v_0 and for all $w \in W$, the trajectories of (1.115) satisfy

$$\lim_{t\to\infty} e(t) = \lim_{t\to\infty} (C_{cw}x_c(t) + D_{cw}v(t)) = 0.$$

In addition to Assumptions 1.6 to 1.8, we need one more assumption as follows.

Assumption 1.9. The pair (C, A) is detectable.

To study the solvability conditions for the robust output regulation problem for discrete-time linear systems (1.112), we first note that the concept of the internal model as defined in Definition 1.22 also applies to the discrete-time linear systems (1.112) with the pair of matrices (G_1, G_2) given by (1.58) and (1.60). Moreover, under Assumptions 1.4 and 1.6, the matrix G_1 with β_i being described by (1.60) has Property 1.5. Thus we can readily obtain the following discrete-time counterparts of Lemmas 1.26 and 1.27 and Theorems 1.30 and 1.31.

Lemma 1.37. *Under Assumptions 1.6 and 1.7, if the pair (G_1, G_2) incorporates a p-copy internal model of the matrix A_1, and G_1 satisfies Property 1.5, then the pair*

$$\left(\begin{bmatrix} A & 0 \\ G_2C & G_1 \end{bmatrix}, \begin{bmatrix} B \\ G_2D \end{bmatrix} \right) \tag{1.116}$$

is stabilizable.

Lemma 1.38. *Under Assumption 1.6, assume $(\mathcal{G}_1, \mathcal{G}_2)$ incorporates a p-copy internal model of A_1. Let*

$$A_c = \begin{bmatrix} \hat{A} & \hat{B} \\ \mathcal{G}_2\hat{C} & \mathcal{G}_1 + \mathcal{G}_2\hat{D} \end{bmatrix} \tag{1.117}$$

be Schur, where $\hat{A}$, $\hat{B}$, $\hat{C}$, $\hat{D}$ are any matrices of appropriate dimensions. Then, for any matrices $\hat{E}$ and $\hat{F}$ of appropriate dimensions, the following matrix equations:

$$\begin{aligned} XA_1 &= \hat{A}X + \hat{B}Z + \hat{E}, \\ ZA_1 &= \mathcal{G}_1 Z + \mathcal{G}_2(\hat{C}X + \hat{D}Z + \hat{F}), \end{aligned} \tag{1.118}$$

have a unique solution X and Z. Moreover, X and Z satisfy

$$0 = \hat{C}X + \hat{D}Z + \hat{F}. \tag{1.119}$$

Theorem 1.39. *Under Assumptions 1.6 and 1.7, the following are equivalent:*

(i) *The transmission zeros condition (1.24) holds.*

(ii) *The discrete-time linear robust output regulation problem is solvable by a dynamic state feedback controller* $(K_1, K_2, \mathcal{G}_1, \mathcal{G}_2)$.

(iii) *There exists an open neighborhood* W *of* $w = 0$ *such that for each* $w \in W$, *the following regulator equations:*

$$\begin{aligned} X_w A_1 &= A_w X_w + B_w U_w + E_w, \\ 0 &= C_w X_w + D_w U_w + F_w, \end{aligned} \tag{1.120}$$

have a solution (X_w, U_w).

Theorem 1.40. *Under Assumptions 1.6, 1.7, and 1.9, the following are equivalent:*

(i) *The transmission zeros condition (1.24) holds.*

(ii) *The discrete-time linear robust output regulation problem is solvable by a dynamic output feedback controller* $(K, \mathcal{G}_1, \mathcal{G}_2)$.

(iii) *There exists an open neighborhood* W *of* $w = 0$ *such that for each* $w \in W$, *the regulator equations (1.120) have a solution* (X_w, U_w).

Remark 1.41. Both the dynamic state and the dynamic output feedback controllers for the discrete-time linear systems can be constructed in the same way as those for continuous-time linear systems. In particular, under Assumptions 1.6 and 1.7, and the transmission zeros condition (1.24), there exists a pair of matrices (G_1, G_2) that incorporates a p-copy internal model of A_1 with G_1 satisfying Property 1.5. By Lemma 1.37, the pair

$$\left(\begin{bmatrix} A & 0 \\ G_2 C & G_1 \end{bmatrix}, \begin{bmatrix} B \\ G_2 D \end{bmatrix} \right)$$

is stabilizable. Thus there exist feedback gains K_1 and K_2 such that the matrix

$$\begin{bmatrix} A + BK_1 & BK_2 \\ G_2(C + DK_1) & G_1 + G_2 D K_2 \end{bmatrix}$$

is Schur. Therefore, the dynamic state feedback control law of the form (1.113) solves the discrete-time robust output regulation problem. Under the additional Assumption 1.9, there exists an L such that $A - LC$ is Schur. Let (K_1, K_2, G_1, G_2) be the dynamic state feedback control law that solves the discrete-time robust output regulation problem. Let $K = (K_1, K_2)$,

$$\mathcal{G}_1 = \begin{bmatrix} A + BK_1 - L(C + DK_1) & (B - LD)K_2 \\ 0 & G_1 \end{bmatrix}, \quad \mathcal{G}_2 = \begin{bmatrix} L \\ G_2 \end{bmatrix}.$$

Then, by exactly the same argument as in the continuous-time case, the dynamic output feedback control law of the form (1.114) solves the discrete-time robust output regulation problem. ∎

Chapter 2

Introduction to Nonlinear Systems

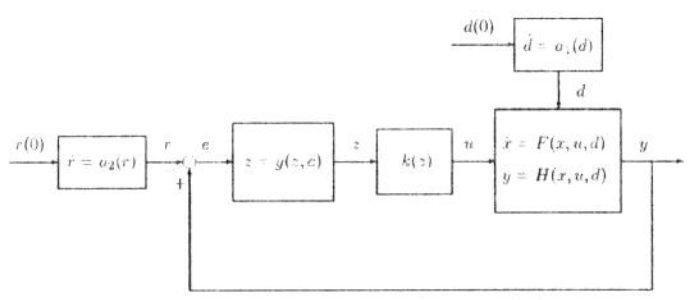

In this chapter, we review some fundamental concepts and results on nonlinear control systems that will be referred to in subsequent chapters. In Section 2.1, we present the descriptions of various nonlinear systems. In Section 2.2, we summarize the Lyapunov stability results for both autonomous and nonautonomous nonlinear systems. Section 2.3 introduces the input-to-state stability of a nonlinear control system. Section 2.4 reviews the center manifold theory. Section 2.5 reviews the discrete-time nonlinear systems and summarizes the center manifold theory for maps. In Sections 2.6 and 2.7, we study the normal form and zero dynamics for single-input, single-output and multi-input, multi-output nonlinear systems, respectively. Finally, in Section 2.8, we close this chapter by introducing some typical nonlinear systems.

The materials presented in this chapter are well known and can be found in many textbooks on nonlinear systems. Thus proofs of almost all results are omitted. For an in-depth treatment of the nonlinear system theory, the reader is referred to books by Carr [7], Khalil [74], Isidori [63], [64], and Nijmeijer and Van der Schaft [88].

2.1 Nonlinear Systems

A general nonlinear dynamic system is described by

$$\dot{x}(t) = f(x(t), t), \quad x(t_0) = x_0, \tag{2.1}$$

where $x \in \mathcal{R}^n$, $t \in [t_0, \infty)$, and $f : \mathcal{R}^n \times \mathcal{R} \to \mathcal{R}^n$. x is called the state of the system, $x_0 \in \mathcal{R}^n$ the initial state, and $t_0 \in \mathcal{R}$ the initial time. The components of x and f are denoted, respectively, by

$$x = \begin{bmatrix} x_1 \\ \vdots \\ x_n \end{bmatrix}, \quad f = \begin{bmatrix} f_1 \\ \vdots \\ f_n \end{bmatrix}.$$

If the function $f(x, t)$ does not explicitly depend on the time t, then (2.1) can be simplified as follows:

$$\dot{x}(t) = f(x(t)), \quad x(t_0) = x_0. \tag{2.2}$$

A dynamic system of the form (2.1) is called a *nonautonomous system*, while (2.2) is called an *autonomous system*.

A general multivariable nonlinear control system is described by the following two equations:

$$\dot{x}(t) = f(x(t), u(t), t), \tag{2.3}$$
$$y(t) = h(x(t), u(t), t), \tag{2.4}$$

where $x \in \mathcal{R}^n$ is the plant state, $u \in \mathcal{R}^m$ the plant input, $y \in \mathcal{R}^p$ the plant output, and $f : \mathcal{R}^n \times \mathcal{R}^m \times \mathcal{R} \to \mathcal{R}^n$, $h : \mathcal{R}^n \times \mathcal{R}^m \times \mathcal{R} \to \mathcal{R}^p$. The components of x, u, y, f, h are denoted, respectively, by

$$x = \begin{bmatrix} x_1 \\ \vdots \\ x_n \end{bmatrix}, \quad u = \begin{bmatrix} u_1 \\ \vdots \\ u_m \end{bmatrix}, \quad y = \begin{bmatrix} y_1 \\ \vdots \\ y_p \end{bmatrix},$$

$$f = \begin{bmatrix} f_1 \\ \vdots \\ f_n \end{bmatrix}, \quad h = \begin{bmatrix} h_1 \\ \vdots \\ h_p \end{bmatrix}.$$

If neither $f(x, u, t)$ nor $h(x, u, t)$ explicitly depends on the time t, then the system (2.3) and (2.4) can be simplified as follows:

$$\dot{x}(t) = f(x(t), u(t)), \tag{2.5}$$
$$y(t) = h(x(t), u(t)). \tag{2.6}$$

We call the system (2.3) and (2.4) a *nonautonomous nonlinear control system* and the system (2.5) and (2.6) an *autonomous nonlinear control system*.

For many autonomous nonlinear control systems, the function $f(x, u)$ is linear in the input u, and the function $h(x, u)$ does not depend on the input u explicitly. In this case, we can write, with some abuse of the notation, $h(x, u) = h(x)$ and $f(x, u) = f(x) + g(x)u$ for some functions $f : \mathcal{R}^n \to \mathcal{R}^n$, $g : \mathcal{R}^n \to \mathcal{R}^{n\times m}$, and $h : \mathcal{R}^n \to \mathcal{R}^p$. Therefore, (2.5) and (2.6) can be further simplified as follows:

$$\begin{aligned} \dot{x}(t) &= f(x(t)) + g(x(t))u(t), \\ y(t) &= h(x(t)). \end{aligned} \tag{2.7}$$

We call (2.7) an *affine nonlinear control system*.

Note that $g(x)$ can be expanded as $g(x) = [g_1(x), \ldots, g_m(x)]$, where $g_i : \mathcal{R}^n \to \mathcal{R}^n$ for $i = 1, \ldots, m$.

The class of nonlinear state feedback control laws takes the following form:

$$u(t) = k(x(t), t), \tag{2.8}$$

where $k : \mathcal{R}^n \times \mathcal{R} \to \mathcal{R}^m$. The composition of the control system (2.3) and the control law (2.8) gives $\dot{x} = f(x, k(x, t), t)$, which is a nonautonomous system of the form (2.1). In particular, when neither the function $f(x, u, t)$ nor the function $k(x, t)$ depends on t explicitly, we obtain an autonomous system of the form (2.2). Other types of nonlinear control laws will be introduced in the subsequent chapters.

2.2 Stability Concepts for Nonlinear Systems

In this section, we review the stability concepts for the system described by (2.1) while viewing (2.2) as a special case of (2.1). Throughout this section, we assume that $f : \mathcal{R}^n \times [t_0, \infty) \to \mathcal{R}^n$ is piecewise continuous in t and *locally Lipschitz* in x; that is, there exists a constant L such that

$$\|f(x, t) - f(y, t)\| \leq L\|x - y\| \tag{2.9}$$

for all (x, t) and (y, t) in some open neighborhood of (x_0, t_0). Under this assumption, given x_0, there exists some $t_1 > t_0$ and a unique continuous function $x : [t_0, t_1] \to \mathcal{R}^n$ that satisfies (2.1). This time function $x(t)$ is called a (local) solution of (2.1) over the interval $[t_0, t_1]$. The solution $x(t)$ is also called the state trajectory or simply the state of (2.1).

A constant vector $x_e \in \mathcal{R}^n$ is said to be an *equilibrium point* of the system (2.1) if

$$f(x_e, t) = 0 \;\; \forall \;\, t \geq t_0. \tag{2.10}$$

If a nonzero vector x_e is an equilibrium point of (2.1), then we can always introduce a new state variable $z = x - x_e$ and define a new system $\dot{z} = f(z + x_e, t)$ which has $z = 0$ as its equilibrium point. Thus, without loss of generality, we can always assume that the origin of $\mathcal{R}^n$ is an equilibrium point of the system (2.1) in this chapter.

Definition 2.1. *The equilibrium point $x_e = 0$ of the system (2.1) is*

(i) Lyapunov stable *at t_0 if for any $R > 0$, there exists an $r(R, t_0) > 0$ such that, for all $\|x(t_0)\| < r(R, t_0)$, $\|x(t)\| < R$ for all $t \geq t_0$.*

(ii) unstable *at t_0 if it is not stable at t_0.*

(iii) asymptotically stable *at t_0 if it is stable at t_0, and there exists a $\delta(t_0) > 0$ such that $\|x(t)\| \to 0$ as $t \to \infty$ for all $\|x(t_0)\| < \delta(t_0)$.*

(iv) globally asymptotically stable *at t_0 if it is stable at t_0 and $\|x(t)\| \to 0$ as $t \to \infty$ for all $x(t_0) \in \mathcal{R}^n$.*

Definition 2.2. *The equilibrium point $x_e = 0$ of the system (2.1) is*

(i) uniformly stable *if for any $R > 0$, there exists $r(R) > 0$, independent of t_0, such that, for all $\|x(t_0)\| < r(R)$, $\|x(t)\| < R$ for all $t \geq t_0$.*

(ii) uniformly asymptotically stable *if it is uniformly stable, and there exists a* $\delta > 0$, *independent of* t_0, *such that, for all* $\|x(t_0)\| < \delta$, $\|x(t)\| \to 0$ *as* $t \to \infty$ *uniformly in* t_0, *that is, for any* $\epsilon > 0$, *there exists a* $T > 0$, *independent of* t_0, *such that, for all* $\|x(t_0)\| < \delta$, $\|x(t)\| < \epsilon$ *whenever* $t > t_0 + T$.

(iii) uniformly globally asymptotically stable *if it is uniformly stable, and for any* $\epsilon > 0$, *and any* $\delta > 0$, *there exists a* $T > 0$, *independent of* t_0, *such that, for all* $\|x(t_0)\| < \delta$, $\|x(t)\| < \epsilon$ *whenever* $t > t_0 + T$.

A typical nonlinear system whose equilibrium point is globally asymptotically stable but not uniformly asymptotically stable is given as follows.

Example 2.3.

$$\dot{x} = -\frac{x}{1+t}, \quad x \in \mathcal{R}. \tag{2.11}$$

It can be verified that, for any initial state $x(t_0)$ with any initial time t_0, the solution of (2.11) is

$$x(t) = x(t_0)\frac{1+t_0}{1+t}, \quad t \geq t_0.$$

It can be seen that the equilibrium point is uniformly stable and globally asymptotically stable. But, given $\epsilon > 0$ and $\delta > 0$, in order to make $\|x(t)\| < \epsilon$ for all $\|x(t_0)\| < \delta$, t must be greater than $T = \frac{\delta(1+t_0)}{\epsilon} - 1$. Since this T cannot be made independent of t_0, the equilibrium point is not uniformly asymptotically stable. ∎

For the autonomous system (2.2), if $x(t)$ is the solution of (2.2) satisfying the initial condition $x(t_0) = x_0$, then $\hat{x}(t) = x(t + t_0)$ is the solution of (2.2) satisfying the initial condition $\hat{x}(0) = x_0$. Thus, we can always assume $t_0 = 0$ for the autonomous system (2.2). Moreover, for the autonomous system (2.2), if the equilibrium point is stable (asymptotically stable, globally asymptotically stable) at t_0, it is also uniformly stable (uniformly asymptotically stable, uniformly globally asymptotically stable).

We now introduce the Lyapunov stability theory to determine the stability of the equilibrium point of the nonlinear systems (2.1) and (2.2), respectively. Let us first focus on the autonomous system (2.2). Assume that $f(x)$ is C^1 (continuously differentiable) in an open neighborhood of the origin of $\mathcal{R}^n$. Define the Jacobian matrix of $f(x)$ at the origin as $F = \frac{\partial f}{\partial x}(0)$. Then we have the following theorem.

Theorem 2.4. *The equilibrium point* 0 *of the system (2.2) is locally asymptotically stable if all the eigenvalues of the matrix* F *have negative real parts, and is unstable if at least one eigenvalue of the matrix* F *has positive real parts.*

Now consider the control system (2.5) and (2.6). Assume $f(x,u)$ and $h(x,u)$ are C^1 in an open neighborhood of $(x,u) = (0,0)$ satisfying $f(0,0) = 0$ and $h(0,0) = 0$. Let

$$A = \frac{\partial f}{\partial x}(0,0), \quad B = \frac{\partial f}{\partial u}(0,0), \quad C = \frac{\partial h}{\partial x}(0,0), \quad D = \frac{\partial h}{\partial u}(0,0). \tag{2.12}$$

Then the system

$$\dot{x} = Ax + Bu, \qquad y = Cx + Du \tag{2.13}$$

is a linear approximation of the system (2.5) and (2.6) and is called the *Jacobian linearization* of system (2.5) and (2.6) at $(x, u) = (0, 0)$.

Suppose the pair (A, B) is stabilizable. Then there exists an $m \times n$ constant matrix K such that all the eigenvalues of the matrix $A + BK$ have negative real parts. Applying a linear state feedback controller

$$u = Kx$$

to the system (2.5) results in an autonomous system

$$\dot{x} = f(x, Kx)$$

with $x = 0$ as an equilibrium point. Clearly, the Jacobian matrix of $f(x, Kx)$ at the origin is given by $A + BK$. Thus Theorem 2.4 concludes that a linear state feedback control is able to (locally) stabilize the control system (2.5) provided that the Jacobian linearization of the system (2.5) at $(x, u) = (0, 0)$ is stabilizable. If, in addition, the pair (C, A) is detectable, there exists a linear output feedback controller of the form

$$u = Kz, \quad \dot{z} = G_1 z + G_2 y \tag{2.14}$$

such that the equilibrium point of the closed-loop system composed of (2.5), (2.6), and (2.14) is locally asymptotically stable.

Remark 2.5. The case in which none of the eigenvalues of the matrix A has positive real parts, but at least one of them has zero real parts, is called the *critical case*. It can be shown that, in the critical case, the equilibrium point of the system (2.2) can be stable, asymptotically stable, or unstable. Thus, the Lyapunov linearization method cannot handle the critical case. But the Lyapunov direct method to be introduced below or the center manifold theory to be introduced in Section 2.4 is sometimes applicable to the critical case. ∎

Definition 2.6. *Let $V : X \to \mathcal{R}$ be a C^1 function with X an open neighborhood of the origin of $\mathcal{R}^n$. V is said to be a (local)* Lyapunov function *of (2.2) if $V(x)$ is positive definite in X, and*

$$\dot{V}(x) \stackrel{def}{=} \sum_{i=1}^{n} \frac{\partial V}{\partial x_i} f_i(x) = \frac{\partial V}{\partial x} f(x) \tag{2.15}$$

is (locally) negative semi-definite. If $X = \mathcal{R}^n$, and $\dot{V}(x)$ is negative semi-definite for all $x \in \mathcal{R}^n$, then $V(x)$ is said to be a global Lyapunov function *for (2.2).*

Theorem 2.7. *If the system (2.2) has a Lyapunov function $V(x)$, then the equilibrium point $x_e = 0$ is Lyapunov stable. If, in addition, $\dot{V}(x)$ is locally negative definite in an open neighborhood of $x_e = 0$, then the equilibrium point $x_e = 0$ is asymptotically stable.*

Theorem 2.8. *Suppose the system (2.2) has a global Lyapunov function $V(x)$, which is* radially unbounded, *that is,*

$$\lim_{\|x\|\to\infty} V(x) = \infty,$$

and further, that $\dot{V}(x)$ is globally negative definite. Then the equilibrium point $x_e = 0$ is globally asymptotically stable.

Theorem 2.9. *Consider an autonomous system of the form*

$$\dot{x}_1 = f_1(x_1, x_2), \qquad \dot{x}_2 = f_2(x_2), \tag{2.16}$$

where $x_1 \in \mathcal{R}^{n_1}$, $x_2 \in \mathcal{R}^{n_2}$, $f_1(0,0) = 0$, and $f_2(0) = 0$. Suppose the equilibrium point $x_1 = 0$ of $\dot{x}_1 = f_1(x_1, 0)$ is asymptotically stable, and the equilibrium point $x_2 = 0$ of $\dot{x}_2 = f_2(x_2)$ is Lyapunov stable. Then the equilibrium point $(x_1, x_2) = (0, 0)$ of (2.16) is Lyapunov stable.

To describe the Lyapunov stability theory for the nonautonomous system (2.1), we introduce the class $\mathcal{K}$ and class $\mathcal{K}_\infty$ functions.

Definition 2.10. *A continuous function $\alpha : [0, a) \to \mathcal{R}^+$ is said to belong to class $\mathcal{K}$ if it is strictly increasing and satisfies $\alpha(0) = 0$, and is said to belong to class $\mathcal{K}_\infty$ if, in addition, $a = \infty$ and $\alpha(r) \to \infty$ as $r \to \infty$.*

Theorem 2.11. *Let $V : \mathcal{R}^n \times \mathcal{R} \to \mathcal{R}^+$ be a C^1 function such that, for some class $\mathcal{K}$ functions $\bar{\alpha}(\cdot)$ and $\underline{\alpha}(\cdot)$, defined on $[0, d)$,*

(i) $\underline{\alpha}(\|x\|) \le V(x,t) \le \bar{\alpha}(\|x\|)$,

(ii) $\dot{V}(x,t) \stackrel{def}{=} \frac{\partial V}{\partial t} + \frac{\partial V}{\partial x} f(x,t) \le 0$ *for all* $\|x\| < d$ *and all* $t \ge t_0$.

Then the origin is uniformly stable. If (ii) *is replaced by*

(iii) $\dot{V}(x,t) \le -\alpha(\|x\|)$ *for all* $\|x\| < d$ *and all* $t \ge t_0$,
where $\alpha(\cdot)$ is some class $\mathcal{K}$ function defined on $[0, d)$,

then the origin is uniformly asymptotically stable.

If $d = \infty$ and $\bar{\alpha}(\cdot)$ and $\underline{\alpha}(\cdot)$ are class $\mathcal{K}_\infty$ functions, then the origin is uniformly globally asymptotically stable.

2.3 Input-to-State Stability

In this section, we will review the concept of the input-to-state stability for the system described by (2.3). This concept was introduced by Sontag in 1989 [100] and has rapidly become an effective tool in the analysis and design of nonlinear control systems. At the beginning, we assume that $f : \mathcal{R}^n \times \mathcal{R}^m \times [0, \infty) \to \mathcal{R}^n$ is piecewise continuous in t and locally Lipschitz in x and satisfies $f(0, 0, t) = 0$ for all $t \ge t_0 \ge 0$.

Definition 2.12. *A continuous function $\beta : [0, a) \times [0, \infty) \to \mathcal{R}^+$ is said to belong to class $\mathcal{KL}$ if, for each fixed s, the function $\beta(\cdot, s)$ is a class $\mathcal{K}$ function defined on $[0, a)$ and, for each fixed r, the function $\beta(r, \cdot) : [0, \infty) \to [0, \infty)$ is decreasing and $\beta(r, s) \to 0$ as $s \to \infty$.*

While the stability of an equilibrium point is a property of the solution of a dynamic system of the form (2.1) excited by an initial state x_0, the input-to-state stability is concerned with a relation between the trajectory of equation (2.3) and the initial state $x(t_0)$ and the input $u(t)$ of (2.3). We will use the notation L_∞^m to denote the set of all piecewise continuous bounded functions $u : [t_0, \infty) \to \mathcal{R}^m$ with the supremum norm

$$\|u(\cdot)\|_\infty = \sup_{t \geq t_0} \|u(t)\|. \tag{2.17}$$

Definition 2.13. *The system (2.3) is said to be* input-to-state stable *(ISS) if there exist a class $\mathcal{KL}$ function β and a class $\mathcal{K}$ function γ such that for any initial state $x(t_0)$ and any input function $u(t) \in L_\infty^m$, the solution $x(t)$ exists and satisfies*

$$\|x(t)\| \leq \beta(\|x(t_0)\|, t - t_0) + \gamma\left(\sup_{t_0 \leq \tau \leq t} \|u(\tau)\|\right), \ t \geq t_0. \tag{2.18}$$

For an ISS system, the solution $x(t)$ is bounded for all initial states $x(t_0)$ and all input functions $u(t) \in L_\infty^m$. In particular, when the input u is held at zero, the solution of (2.3) starting from any initial state $x(t_0)$ for any initial time t_0 satisfies

$$\|x(t)\| \leq \beta(\|x(t_0)\|, t - t_0). \tag{2.19}$$

Thus, the equilibrium point 0 of the unforced system $\dot{x} = f(x, 0, t)$ is uniformly globally asymptotically stable.

On the other hand, for any $x(t_0)$ and any t_0, $\beta(\|x(t_0)\|, t - t_0) \to 0$ as $t \to \infty$. Thus

$$\limsup_{t \to \infty} \|x(t)\| \leq \gamma(\|u(\cdot)\|_\infty). \tag{2.20}$$

That is, as t goes to ∞, the solution $x(t)$ will ultimately be bounded by a class $\mathcal{K}$ function of $\|u(\cdot)\|$. Thus, the class $\mathcal{K}$ function γ will be called a gain function of (2.3).

Remark 2.14. Since $\max\{\beta, \gamma\} \leq \beta + \gamma \leq \max\{2\beta, 2\gamma\}$ for any pair $\beta \geq 0$, $\gamma \geq 0$, an equivalent way to characterize the input-to-state stability of (2.3) is that there exist a class $\mathcal{KL}$ function β and a class $\mathcal{K}$ function γ such that for any initial state $x(t_0)$ and any input function $u(\cdot) \in L_\infty^m$, the solution $x(t)$ exists and satisfies

$$\|x(t)\| \leq \max\left\{\beta(\|x(t_0)\|, t - t_0), \gamma\left(\sup_{t_0 \leq \tau \leq t} \|u(\tau)\|\right)\right\}, \ t \geq t_0. \quad \blacksquare \tag{2.21}$$

Definition 2.15. *Let $V : \mathcal{R}^n \times \mathcal{R} \to \mathcal{R}^+$ be a C^1 function. It is called an* ISS-Lyapunov function *for system (2.3) if there exist class $\mathcal{K}_\infty$ functions $\bar{\alpha}(\cdot)$, $\underline{\alpha}(\cdot)$, and $\alpha(\cdot)$ and a class $\mathcal{K}$ function $\chi(\cdot)$ such that*

$$\text{(i)} \qquad \underline{\alpha}(\|x\|) \leq V(x, t) \leq \bar{\alpha}(\|x\|),$$

$$\text{(ii)} \qquad \frac{\partial V}{\partial t} + \frac{\partial V}{\partial x} f(x, u, t) \leq -\alpha(\|x\|)$$

for all $\|x\| \geq \chi(\|u\|)$, $x \in \mathcal{R}^n$, $u \in L_\infty^m$, and $t \geq t_0$.

Theorem 2.16. *If the system (2.3) has an ISS-Lyapunov function, then it is ISS with a gain function* $\underline{\alpha}^{-1} \circ \bar{\alpha} \circ \chi$; *that is, there exist a class* $\mathcal{KL}$ *function* β *and a class* $\mathcal{K}$ *function* $\gamma = \underline{\alpha}^{-1} \circ \bar{\alpha} \circ \chi$ *such that for any initial state* $x(t_0)$ *and any input function* $u(\cdot) \in L_\infty^m$, *the solution* $x(t)$ *of (2.3) exists and satisfies (2.21).*

Now assume $V : \mathcal{R}^n \times \mathcal{R} \to \mathcal{R}^+$ is a C^1 function and satisfies

$$\frac{\partial V}{\partial t} + \frac{\partial V}{\partial x} f(x,u,t) \leq -\alpha(\|x\|) + \sigma(\|u\|) \tag{2.22}$$

for all $x \in \mathcal{R}^n$, $u \in L_\infty^m$, and $t \geq t_0$, where $\alpha(\cdot)$ is some class $\mathcal{K}_\infty$ function and $\sigma(\cdot)$ is some class $\mathcal{K}$ function. Let

$$\chi(r) = \alpha^{-1}\left(\frac{\sigma(r)}{\epsilon}\right) \tag{2.23}$$

with $0 < \epsilon < 1$. Noting the fact that

$$\|x\| \geq \chi(\|u\|) \quad \Rightarrow \quad \sigma(\|u\|) \leq \epsilon\alpha(\|x\|)$$

and using (2.22) gives

$$\frac{\partial V}{\partial t} + \frac{\partial V}{\partial x} f(x,u,t) \leq -(1-\epsilon)\alpha(\|x\|) \tag{2.24}$$

for all $\|x\| \geq \chi(\|u\|)$, $x \in \mathcal{R}^n$, $u \in L_\infty^m$, and $t \geq t_0$. Thus, $V(x)$ is an ISS-Lyapunov function of (2.3). As a result, we obtain the following theorem.

Theorem 2.17. *Let* $V : \mathcal{R}^n \times \mathcal{R} \to \mathcal{R}^+$ *be a* C^1 *function satisfying, for some class* $\mathcal{K}_\infty$ *functions* $\bar{\alpha}(\cdot)$, $\underline{\alpha}(\cdot)$, *and* $\alpha(\cdot)$ *and a class* $\mathcal{K}$ *function* $\sigma(\cdot)$,

$$\text{(i)} \qquad \underline{\alpha}(\|x\|) \leq V(x,t) \leq \bar{\alpha}(\|x\|),$$

$$\text{(ii)} \qquad \frac{\partial V}{\partial t} + \frac{\partial V}{\partial x} f(x,u,t) \leq -\alpha(\|x\|) + \sigma(\|u\|)$$

for all $x \in \mathcal{R}^n$, $u \in L_\infty^m$, *and* $t \geq t_0$. *Then the system (2.3) is ISS with a gain function* $\underline{\alpha}^{-1} \circ \bar{\alpha} \circ \alpha^{-1} \circ \left(\frac{\sigma}{\epsilon}\right)$ *for any* $0 < \epsilon < 1$.

Theorem 2.18 (Small Gain Theorem). *Consider the following system:*

$$\dot{x}_1 = f_1(x_1, x_2, u, t), \quad t \geq t_0 \geq 0, \tag{2.25}$$

$$\dot{x}_2 = f_2(x_1, x_2, u, t), \quad t \geq t_0 \geq 0, \tag{2.26}$$

where, for $i = 1, 2$ *and* $x_i \in \mathcal{R}^{n_i}$, f_i *are locally Lipschitz in* $\mathrm{col}(x_1, x_2, u)$ *and piecewise continuous in* t, $u \in \mathcal{R}^m$, *and, for all* $t \geq t_0 \geq 0$, $f_1(0,0,0,t) = 0$ *and* $f_2(0,0,0,t) = 0$.

Assume that the subsystem (2.25) is ISS viewing x_1 *as state and* $\mathrm{col}(x_2, u)$ *as input, and the subsystem (2.26) is ISS viewing* x_2 *as state and* $\mathrm{col}(x_1, u)$ *as input; that is, there exist class* $\mathcal{KL}$ *functions* $\beta_1(\cdot,\cdot)$, $\beta_2(\cdot,\cdot)$ *and class* $\mathcal{K}$ *functions* $\gamma_1^x(\cdot)$, $\gamma_2^x(\cdot)$, $\gamma_1^u(\cdot)$, $\gamma_2^u(\cdot)$

such that for any initial state $x_1(t_0)$, *and any input functions* $x_2(\cdot) \in L_\infty^{n_2}$, $u(\cdot) \in L_\infty^m$, *the solution of (2.25) exists and satisfies, for all* $t \geq t_0 \geq 0$,

$$\|x_1(t)\| \leq \max\left\{\beta_1(\|x_1(t_0)\|, t - t_0), \gamma_1^x\left(\sup_{t_0 \leq \tau \leq t} \|x_2(\tau)\|\right), \gamma_1^u\left(\sup_{t_0 \leq \tau \leq t} \|u(\tau)\|\right)\right\}, \quad (2.27)$$

and for any initial state $x_2(t_0)$, *and any input functions* $x_1(\cdot) \in L_\infty^{n_1}$, $u(\cdot) \in L_\infty^m$, *the solution of (2.26) exists and satisfies, for all* $t \geq t_0 \geq 0$,

$$\|x_2(t)\| \leq \max\left\{\beta_2(\|x_2(t_0)\|, t - t_0), \gamma_2^x\left(\sup_{t_0 \leq \tau \leq t} \|x_1(\tau)\|\right), \gamma_2^u\left(\sup_{t_0 \leq \tau \leq t} \|u(\tau)\|\right)\right\}. \quad (2.28)$$

Further assume that

$$\gamma_1^x(\gamma_2^x(r)) < r, \quad \forall r > 0. \quad (2.29)$$

Then the system (2.25) and (2.26) is ISS viewing $\mathrm{col}(x_1, x_2)$ *as state and* u *as input; that is, there exist class* $\mathcal{KL}$ *functions* $\beta(\cdot, \cdot)$ *and class* $\mathcal{K}$ *functions* $\gamma(\cdot)$ *such that for any initial state* $x(t_0)$ *and any input function* $u(\cdot) \in L_\infty^m$, *the solution of (2.25) and (2.26) exists and satisfies, for all* $t \geq t_0 \geq 0$,

$$\|x(t)\| \leq \max\left\{\beta(\|x(t_0)\|, t - t_0), \gamma\left(\sup_{t_0 \leq \tau \leq t} \|u(\tau)\|\right)\right\} \quad (2.30)$$

with the gain function given by any class $\mathcal{K}$ *function satisfying*

$$\gamma(r) \geq \max\left\{2\gamma_1^u(r), 2\gamma_1^x \circ \gamma_2^u(r), 2\gamma_2^u(r), 2\gamma_2^x \circ \gamma_1^u(r)\right\}, \quad \forall r > 0. \quad (2.31)$$

Two special cases of Theorem 2.18 are worth mentioning, namely, the case where f_1 does not depend on x_2 explicitly and the case where f_1 does not depend on u explicitly. Specializing Theorem 2.18 to these two cases gives the following corollary.

Corollary 2.19. *Consider the following system:*

$$\dot{x}_1 = f_1(x_1, x_2, t), \quad t \geq t_0 \geq 0, \quad (2.32)$$

$$\dot{x}_2 = f_2(x_1, x_2, u, t), \quad t \geq t_0 \geq 0. \quad (2.33)$$

Assume that the subsystem (2.32) is ISS viewing x_1 *as state and* x_2 *as input, and that the subsystem (2.33) is ISS viewing* x_2 *as state and* $\mathrm{col}(x_1, u)$ *as input; that is, there exist class* $\mathcal{KL}$ *functions* $\beta_1(\cdot, \cdot)$, $\beta_2(\cdot, \cdot)$ *and class* $\mathcal{K}$ *functions* $\gamma_1^x(\cdot)$, $\gamma_2^x(\cdot)$, $\gamma_2^u(\cdot)$, *such that, for all* $t \geq t_0 \geq 0$,

$$\|x_1(t)\| \leq \max\left\{\beta_1(\|x_1(t_0)\|, t - t_0), \gamma_1^x\left(\sup_{t_0 \leq \tau \leq t} \|x_2(\tau)\|\right)\right\}, \quad \forall x_2(\cdot) \in L_\infty^{n_2}, \quad (2.34)$$

$$\|x_2(t)\| \leq \max\left\{\beta_2(\|x_2(t_0)\|, t - t_0), \gamma_2^x\left(\sup_{t_0 \leq \tau \leq t} \|x_1(\tau)\|\right), \gamma_2^u\left(\sup_{t_0 \leq \tau \leq t} \|u(\tau)\|\right)\right\},$$
$$\forall x_1(\cdot) \in L_\infty^{n_1} \text{ and } \forall u(\cdot) \in L_\infty^m. \quad (2.35)$$

Further assume that (2.29) holds. Then the system (2.32) and (2.33) is ISS viewing $\mathrm{col}(x_1, x_2)$ *as state and u as input, with its gain function given by any class* $\mathcal{K}$ *function satisfying*

$$\gamma(r) \geq \max\left\{2\gamma_1^x \circ \gamma_2^u(r), 2\gamma_2^u(r)\right\}, \quad \forall r > 0. \tag{2.36}$$

Corollary 2.20. *Consider the following system:*

$$\dot{x}_1 = f_1(x_1, u, t), \quad t \geq t_0 \geq 0, \tag{2.37}$$

$$\dot{x}_2 = f_2(x_1, x_2, u, t), \quad t \geq t_0 \geq 0. \tag{2.38}$$

Assume that the subsystem (2.37) is ISS viewing x_1 as state and u as input, and the subsystem (2.38) is ISS viewing x_2 as state and $\mathrm{col}(x_1, u)$ *as input; that is, there exist class* $\mathcal{KL}$ *functions* $\beta_1(\cdot,\cdot)$, $\beta_2(\cdot,\cdot)$ *and class* $\mathcal{K}$ *functions* $\gamma_2^x(\cdot)$, $\gamma_1^u(\cdot)$, $\gamma_2^u(\cdot)$, *such that, for all* $t \geq t_0 \geq 0$,

$$\|x_1(t)\| \leq \max\left\{\beta_1(\|x_1(t_0)\|, t - t_0), \gamma_1^u\left(\sup_{t_0 \leq \tau \leq t} \|u(\tau)\|\right)\right\}, \quad \forall u(\cdot) \in L_\infty^n, \tag{2.39}$$

$$\|x_2(t)\| \leq \max\left\{\beta_2(\|x_2(t_0)\|, t - t_0), \gamma_2^x\left(\sup_{t_0 \leq \tau \leq t} \|x_1(\tau)\|\right), \gamma_2^u\left(\sup_{t_0 \leq \tau \leq t} \|u(\tau)\|\right)\right\},$$
$$\forall x_1(\cdot) \in L_\infty^{n_1} \text{ and } \forall u(\cdot) \in L_\infty^n. \tag{2.40}$$

Then the system (2.37) and (2.38) is ISS viewing $\mathrm{col}(x_1, x_2)$ *as state and u as input with its gain function given by any class* $\mathcal{K}$ *function satisfying*

$$\gamma(r) \geq \max\left\{2\gamma_1^u(r), 2\gamma_2^u(r), 2\gamma_2^x \circ \gamma_1^u(r)\right\}, \quad \forall r > 0. \tag{2.41}$$

Proof. If the inequality (2.39) holds, then the inequality (2.27) also holds for any class $\mathcal{K}$ functions $\gamma_1^x(\cdot)$. In particular, when $\gamma_1^x(r) = \min\left\{\gamma_2^{x-1}\left(\frac{r}{2}\right), \frac{\gamma_2^{x-1}(r)}{2}, r\right\}$, (2.29) holds, and (2.31) becomes (2.41). □

Remark 2.21. In Chapter 7, we will encounter systems of the following form:

$$\dot{x} = f\left(x, u, \mu(t)\right), \tag{2.42}$$

where $x \in \mathcal{R}^n$ is the state, $u \in \mathcal{R}^m$ is input, and $\mu : [t_0, \infty) \to \Sigma \subset \mathcal{R}^{n_\mu}$ is a piecewise continuous function with Σ being a prescribed compact set of $\mathcal{R}^{n_\mu}$. The function μ is used to model the system's uncertainty or disturbance. It is assumed that the function f is locally Lipschitz with respect to x and satisfies $f(0, 0, \mu) = 0$ for all $\mu \in \mathcal{R}^p$. For each given $\mu(t)$, system (2.42) can be viewed as a special form of (2.3). Thus we can still apply the input-to-state stability concept to system (2.42). Moreover, if we let Σ_μ be the class of piecewise functions from $[t_0, \infty)$ to Σ with Σ, being a prescribed compact set of $\mathcal{R}^{n_\mu}$, it is possible to define the concept of robust input-to-state stability on (2.42) as follows.

Definition 2.22. *Given Σ_μ, the system (2.42) is said to be* robust input-to-state stable *(RISS) with respect to μ if there exist a class $\mathcal{KL}$ function β and a class $\mathcal{K}$ function γ, both of which are independent of any $\mu \in \Sigma_\mu$ such that for any initial state $x(t_0)$, any input function $u(t) \in L_\infty^m$, and any piecewise continuous function $\mu \in \Sigma_\mu$, the solution $x(t)$ exists and satisfies inequality (2.18) or, equivalently, inequality (2.21).*

The ISS-Lyapunov function for (2.3) defined in Definition 2.15 can also be extended to the RISS-Lyapunov function for (2.42) if, in Definition 2.15, $f(x, u, t)$ is viewed as $f(x, u, \mu(t))$, and all the class $\mathcal{K}_\infty$ functions $\bar{\alpha}(\cdot)$, $\underline{\alpha}(\cdot)$, $\alpha(\cdot)$ and the class $\mathcal{K}$ function $\chi(\cdot)$ are assumed to be independent of $\mu \in \Sigma_\mu$. Similarly, Theorems 2.16 to 2.18 also apply to system (2.42) if input-to-state stability is replaced everywhere by robust input-to-state stability, and all class $\mathcal{KL}$ functions, all class $\mathcal{K}_\infty$ functions, and class $\mathcal{K}$ functions in these theorems are assumed to be independent of $\mu \in \Sigma_\mu$. ∎

2.4 Center Manifold Theory

The center manifold theory will play a crucial role in establishing the solvability of the nonlinear output regulation problem. In this section, we will present a few results from the center manifold theory for the autonomous system (2.2) with the assumption that $f(\cdot)$ is a locally defined sufficiently smooth function vanishing at the origin; that is, $f(\cdot)$ is a C^k function for some sufficiently large integer k defined in an open neighborhood of the origin of $\mathcal{R}^n$ and $f(0) = 0$. Readers are referred to Carr [7] for the proofs of these results.

Definition 2.23. *Let X be an open set of $\mathcal{R}^n$. A set of the form*

$$M = \{x \in X \mid H(x) = 0\}, \tag{2.43}$$

where $H : \mathcal{R}^n \to \mathcal{R}^{n_1}$ is a sufficiently smooth function and rank $\frac{\partial H}{\partial x}(x) = n_1$ *for all $x \in M$ is called an $(n - n_1)$-dimensional* hypersurface *in $\mathcal{R}^n$.*

A hypersurface is a special type of a *manifold* in $\mathcal{R}^n$. A set M as described in (2.43) is called a (locally) *invariant manifold* of (2.2) if the solution of (2.2) starting from $x_0 \in M$ remains in M for sufficiently small $t > 0$.

Remark 2.24. If the system (2.2) has an invariant manifold M which contains the origin, then by the Implicit Function Theorem [93], there exist some partition $x = \text{col}(x^1, x^2)$ with $x^1 \in \mathcal{R}^{n_1}$ and $x^2 \in \mathcal{R}^{n_2}$ with $n_2 = n - n_1$ and a locally defined sufficiently smooth function $x^1 = \sigma(x^2)$ satisfying $\sigma(0) = 0$ such that $H(\sigma(x^2), x^2) = 0$. Corresponding to the partition $x = \text{col}(x^1, x^2)$, we can decompose the system (2.2) as follows:

$$\dot{x}^1 = f^1(x^1, x^2), \qquad \dot{x}^2 = f^2(x^1, x^2). \tag{2.44}$$

Let $\text{col}(x^1(t), x^2(t))$ be a solution of (2.44) starting from an initial state $\text{col}(x^1(0), x^2(0)) \in M$. Then the fact that M is an invariant manifold for (2.2) implies that $x^1(t) = \sigma(x^2(t))$ for sufficiently small $t > 0$. Differentiating $x^1(t) = \sigma(x^2(t))$ with respect to t gives

$$\dot{x}^1(t) = f^1(\sigma(x^2(t)), x^2(t)) = \frac{\partial \sigma}{\partial x^2}\dot{x}^2(t) = \frac{\partial \sigma}{\partial x^2} f^2(\sigma(x^2(t)), x^2(t)). \tag{2.45}$$

The function $\sigma(\cdot)$ must satisfy (2.45) for all solutions of (2.44) contained in M. Thus the function $\sigma(\cdot)$ must satisfy the following partial differential equation:

$$\frac{\partial \sigma}{\partial x^2} f^2(\sigma(x^2), x^2) = f^1(\sigma(x^2), x^2). \tag{2.46}$$

In what follows, (2.46) will be called an *invariant manifold equation*. On the other hand, suppose (2.44) is a decomposition of (2.2) with $x^1 \in \mathcal{R}^{n_1}$ and $x^2 \in \mathcal{R}^{n_2}$. Let $\sigma : \mathcal{R}^{n_2} \to \mathcal{R}^{n_1}$ be any sufficiently smooth function satisfying (2.46) for all x^2 in an open neighborhood of the origin of $\mathcal{R}^{n_2}$. Then it can be easily verified that the solution $(x^1(t), x^2(t))$ of (2.44) starting from any sufficiently small initial state $(x^1(0), x^2(0))$ satisfying $x^1(0) = \sigma(x^2(0))$ will satisfy $x^1(t) = \sigma(x^2(t))$ for sufficiently small $t > 0$. Thus, the hypersurface $H(x) = x^1 - \sigma(x^2) = 0$ defines an invariant manifold for (2.2). ∎

Now consider the nonlinear system (2.2), and let $F \in \mathcal{R}^{n \times n}$ be the Jacobian matrix of $f(x)$ at the origin. Assume F has $0 < n_1 < n$ eigenvalues with nonzero real parts and $n_2 = n - n_1$ eigenvalues with zero real parts. Then there exists a nonsingular matrix T such that, in the new coordinates $\text{col}(y, z) = Tx$ where $y \in \mathcal{R}^{n_1}$ and $z \in \mathcal{R}^{n_2}$, (2.2) can be written as follows:

$$\dot{y} = f_1(y, z), \qquad \dot{z} = f_2(y, z) \tag{2.47}$$

with

$$\begin{bmatrix} \frac{\partial f_1}{\partial(y,z)}(0, 0) \\ \frac{\partial f_2}{\partial(y,z)}(0, 0) \end{bmatrix} = \begin{bmatrix} A & B \\ 0 & A_1 \end{bmatrix},$$

where all the eigenvalues of A have nonzero real parts and all the eigenvalues of A_1 have zero real parts.

Theorem 2.25 (Center Manifold Theorem). *Consider the system (2.47). There exist an open neighborhood $Z \subset \mathcal{R}^{n_2}$ of $z = 0$ and a C^{k-1} function $\mathbf{y} : Z \to \mathcal{R}^{n_1}$ with $\mathbf{y}(0) = 0$, such that, for all $z \in Z$,*

$$\frac{\partial \mathbf{y}}{\partial z} f_2(\mathbf{y}(z), z) = f_1(\mathbf{y}(z), z). \tag{2.48}$$

Let

$$M = \{(y, z) \in \mathcal{R}^{n_1} \times Z \mid y = \mathbf{y}(z)\}.$$

By Remark 2.24, M is an n_2-dimensional invariant manifold for (2.47) passing through the origin. Moreover, (2.48) implies that $\frac{\partial \mathbf{y}}{\partial z}(0)$ satisfies the following Sylvester equation:

$$\frac{\partial \mathbf{y}}{\partial z}(0) A_1 = A \frac{\partial \mathbf{y}}{\partial z}(0) + B,$$

which yields

$$\begin{bmatrix} \frac{\partial \mathbf{y}}{\partial z}(0) \\ I_{n_2} \end{bmatrix} A_1 = \begin{bmatrix} A & B \\ 0 & A_1 \end{bmatrix} \begin{bmatrix} \frac{\partial \mathbf{y}}{\partial z}(0) \\ I_{n_2} \end{bmatrix}. \tag{2.49}$$

That is, the tangent space to the manifold $y = \mathbf{y}(z)$ at the origin is the invariant subspace of the linear mapping F spanned by all generalized eigenvectors of F associated with all eigenvalues of F with zero real parts. For this reason, the manifold M is called a center manifold for (2.47) at the origin.

Theorem 2.26. *Consider the system (2.47). Let $\mathbf{y}^{(l)} : \mathcal{R}^{n_2} \to \mathcal{R}^{n_1}$ be a C^1 function with $\mathbf{y}^{(l)}(0) = 0$ and*

$$\frac{\partial \mathbf{y}^{(l)}}{\partial z} f_2(\mathbf{y}^{(l)}(z), z) = f_1(\mathbf{y}^{(l)}(z), z) + O(\|z\|^{l+1}), \tag{2.50}$$

where $O(\|z\|^l) : \mathcal{R}^{n_2} \to \mathcal{R}^{n_1}$ is a sufficiently smooth function such that

$$\lim_{z \to 0} \frac{\|O(\|z\|^l)\|}{\|z\|^l}$$

is a finite constant for some integer $l > 1$. Then

$$\mathbf{y}(z) = \mathbf{y}^{(l)}(z) + O(\|z\|^{l+1}), \tag{2.51}$$

where $\mathbf{y}(z)$ is any solution of equation (2.48) satisfying $\mathbf{y}(0) = 0$.

Theorem 2.27 (Reduction Theorem). *Consider the system (2.47). Suppose all the eigenvalues of the matrix A have negative real parts. Let $\mathbf{y}(z)$ be a solution of equation (2.48) satisfying $\mathbf{y}(0) = 0$. Then the equilibrium point of the system (2.47) at the origin is Lyapunov stable (asymptotically stable) (unstable) if and only if the equilibrium point $v = 0$ of the system*

$$\dot{v} = f_2(\mathbf{y}(v), v), \quad t \geq 0, \tag{2.52}$$

is Lyapunov stable (asymptotically stable) (unstable).

Theorem 2.28. *Consider the system (2.47). Suppose all the eigenvalues of the matrix A have negative real parts and the equilibrium point of the system (2.52) at $v = 0$ is stable. Let $\mathrm{col}(y(t),\ z(t))$ be a solution of equation (2.47) with $\mathrm{col}(y(0),\ z(0))$ sufficiently small. Then, there exist positive constants δ and λ such that, for all $t > 0$,*

$$\|y(t) - \mathbf{y}(z(t))\| \leq \delta e^{-\lambda t} \|y(0) - \mathbf{y}(z(0))\|. \tag{2.53}$$

The center manifold described in Theorem 2.28 is called a stable center manifold.

2.5 Discrete-Time Nonlinear Systems and Center Manifold Theory for Maps

A discrete-time autonomous nonlinear dynamic system is described by the following equation:

$$x(t+1) = f(x(t)), \quad x(t_0) = x_0, \tag{2.54}$$

where $x \in \mathcal{R}^n$ is called the state of the system, $f : \mathcal{R}^n \to \mathcal{R}^n$, $x_0 \in \mathcal{R}^n$ is the initial state, t_0 is an integer called the initial time, and $t = t_0, t_0 + 1, t_0 + 2, \ldots$. Without loss of generality, we assume $t_0 = 0$ throughout this book.

A constant vector $x_e \in \mathcal{R}^n$ is said to be an equilibrium point of the system (2.54) if

$$f(x_e) = x_e. \tag{2.55}$$

If a nonzero vector x_e is an equilibrium point of (2.54), then we can always introduce a new state variable $z = x - x_e$ and a new system $z(t+1) = f(z(t) + x_e) - f(x_e)$ that has $z_e = 0$ as its equilibrium point. Thus, without loss of generality, we can always assume that the origin of $\mathcal{R}^n$ is an equilibrium point of the system (2.54).

Definition 2.29. *The equilibrium point $x_e = 0$ of the system (2.54) is*

(i) Lyapunov stable *if for any $R > 0$, there exists an $r(R) > 0$ such that, for all $\|x(0)\| < r(R)$, $\|x(t)\| < R$ for all $t \geq 0$.*

(ii) unstable *if it is not stable.*

(iii) asymptotically stable *if it is stable, and there exists a $\delta > 0$ such that $\|x(t)\| \to 0$ as $t \to \infty$ for all $\|x(0)\| < \delta$.*

(iv) globally asymptotically stable *if it is stable and $\|x(t)\| \to 0$ as $t \to \infty$ for all $x(0) \in \mathcal{R}^n$.*

Theorem 2.30. *Assume that the function $f(x)$ is C^1 in an open neighborhood of the origin of $\mathcal{R}^n$ and $f(0) = 0$. Let $F \in \mathcal{R}^{n\times n}$ be the Jacobian matrix of $f(x)$ at the origin. The equilibrium point 0 of the system (2.54) is asymptotically stable if all the eigenvalues of the matrix F have modulus smaller than 1 and is unstable if at least one eigenvalue of the matrix F has modulus greater than 1.*

In the following, we will introduce four basic theorems of the center manifold theory for maps that are parallel to Theorems 2.25 to 2.28. These theorems will play the same role to discrete-time nonlinear systems as Theorems 2.25 to 2.28 do to continuous-time nonlinear systems. We will assume that the function f that defines the nonlinear system (2.54) is C^k for some integer $k \geq 2$, and vanishes at the origin. Also assume the Jacobian matrix F of $f(x)$ at the origin has $0 < n_1 < n$ eigenvalues with modulus not equal to 1 and $n_2 = n - n_1$ eigenvalues with modulus equal to 1. Then there exists a nonsingular matrix T such that, in the new coordinates $\mathrm{col}(y, z) = Tx$, where $y \in \mathcal{R}^{n_1}$ and $z \in \mathcal{R}^{n_2}$, (2.54) can be written as follows:

$$\begin{aligned} y(t+1) &= f_1(y(t), z(t)), \\ z(t+1) &= f_2(y(t), z(t)), \end{aligned} \tag{2.56}$$

with

$$\begin{bmatrix} \frac{\partial f_1}{\partial(y,z)}(0,0) \\ \frac{\partial f_2}{\partial(y,z)}(0,0) \end{bmatrix} = \begin{bmatrix} A & B \\ 0 & A_1 \end{bmatrix}$$

where all the eigenvalues of A have modulus not equal to 1, and all the eigenvalues of A_1 have modulus equal to 1.

Theorem 2.31 (Center Manifold Theorem for Maps). *Consider the system (2.56). There exist an open neighborhood $Z \subset \mathcal{R}^{n_2}$ of $z = 0$ and a C^{k-1} with $k \geq 2$ function $\mathbf{y} : Z \to \mathcal{R}^{n_1}$ with $\mathbf{y}(0) = 0$, such that, for all $z \in Z$,*

$$\mathbf{y}(f_2(\mathbf{y}(z), z)) = f_1(\mathbf{y}(z), z). \tag{2.57}$$

It can be easily verified that the function $\mathbf{y}$ has the property that the solution $\mathrm{col}(y(t), z(t))$ of (2.56) starting from any sufficiently small initial state $\mathrm{col}(y(0), z(0))$ satisfying $y(0) = \mathbf{y}(z(0))$ will satisfy $y(t) = \mathbf{y}(z(t))$ for all t as long as $z(t) \in Z$. In other words, let

$$M_d = \{(y, z) \in \mathcal{R}^{n_1} \times Z \mid y = \mathbf{y}(z)\}.$$

Then M_d is a (locally) invariant manifold of (2.56) in the sense that the solution of (2.56) starting from this manifold will remain in this manifold for all t as long as $z(t) \in Z$. Moreover, a relation similar to equation (2.49) holds. For this reason, we call M_d a center manifold at the origin of the map $\mathrm{col}(f_1, f_2) : \mathcal{R}^n \to \mathcal{R}^n$, or a center manifold of (2.56) passing through the origin.

Theorem 2.32. *Consider the system (2.56). Let $\mathbf{y}^{(l)} : \mathcal{R}^{n_2} \to \mathcal{R}^{n_1}$ be a C^1 map with $\mathbf{y}^{(l)}(0) = 0$ and*

$$\mathbf{y}^{(l)}(f_2(\mathbf{y}^{(l)}(z), z)) = f_1(\mathbf{y}^{(l)}(z), z) + O(\|z\|^{l+1}). \tag{2.58}$$

Then

$$\mathbf{y}(z) = \mathbf{y}^{(l)}(z) + O(\|z\|^{l+1}), \tag{2.59}$$

where $\mathbf{y}(z)$ is any solution of equation (2.57) satisfying $\mathbf{y}(0) = 0$.

Theorem 2.33 (Reduction Theorem). *Consider the system (2.56). Suppose all the eigenvalues of the matrix A have modulus smaller than 1. Let $\mathbf{y}(z)$ be a solution of equation (2.57) satisfying $\mathbf{y}(0) = 0$. Then the equilibrium point of the system (2.56) at the origin is Lyapunov stable (asymptotically stable) (unstable) if and only if the equilibrium point $v = 0$ of the following system:*

$$v(t+1) = f_2(\mathbf{y}(v(t)), v(t)), \quad t = 0, 1, \ldots, \tag{2.60}$$

is Lyapunov stable (asymptotically stable) (unstable).

Theorem 2.34. *Consider the system (2.56). Suppose all the eigenvalues of the matrix A have modulus smaller than 1 and the equilibrium point of the system (2.60) at $v = 0$ is stable. Let $\mathrm{col}(y(t),\ z(t))$ be a solution of equation (2.56) with $\mathrm{col}(y(0),\ z(0))$ sufficiently small. Then, there exists a solution $v(t)$ of the system (2.60) such that, for all $t = 0, 1, \ldots,$*

$$\begin{aligned} \|z(t) - v(t)\| &\leq \delta\lambda^t, \\ \|y(t) - \mathbf{y}(v(t))\| &\leq \delta\lambda^t, \end{aligned} \tag{2.61}$$

where δ and λ are positive constants with $\lambda < 1$.

The center manifold described in Theorem 2.34 is called a stable center manifold.

2.6 Normal Form and Zero Dynamics of SISO Nonlinear Systems

In this and subsequent sections, we will review the concepts of the normal form and zero dynamics for the class of affine nonlinear systems (2.7). This section will focus on the single-input, single-output (SISO) systems while multiple-input, multiple-output (MIMO) systems will be covered in the next section. Normal form and zero dynamics provide structural information on the nonlinear systems and will be used in Chapter 3 for studying the solvability of the nonlinear regulator equations. We will use a rather casual manner to present these concepts while referring readers to Isidori [63] for all the proofs.

Throughout this section, we will call a sufficiently smooth function $f : \mathcal{R}^n \to \mathcal{R}^n$ a vector field in $\mathcal{R}^n$. We begin by introducing some notations and terminology.

Definition 2.35. *Let $h : \mathcal{R}^n \to \mathcal{R}$ be a sufficiently smooth scalar function, and $f : \mathcal{R}^n \to \mathcal{R}^n$ a vector field. Then*

$$
\begin{aligned}
\frac{\partial h}{\partial x} &\overset{def}{=} \left[\frac{\partial h}{\partial x_1}, \dots, \frac{\partial h}{\partial x_n} \right], \\
L_f^0 h(x) &\overset{def}{=} h(x), \\
L_f h(x) &\overset{def}{=} \sum_{i=1}^{n} \frac{\partial h}{\partial x_i} f_i(x) = \frac{\partial h}{\partial x} f(x), \\
&\vdots \\
L_f^k h(x) &\overset{def}{=} L_f(L_f^{k-1} h)(x) = \frac{\partial L_f^{k-1} h}{\partial x} f(x).
\end{aligned}
$$

Also, let $g : \mathcal{R}^n \to \mathcal{R}^n$ be a vector field; then, for $k = 0, 1, \dots,$

$$
L_g L_f^k h(x) \overset{def}{=} L_g(L_f^k h)(x) = \frac{\partial L_f^k h}{\partial x} g(x).
$$

We will call $\frac{\partial h}{\partial x}$ the *gradient* of $h(x)$ and $L_f h(x)$ the *Lie derivative* of the function h along the vector field f.

Definition 2.36. *The system (2.7) is said to have a* relative degree r *at x^0 if*

(i)

$$
L_g L_f^k h(x) = 0 \tag{2.62}
$$

for all $k < r - 1$ and for all x in an open neighborhood of x^0, and

(ii)

$$
L_g L_f^{r-1} h(x^0) \neq 0. \tag{2.63}
$$

Example 2.37. Consider a three-dimensional system of the form (2.7) with $x = \text{col}(x_1, x_2, x_3)$, and

$$f(x) = \begin{bmatrix} 0 \\ x_1 + x_2 \\ x_1^2 + x_2 + \theta x_3 \end{bmatrix}, \quad g(x) = \begin{bmatrix} \exp(x_2) \\ 0 \\ 0 \end{bmatrix}, \quad h(x) = x_2, \tag{2.64}$$

where θ is any real number. Simple calculation gives

$$L_g h(x) = 0, \quad L_g L_f h(x) = \exp(x_2).$$

Thus, by Definition 2.36, this system has a relative degree 2 at any point $x^0 \in \mathcal{R}^3$. ∎

Remark 2.38. The system may not have a well-defined relative degree at some point x^0 when there exists a positive integer r such that

(i)

$$L_g L_f^k h(x) = 0$$

for all $k < r - 1$ and for all x in an open neighborhood of x^0, and

(ii)

$$L_g L_f^{r-1} h(x^0) = 0.$$

However, there exists no open neighborhood of x^0 such that

$$L_g L_f^{r-1} h(x) = 0$$

in this neighborhood. For instance, in Example 2.37, if the function $\exp(x_2)$ is replaced by $\sin x_2$, then the system does not have a well-defined relative degree at $x^0 = 0$. It will be seen later that the ball and beam system to be described in Section 2.8 does not have a well-defined relative degree at $x^0 = 0$, either. ∎

Assume the system has a relative degree r at x^0. Then it can be verified that the trajectory of the system starting from any $x(0)$ sufficiently close to x^0 is such that, for sufficiently small t,

$$\begin{aligned} \dot{y}(t) &= L_f h(x(t)), \\ &\vdots \\ y^{(r-1)}(t) &= L_f^{r-1} h(x(t)), \\ y^{(r)}(t) &= L_f^r h(x(t)) + L_g L_f^{r-1} h(x(t)) u(t), \end{aligned} \tag{2.65}$$

with

$$L_g L_f^{r-1} h(x^0) \neq 0.$$

Solving the equation

$$\hat{u}(t) = L_f^r h(x(t)) + L_g L_f^{r-1} h(x(t)) u(t),$$

where $\hat{u} \in \mathcal{R}$ is viewed as a new input to the system (2.7), gives a state feedback controller of the form

$$u(t) = \frac{-L_f^r h(x(t)) + \hat{u}(t)}{L_g L_f^{r-1} h(x(t))}. \tag{2.66}$$

Applying (2.66) to system (2.7) results in a new system whose input–output relation obeys, for all sufficiently small t,

$$y^{(r)}(t) = \hat{u}(t).$$

Returning to Example 2.37, a direct calculation gives

$$\dot{y} = x_1 + x_2, \qquad \ddot{y} = x_1 + x_2 + \exp(x_2)u.$$

So the controller

$$u = \frac{-x_1 - x_2 + \hat{u}}{\exp(x_2)}$$

gives the relationship

$$\ddot{y} = \hat{u}.$$

Remark 2.39. The control law (2.66) is called the *input–output linearizing control law*, as it results in a linear input–output relation between the new input $\hat{u}$ and the output y. A further linear feedback control of the form

$$\hat{u}(t) = -\alpha_0 y - \alpha_1 \dot{y} - \cdots - \alpha_{r-1} y^{(r-1)}, \tag{2.67}$$

where $\alpha_0, \alpha_1, \ldots, \alpha_{r-1}$ are such that

$$\lambda^r + \alpha_{r-1}\lambda^{r-1} + \cdots + \alpha_1\lambda + \alpha_0$$

is a Hurwitz polynomial, will make the output y satisfy a stable linear differential equation as follows:

$$y^{(r)} + \alpha_{r-1} y^{(r-1)} + \cdots + \alpha_1 \dot{y} + \alpha_0 y = 0.$$

Thus, the output $y(t)$ will approach 0 as $t \to \infty$. The composition of (2.66) and (2.67) yields a state feedback control law of the form

$$u = \frac{-L_f^r h(x) - \sum_{i=0}^{r-1} \alpha_i L_f^i h(x)}{L_g L_f^{r-1} h(x)}, \tag{2.68}$$

which will be called an input–output linearization–based control law. It should be noted that such a control law may not guarantee the asymptotic stability of the equilibrium point of the closed-loop system. In fact, the closed-loop system composed of (2.64) and the control law (2.68) with $r = 2$ is

$$\begin{bmatrix} \dot{x}_1 \\ \dot{x}_2 \\ \dot{x}_3 \end{bmatrix} = \begin{bmatrix} -x_1 - x_2 - \alpha_0 x_2 - \alpha_1(x_1 + x_2) \\ x_1 + x_2 \\ x_1^2 + x_2 + \theta x_3 \end{bmatrix}. \tag{2.69}$$

The Jacobian matrix of (2.69) at the origin is given by

$$\begin{bmatrix} -(1+\alpha_1) & -(1+\alpha_0+\alpha_1) & 0 \\ 1 & 1 & 0 \\ 0 & 1 & \theta \end{bmatrix},$$

which has a characteristic polynomial

$$(\lambda^2+\alpha_1\lambda+\alpha_0)(\lambda-\theta).$$

By the Lyapunov linearization method, when $\theta > 0$, the equilibrium point of the closed-loop system is unstable regardless of the choice of α_0 and α_1. It will be seen later that when $\theta > 0$, the system is a nonminimum phase system, and the input–output linearization–based control law can only stabilize a minimum phase system. ∎

Next we will introduce the normal form and the zero dynamics for the system (2.7) with $m = p = 1$.

Definition 2.40. *Let $T(x)$ be a sufficiently smooth vector field defined on some open set $X \subset \mathcal{R}^n$. $T(x)$ is said to be a local* diffeomorphism *on $X \subset \mathcal{R}^n$ if there exists a sufficiently smooth vector field $T^{-1}(z)$ defined on X such that $T^{-1}(T(x)) = x$ for all $x \in X$. If $X = \mathcal{R}^n$, then $T(x)$ is said to be a global* diffeomorphism *on $\mathcal{R}^n$.*

If $T(x)$ is a diffeomorphism on $X \subset \mathcal{R}^n$, then we can define a coordinate transformation $z = T(x)$ for (2.7). Under the new state vector z, the system (2.7) can be expressed as follows:

$$\dot{z} = \left(\frac{\partial T}{\partial x}(f(x)+g(x)u)\right)\bigg|_{x=T^{-1}(z)}. \tag{2.70}$$

Moreover, if $T(0) = 0$, then $T^{-1}(0) = 0$. Thus, the origin $z = 0$ is also an equilibrium point of (2.70) when $T(0) = 0$. We will say that (2.7) is diffeomorphic to (2.70) on $X \subset \mathcal{R}^n$.

Remark 2.41. It can be shown that, if the system (2.7) has a relative degree r at x^0, then the following row vectors:

$$\frac{\partial h}{\partial x}(x^0), \quad \frac{\partial L_f h}{\partial x}(x^0), \quad \ldots, \quad \frac{\partial L_f^{r-1} h}{\partial x}(x^0)$$

are linearly independent [63]. As a result, if, at a point x^0, the relative degree r of (2.7) is well defined, then $r \le n$. For convenience, let

$$H(x) = \begin{bmatrix} h(x) \\ L_f h(x) \\ \vdots \\ L_f^{r-1} h(x) \end{bmatrix}. \tag{2.71}$$

We will call $H(x)$ the H-vector of (2.7). Clearly, if the system (2.7) has a relative degree r at x^0, then the rows of $\frac{\partial H}{\partial x}(x^0)$ are linearly independent. ∎

Now assume that the system (2.7) has a relative degree r at $x^0 = 0$. Let

$$\begin{aligned} T_1(x) &= h(x), \\ T_2(x) &= L_f h(x), \\ &\vdots \\ T_r(x) &= L_f^{r-1} h(x). \end{aligned}$$

By Remark 2.41, there exist $n - r$ sufficiently smooth functions $T_{r+1}(x), \ldots, T_n(x)$ such that the vector field

$$T(x) = \begin{bmatrix} T_1(x) \\ T_2(x) \\ \vdots \\ T_r(x) \\ T_{r+1}(x) \\ \vdots \\ T_n(x) \end{bmatrix} \tag{2.72}$$

is a diffeomorphism on an open neighborhood X of 0 and satisfies $T(0) = 0$. Let $z_i = T_i(x)$, $i = 1, \cdots, n$. Then z_i's satisfy the following equations:

$$\begin{aligned} \dot{z}_1 &= z_2, \\ &\vdots \\ \dot{z}_{r-1} &= z_r, \\ \dot{z}_r &= \left(L_f^r h(x) + L_g L_f^{r-1} h(x) u \right)\Big|_{x=T^{-1}(z)}, \\ \dot{z}_{r+1} &= \left(L_f T_{r+1}(x) + L_g T_{r+1}(x) u \right)\big|_{x=T^{-1}(z)}, \\ &\vdots \\ \dot{z}_n &= \left(L_f T_n(x) + L_g T_n(x) u \right)\big|_{x=T^{-1}(z)}, \\ y &= z_1. \end{aligned} \tag{2.73}$$

We call (2.73) the *normal form* of the system (2.7).

Remark 2.42. By the Frobenius Theorem, it is possible to choose $T_i(x)$, $i = r + 1, \ldots, n$, such that $T(x)$ is locally invertible and

$$L_g T_i(x) = 0, \quad i = r + 1, \ldots, n, \tag{2.74}$$

for x in an open neighborhood of x^0. It is clear from (2.73) that this set of choices will render the equations (2.73) a more special expression as follows:

$$
\begin{aligned}
\dot{z}_1 &= z_2, \\
&\vdots \\
\dot{z}_{r-1} &= z_r, \\
\dot{z}_r &= \left(L_f^r h(x) + L_g L_f^{r-1} h(x) u\right)\Big|_{x=T^{-1}(z)}, \\
\dot{z}_{r+1} &= L_f T_{r+1}(x)\big|_{x=T^{-1}(z)}, \\
&\vdots \\
\dot{z}_n &= L_f T_n(x)\big|_{x=T^{-1}(z)}, \\
y &= z_1. \quad \blacksquare
\end{aligned}
\tag{2.75}
$$

Next, we will introduce the notion of a (local) output zeroing manifold for the general nonlinear system described by (2.5) and (2.6).

Definition 2.43. *Let M be a manifold containing the origin of $\mathcal{R}^n$. M is called a (local)* control invariant manifold *of the system described by (2.5) if there exists a sufficiently smooth state feedback control of the form $u = k(x)$ with $k(0) = 0$ such that M is a (local) invariant manifold of $\dot{x} = f(x, k(x))$, and it is called a (local)* output zeroing manifold *of the system (2.5) and (2.6) if it is a (local) control invariant manifold of (2.5) and is contained in the kernel of the mapping $h(x, k(x))$; that is, for all $x \in M$, $h(x, k(x)) = 0$.*

Returning to the affine nonlinear system (2.7), assume that the system (2.7) has a relative degree r at $x^0 = 0$ and let the function $H(x)$ be defined as in (2.71). Then there exists an open neighborhood X of the origin of $\mathcal{R}^n$ such that $M = \{x \in X \mid H(x) = 0\}$ is a manifold of dimension $n - r$. We will show that the set M is a (local) output zeroing manifold of (2.7). In fact, by the definition of $H(x)$, M is contained in the kernel of the output mapping $h(x)$. Now assume that the normal form of (2.7) is given by (2.73). Define a state feedback control law as follows:

$$
u = \left(\frac{-L_f^r h(x) + \hat{u}}{L_g L_f^{r-1} h(x)}\right)\Bigg|_{x=T^{-1}(z)}. \tag{2.76}
$$

Then the closed-loop system is given by

$$
\begin{aligned}
\dot{z}_1 &= z_2, \\
&\vdots \\
\dot{z}_{r-1} &= z_r, \\
\dot{z}_r &= \hat{u}, \\
\dot{z}_i &= \left(L_f T_i(x) + L_g T_i(x)\left(\frac{-L_f^r h(x) + \hat{u}}{L_g L_f^{r-1} h(x)}\right)\right)\Bigg|_{x=T^{-1}(z)}, \\
& \quad i = r+1, \ldots, n, \\
y &= z_1.
\end{aligned}
\tag{2.77}
$$

Then it is clear from (2.77) that, under the state feedback control $\hat{u} = 0$, for all initial states $z(0) = \text{col}(z_1(0), \dots, z_n(0))$ satisfying $z_1(0) = z_2(0) = \cdots = z_r(0) = 0$, the first r components of the solution $z(t)$ of (2.77) starting from $z(0)$ are identically zero for sufficiently small t. This is the same as saying that, in the original coordinates x, under the state feedback control $u = (-L_f^r h(x))/(L_g L_f^{r-1} h(x))$, for all sufficiently small t, the solution $x(t)$ of (2.7) starting from any initial state $x(0) \in M$ belongs to M. Thus, M is an invariant manifold of (2.7). Moreover, by the definition of M, $h(x) = 0$ for all $x \in M$, and thus M is an output zeroing manifold of (2.7).

Remark 2.44. A system may have several output zeroing manifolds of different dimensions. An output zeroing manifold M is *locally maximal* if, for any other (local) output zeroing manifold M', there exists an open neighborhood X of the origin of $\mathcal{R}^n$ such that $X \cap M' \subseteq X \cap M$. It can be shown that if the system (2.7) has a relative degree r at the origin, then the manifold defined by the hypersurface $H(x) = 0$ with $H(x)$ being given by (2.71) is the (locally) maximal output zeroing manifold of (2.7). In fact, assume that M' is any other (local) output zeroing manifold of (2.7) under a sufficiently smooth state feedback control $u = k'(x)$; then, the closed-loop system has the property that, for any sufficiently small $x(0) \in M'$, the solution $x(t)$ of the closed-loop system starting from any $x(0)$ sufficiently close to x^0 is such that $y(t) = h(x(t)) = 0$ for all sufficiently small $t \geq 0$. Therefore, the derivatives of $y(t)$ up to any orders are identically zero for all sufficiently small $t \geq 0$. It follows from (2.65) that $x(t) \in M$ for all sufficiently small $t \geq 0$. ■

We can put equation (2.77) into a more compact form. To this end, let

$$a(z) = L_f^r h(T^{(-1)}(z)), \qquad b(z) = L_g L_f^{r-1} h(T^{(-1)}(z)),$$

$$c(z) = \begin{bmatrix} L_f T_{r+1}\left(T^{-1}(z)\right) \\ \vdots \\ L_f T_n\left(T^{-1}(z)\right) \end{bmatrix}, \quad d(z) = \begin{bmatrix} L_g T_{r+1}\left(T^{-1}(z)\right) \\ \vdots \\ L_g T_n\left(T^{-1}(z)\right) \end{bmatrix},$$

$$\xi = \begin{bmatrix} z_1 \\ \vdots \\ z_r \end{bmatrix}, \quad \eta = \begin{bmatrix} z_{r+1} \\ \vdots \\ z_n \end{bmatrix}.$$

Then equation (2.77) becomes the following:

$$\begin{aligned} \dot{\xi} &= A_r \xi + B_r \hat{u}, \\ \dot{\eta} &= c(\xi, \eta) + d(\xi, \eta) \left(\frac{-a(\xi, \eta) + \hat{u}}{b(\xi, \eta)} \right) \stackrel{def}{=} \gamma(\xi, \eta, \hat{u}), \\ y &= C_r \xi, \end{aligned} \tag{2.78}$$

where

$$A_r = \begin{bmatrix} 0 & 1 & 0 & \cdots & 0 \\ 0 & 0 & 1 & \cdots & 0 \\ \vdots & \vdots & \vdots & \ddots & \vdots \\ 0 & 0 & 0 & \cdots & 1 \\ 0 & 0 & 0 & \cdots & 0 \end{bmatrix}, \quad B_r = \begin{bmatrix} 0 \\ 0 \\ \vdots \\ 0 \\ 1 \end{bmatrix}, \quad C_r = [1 \ 0 \ \cdots \ 0].$$

From (2.78), we can define an $(n-r)$-dimensional subsystem as follows:

$$\dot{\eta}(t) = \gamma(0, \eta(t), 0), \tag{2.79}$$

which has an equilibrium point at $\eta = 0$. This system is precisely the system that governs the motion of the last $n-r$ components of z when the motion of the system (2.77) is restricted to the manifold M. For this reason we call the subsystem (2.79) *the zero dynamics* of (2.7).

Remark 2.45.

(i) If a feedback control $u = k(x)$ is required to render the output $y(t)$ of the system (2.7) zero for all sufficiently small t, then, necessarily, the solution of the system (2.7) must be on the manifold M and the feedback control $u = k(x)$ must take the form (2.76) with $\hat{u} = 0$. Thus, requiring the output $y(t)$ of the system (2.7) to be zero for all sufficiently small $t > 0$ uniquely identifies the zero dynamics (2.79) module coordinate transformations.

(ii) The subsystem (2.79) is identified from the normal form (2.73). Thus the representation of the function γ also depends on the way that $T(x)$ is chosen. Nevertheless, for different choices of $T(x)$, the resulting zero dynamics are locally diffeomorphic to each other.

(iii) Let the Jacobian linearization of system (2.7) be

$$\dot{x} = Ax + Bu, \qquad y = Cx. \tag{2.80}$$

Then the transfer function of (2.80) is

$$P(s) = \frac{\det \begin{bmatrix} sI - A & -B \\ C & 0 \end{bmatrix}}{\det(sI - A)}.$$

On the other hand, it can be verified that the transfer function of (2.80) is also given by

$$P(s) = CA^{r-1}B\frac{\det(sI - Q)}{\det(sI - A)},$$

where Q is the Jacobian matrix of $\gamma(0, \eta, 0)$ at $\eta = 0$. Thus, if the triple (A, B, C) is controllable and observable, then the eigenvalues of Q coincide with the zeros of (2.80). Therefore, naturally, we call the system (2.7) *minimum phase* if all the eigenvalues of Q have negative real parts or *nonminimum phase* if at least one eigenvalue of Q has positive real parts. In the critical case when none of the eigenvalues of Q have positive real parts but at least one eigenvalue of Q has zero real parts, we define (2.7) to be *minimum phase* if the equilibrium point $\eta = 0$ of the zero dynamics (2.79) is asymptotically stable and *nonminimum phase* if the equilibrium point $\eta = 0$ of the zero dynamics (2.79) is unstable. Returning to Example 2.37, it can be verified that the zero dynamics of the system is $\dot{x}_3 = \theta x_3$. Therefore, the system is nonminumum phase when $\theta > 0$.

(iv) It can also be verified that the matrix Q is unaffected under the class of input–output linearization–based control laws (2.68). Therefore, the input–output linearization–based control laws can only stabilize a minimum phase nonlinear system.

(v) The equilibrium point of the zero dynamics is called *hyperbolic* if all the eigenvalues of Q have nonzero real parts. Otherwise it is called *nonhyperbolic*. In Chapter 3, we will see that nonlinear systems whose zero dynamics has a nonhyperbolic equilibrium point present a hurdle to the solvability of the output regulation problem. ∎

Remark 2.46. We can always choose the functions $T_{r+1}(x), \ldots, T_n(x)$ to be some $n-r$ components of x. In this case, the zero dynamics of (2.7) can be represented using these $n-r$ components of x. This procedure can be detailed as follows:

(i) By Remark 2.41, there exist r components of x denoted by $x_{j_1}, \ldots, x_{j_r}$ such that

$$\text{rank}\begin{bmatrix} \frac{\partial h}{\partial x_{j_1}}(0) & \cdots & \frac{\partial h}{\partial x_{jr}}(0) \\ \frac{\partial L_f h}{\partial x_{j_1}}(0) & \cdots & \frac{\partial L_f h}{\partial x_{jr}}(0) \\ & \vdots & \\ \frac{\partial L_f^{r-1} h}{\partial x_{j_1}}(0) & \cdots & \frac{\partial L_f^{r-1} h}{\partial x_{jr}}(0) \end{bmatrix} = r.$$

Denote the remaining $n-r$ components of x by $x_{j_{r+1}}, \ldots, x_{j_n}$; then, by the Implicit Function Theorem, there exist an open neighborhood X_0 of the origin of $\mathcal{R}^{(n-r)}$ and a function $\sigma : X_0 \to \mathcal{R}^r$ satisfying $\sigma(0) = 0$ such that

$$\begin{aligned} h(x)\Big|_{(x_{j_1},\ldots,x_{j_r})=\sigma(x_{j_{r+1}},\ldots,x_{jn})} &= 0, \\ L_f h(x)\Big|_{(x_{j_1},\ldots,x_{jr})=\sigma(x_{j_{r+1}},\ldots,x_{jn})} &= 0, \\ &\vdots \\ L_f^{r-1} h(x)\Big|_{(x_{j_1},\ldots,x_{j_r})=\sigma(x_{j_{r+1}},\ldots,x_{jn})} &= 0. \end{aligned}$$

Clearly, the function defined by

$$T(x) = \begin{bmatrix} h(x) \\ \vdots \\ L_f^{r-1} h(x) \\ x_{j_{r+1}} \\ \vdots \\ x_{jn} \end{bmatrix}$$

is invertible in an open neighborhood of the origin of $x^0 = 0$.

(ii) Let $\xi = \text{col}(z_1, z_2, \ldots, z_r) = \text{col}(h(x), \ldots, L_f^{r-1} h(x))$, $\eta = \text{col}(z_{r+1}, \ldots, z_n) = \text{col}(x_{j_{r+1}}, \ldots, x_{j_n})$, and

$$u_e(x) = -\frac{L_f^r h(x)}{L_g L_f^{r-1} h(x)}.$$

Then the zero dynamics as defined in (2.79) has the following representation:

$$\dot{x}_{j_{r+1}} = \left(f_{j_{r+1}}(x) + g^{j_{r+1}}(x)u_e(x)\right)\Big|_{(x_{j_1},\dots,x_{j_r})=\sigma(x_{j_{r+1}},\dots,x_{j_n})},$$
$$\vdots$$
$$\dot{x}_{j_n} = \left(f_{j_n}(x) + g^{j_n}(x)u_e(x)\right)\Big|_{(x_{j_1},\dots,x_{j_r})=\sigma(x_{j_{r+1}},\dots,x_{j_n})},$$

where, for $j = 1, \dots, n$, g^j is the jth component of g. It is noted that, in deriving the above representation of the zero dynamics, there is no need to resort to the normal form of system (2.7). ∎

2.7 Normal Form and Zero Dynamics of MIMO Nonlinear Systems

In this section, we will further extend such notions as the relative degree, normal form, and zero dynamics to MIMO affine nonlinear systems (2.7) with $m \geq p \geq 1$.

Definition 2.47. *For each $i = 1, \dots, p$, the ith output y_i of the system (2.7) is said to have a* relative degree *r_i at a point x^0 if*

(i)

$$L_g L_f^k h_i(x) \stackrel{def}{=} \left[L_{g_1} L_f^k h_i(x), L_{g_2} L_f^k h_i(x), \dots, L_{g_m} L_f^k h_i(x)\right] = 0_{1\times m} \tag{2.81}$$

for all $k < r_i - 1$, and if for all x in an open neighborhood of x^0,

(ii)

$$L_g L_f^{r_i-1} h_i(x^0) \neq 0_{1\times m}. \tag{2.82}$$

The system (2.7) is said to have a vector relative degree *$\{r_1, \dots, r_p\}$ at a point x^0 if*

(i) *for all $1 \leq i \leq p$, the ith output $h_i(x)$ has a relative degree r_i at x^0, and*
(ii) *the $p \times m$ matrix*

$$D(x) = \begin{bmatrix} L_g L_f^{r_1-1} h_1(x) \\ L_g L_f^{r_2-1} h_2(x) \\ \vdots \\ L_g L_f^{r_p-1} h_p(x) \end{bmatrix} \tag{2.83}$$

has full row rank at $x = x^0$.

Suppose, for each $i = 1, \dots, p$, the output y_i of the system (2.7) has a scalar relative degree r_i at $x = x^0$. Then the trajectory starting from any $x(0)$ sufficiently close to x^0 is such that

$$\begin{aligned} \dot{y}_i(t) &= L_f h_i(x), \\ &\vdots \\ y_i^{(r_i-1)}(t) &= L_f^{r_i-1} h_i(x), \\ y_i^{(r_i)}(t) &= L_f^{r_i} h_i(x) + L_g L_f^{r_i-1} h_i(x) u, \end{aligned} \tag{2.84}$$

where

$$L_g L_f^{r_i-1} h_i(x^0) \neq 0_{1\times m}. \tag{2.85}$$

Let

$$E(x) = \begin{bmatrix} L_f^{r_1} h_1(x) \\ L_f^{r_2} h_2(x) \\ \vdots \\ L_f^{r_p} h_p(x) \end{bmatrix} \quad \text{and} \quad Y^{(r)} = \begin{bmatrix} y_1^{(r_1)} \\ y_2^{(r_2)} \\ \vdots \\ y_p^{(r_p)} \end{bmatrix}. \tag{2.86}$$

Then $Y^{(r)}$ and the input u can be related by the following equation:

$$Y^{(r)} = E(x) + D(x)u. \tag{2.87}$$

Further, if the system has vector relative degree at x^0, then $D(x^0)$ has full row rank; hence $(D(x)D^T(x))$ is invertible in an open neighborhood of x^0. Thus, the following equation:

$$\hat{u} = E(x) + D(x)u \tag{2.88}$$

is solvable for u. When $p = m$, the solution of (2.88) is unique. When $p < m$, the solution of (2.88) is not unique. One of the solutions of (2.88) is given by

$$u = D^T(x)(D(x)D^T(x))^{-1}(-E(x) + \hat{u}). \tag{2.89}$$

Under this control law, the trajectory starting from any $x(0)$ sufficiently close to x^0 is such that, for all sufficiently small $t \geq 0$,

$$Y^{(r)}(t) = \hat{u}(t). \tag{2.90}$$

Thus the control law (2.89), which is an extension of (2.66), achieves the input–output linearization for the system (2.7) for the general case when $m \geq p \geq 1$.

Remark 2.48. For convenience of later reference, we will call $D(x)$ and $E(x)$ the *decoupling matrix* and the E-vector of (2.7), respectively. Also, we extend the H-vector defined in (2.71) for single-output systems to multi-output systems as follows:

$$H(x) = \begin{bmatrix} h_1(x) \\ L_f h_1(x) \\ \vdots \\ L_f^{r_1-1} h_1(x) \\ \vdots \\ h_p(x) \\ L_f h_p(x) \\ \vdots \\ L_f^{r_p-1} h_p(x) \end{bmatrix}. \tag{2.91}$$

We will still call this vector H-vector of (2.7). Again, it can be shown that if the system (2.7) has a vector relative degree $\{r_1, \dots, r_p\}$ at x^0, then the rows of $\frac{\partial H}{\partial x}(x^0)$ are linearly independent [63]. ∎

Example 2.49. Consider a two-input, two-output system of the form (2.7) with $x = \text{col}(x_1, x_2, x_3, x_4)$:

$$f(x) = \begin{bmatrix} 0 \\ x_3 + x_4 \\ 0 \\ x_1 \end{bmatrix}, \quad g(x) = \begin{bmatrix} 0 & \exp(x_2) \\ 0 & 0 \\ 1 & 0 \\ 0 & 0 \end{bmatrix}, \quad h(x) = \begin{bmatrix} x_1 + x_3 + x_4 \\ x_2 \end{bmatrix}.$$

Simple calculation gives

$$\begin{array}{lll} h_1(x) = x_1 + x_3 + x_4, & L_f h_1(x) = x_1, & L_g h_1(x) = [1 \ \exp(x_2)], \\ h_2(x) = x_2, & L_f h_2(x) = x_3 + x_4, & L_f^2 h_2(x) = x_1, \\ L_g h_2(x) = [0 \ 0], & L_g L_f h_2(x) = [1 \ 0]. & \end{array}$$

Thus, the system has well-defined scalar relative degree $\{r_1, r_2\} = \{1, 2\}$ at any x^0. Also, we have

$$E(x) = \begin{bmatrix} x_1 \\ x_1 \end{bmatrix}, \quad D(x) = \begin{bmatrix} 1 & \exp(x_2) \\ 1 & 0 \end{bmatrix}.$$

Since rank $D(x) = 2$ for all x^0, the system has a vector relative degree at any x^0. Using (2.89) gives an input–output linearizing controller

$$u = \begin{bmatrix} \hat{u}_2 - x_1 \\ \frac{\hat{u}_1 - \hat{u}_2}{\exp(x_2)} \end{bmatrix},$$

which results in

$$\dot{y}_1 = \hat{u}_1, \quad \ddot{y}_2 = \hat{u}_2. \quad \blacksquare$$

To describe the normal form and zero dynamics for MIMO systems, assume the system (2.7) has a vector relative degree $\{r_1, \dots, r_p\}$ at $x^0 = 0$, and by Remark 2.48, if

$r \overset{def}{=} r_1 + \cdots + r_p$ is less than n, then there exist $n-r$ scalar functions $T_{r+1}(x), \ldots, T_n(x)$ such that

$$T(x) = \begin{bmatrix} H(x) \\ T_{r+1}(x) \\ \vdots \\ T_n(x) \end{bmatrix} \tag{2.92}$$

is invertible in an open neighborhood of $x^0 = 0$ and satisfies $T(0) = 0$. Consider the coordinates transformation

$$z = T(x), \tag{2.93}$$

where z is an n-dimensional vector whose components are denoted by

$$z = \begin{bmatrix} z_1^1 \\ z_2^1 \\ \vdots \\ z_{r_1}^1 \\ \vdots \\ z_1^p \\ z_2^p \\ \vdots \\ z_{r_p}^p \\ z_{r+1} \\ \vdots \\ z_n \end{bmatrix}.$$

In terms of z, (2.7) can be represented as follows:

$$\begin{aligned} \dot{z}_1^i &= z_2^i, \\ &\vdots \\ \dot{z}_{r_i-1}^i &= z_{r_i}^i, \\ \dot{z}_{r_i}^i &= \left(L_f^{r_i} h_i(x) + L_g L_f^{r_i-1} h_i(x) u\right)\Big|_{x=T^{-1}(z)}, \\ \dot{z}_{r+1} &= \left(L_f T_{r+1}(x) + L_g T_{r+1}(x) u\right)\Big|_{x=T^{-1}(z)}, \\ &\vdots \\ \dot{z}_n &= \left(L_f T_n(x) + L_g T_n(x) u\right)\Big|_{x=T^{-1}(z)}, \\ y_i &= z_1^i, \quad i = 1, \ldots, p. \end{aligned} \tag{2.94}$$

Equation (2.94) can be viewed as an extension of (2.73) to the MIMO system and is called the *normal form* of the MIMO system (2.7). If $T_j(x)$, $j = r+1, \ldots, n$, can be chosen such that $L_gT_j(x) = 0, \ j = r+1, \ldots, n$, then the last $n-r$ equations of (2.94) can be made independent of u. Unfortunately, for MIMO systems, it is in general impossible to make $L_gT_j(x) = 0, \ j = r+1, \ldots, n$. Nevertheless, it is possible to show the existence of $n-r$ sufficiently smooth functions $T_j(x), \ j = r+1, \ldots, n$, such that $L_gT_j(x) = 0, \ j = r+1, \ldots, n$, under the assumption that the distribution span$\{g_1, \ldots, g_m\}$ is involutive near $x = 0$.[2]

Next, let $k(x, \hat{u})$ be any solution of (2.88), for example,

$$k(x, \hat{u}) = D^T(x)(D(x)D^T(x))^{-1}(-E(x) + \hat{u}). \tag{2.95}$$

Then, applying the input transformation $u = k(x, \hat{u})$ to (2.94) gives

$$\begin{aligned}
\dot{z}_1^i &= z_2^i, \\
&\vdots \\
\dot{z}_{r_i-1}^i &= z_{r_i}^i, \\
\dot{z}_{r_i}^i &= \hat{u}_i, \\
\dot{z}_j &= \left(L_fT_j(x) + L_gT_j(x)k(x, \hat{u})\right)\big|_{x=T^{-1}(z)}, \quad j = r+1, \ldots, n, \\
y_i &= z_1^i, \quad i = 1, \ldots, p.
\end{aligned} \tag{2.96}$$

It can be seen that system (2.96) exhibits a linear input–output relation.

From system (2.96), it can be seen that, under the state feedback control $\hat{u} = 0$, for all initial states $z(0) = (z_1(0), \ldots, z_n(0))$ satisfying $z_1(0) = z_2(0) = \cdots = z_r(0) = 0$, the first r components of the solution $z(t)$ of (2.96) starting from $z(0)$ are identically zero for sufficiently small t. This is the same as saying that, in the original coordinates x, under the state feedback control $u = u_e(x) = k(x, 0)$, for all sufficiently small t, the solution $x(t)$ of (2.7) starting from any initial state $x(0) \in M$ belongs to M, where $M \stackrel{def}{=} \{x \in X \mid H(x) = 0\}$ with X an open neighborhood of the origin of $\mathcal{R}^n$. Thus, M is an output zeroing manifold of the MIMO system (2.7). Note that though $u_e(x)$ may not be unique when $p < m$, this manifold is uniquely defined by $H(x) = 0$.

Next, we can define the zero dynamics of the MIMO system (2.7) similarly to that of the SISO system. Let $z = \text{col}(\xi, \eta)$, where

$$\xi = \begin{bmatrix} z_1^1 \\ z_2^1 \\ \vdots \\ z_{r_1}^1 \\ \vdots \\ z_1^p \\ z_2^p \\ \vdots \\ z_{r_p}^p \end{bmatrix}, \quad \eta = \begin{bmatrix} z_{(r+1)} \\ \vdots \\ z_n \end{bmatrix}.$$

[2]See Chapter 5 of [63] for details.

Then the $n - r$ equations of (2.94) governing $z_{r+1}, \ldots, z_n$ can be put into the following compact form:

$$\dot{\eta} = \gamma(\xi, \eta, u). \tag{2.97}$$

From (2.97), we can identify an $(n - r)$-dimensional subsystem

$$\dot{\eta}(t) = \gamma(0, \eta, u_e(T^{-1}(0, \eta))). \tag{2.98}$$

Similar to the SISO case, this subsystem can be viewed as being induced by the requirement of rendering the output $y(t) = 0$ for all sufficiently small $t \geq 0$ under the state feedback control $u = u_e(x)$, and is thus called the *zero dynamics* of (2.7).

Remark 2.50. When $p = m$, $u_e(x)$ is uniquely defined by $u_e(x) = -D^{-1}(x)E(x)$. Hence, the zero dynamics (2.98) is also unique within the coordinate transformations. When $p < m$, the zero dynamics (2.98) is not unique because $u_e(x)$ is not. In particular, the stability property of the equilibrium of (2.98) at the origin may depend on the particular function $u_e(x)$. To better illustrate this point, perform a partition $u = \mathrm{col}(u^1, u^2)$ with $u^1 \in \mathcal{R}^p$, $u^2 \in \mathcal{R}^{m-p}$. Then there exists a function $k_u : \mathcal{R}^{n+m-p} \to \mathcal{R}^m$ such that

$$E(x) + D(x)k_u(x, u^2) = 0 \tag{2.99}$$

regardless of the values of u^2. Substituting $u = k_u(x, u^2)$ into (2.97) gives

$$\dot{\eta} = \gamma(\xi, \eta, k_u(T^{-1}(\xi, \eta), u^2)).$$

Let $\theta(\xi, \eta)$ be any sufficiently smooth function satisfying $\theta(0, 0) = 0$. It can be seen that, under the state feedback control $u = k_u(x, \theta(0, \eta))|_{x=T^{-1}(0,\eta)}$, when $\mathrm{col}(\xi(0), \eta(0)) \in M$, $\xi(t)$ will be identically 0 for sufficiently small $t > 0$, and $\eta(t)$ will be governed by the system

$$\dot{\eta} = \gamma(0, \eta, k_u(T^{-1}(0, \eta), \theta(0, \eta))). \tag{2.100}$$

Thus, (2.100) can be viewed as a family of the zero dynamics of (2.7) parameterized by function $\theta(0, \eta)$. It is interesting to note that $\theta(0, \eta)$ can be used to modify the zero dynamics of system (2.7). ∎

Example 2.51. To find the normal form and zero dynamics of the system in Example 2.49, note that

$$\begin{aligned} z_1^1 &= h_1(x) = x_1 + x_3 + x_4, \\ z_1^2 &= h_2(x) = x_2, \\ z_2^2 &= L_f h_2(x) = x_3 + x_4. \end{aligned}$$

Choose $z_4 = x_4$. Then

$$T(x) = \begin{bmatrix} x_1 + x_3 + x_4 \\ x_2 \\ x_3 + x_4 \\ x_4 \end{bmatrix}$$

is invertible for all $x \in \mathcal{R}^4$. The inverse mapping of $T(x)$ can be obtained as follows:

$$\begin{bmatrix} x_1 \\ x_2 \\ x_3 \\ x_4 \end{bmatrix} = T^{-1}(z) = \begin{bmatrix} z_1^1 - z_2^2 \\ z_1^2 \\ z_2^2 - z_4 \\ z_4 \end{bmatrix}.$$

In terms of z, we can obtain the normal form of the system as follows:

$$\begin{aligned} \dot{z}_1^1 &= z_1^1 - z_2^2 + u_1 + \exp(z_1^2)u_2, \\ \dot{z}_1^2 &= z_2^2, \\ \dot{z}_2^2 &= z_1^1 - z_2^2 + u_1, \\ \dot{z}_4 &= z_1^1 - z_2^2, \\ y_1 &= z_1^1, \\ y_2 &= z_1^2. \end{aligned} \tag{2.101}$$

Further, let

$$\begin{aligned} u_1 &= \hat{u}_2 - z_1^1 + z_2^2, \\ u_2 &= \frac{\hat{u}_1 - \hat{u}_2}{\exp(z_1^2)}. \end{aligned}$$

Then equation (2.96) takes the form

$$\begin{aligned} \dot{z}_1^1 &= \hat{u}_1, \\ \dot{z}_1^2 &= z_2^2, \\ \dot{z}_2^2 &= \hat{u}_2, \\ \dot{z}_4 &= z_1^1 - z_2^2, \\ y_1 &= z_1^1, \\ y_2 &= z_1^2. \end{aligned} \tag{2.102}$$

From equation (2.102), it is clear that (2.97) becomes

$$\dot{z}_4 = z_1^1 - z_2^2$$

so that the zero dynamics of (2.102) is given by, according to (2.98),

$$\dot{z}_4 = 0. \quad \blacksquare$$

Remark 2.52. Let $Q = \frac{\partial \gamma}{\partial \eta}(0, 0, 0)$. Then, if the Jacobian linearization of system (2.7) is controllable and observable, the eigenvalues of Q coincide with the transmission zeros of the Jacobian linearization of system (2.7). As in the SISO case, we will call system (2.7) *minimum phase* if all the eigenvalues of Q have negative real parts, and *nonminimum phase* if at least one of the eigenvalues of Q has positive real part. The critical case can also be classified in a way similar to the SISO case. $\blacksquare$

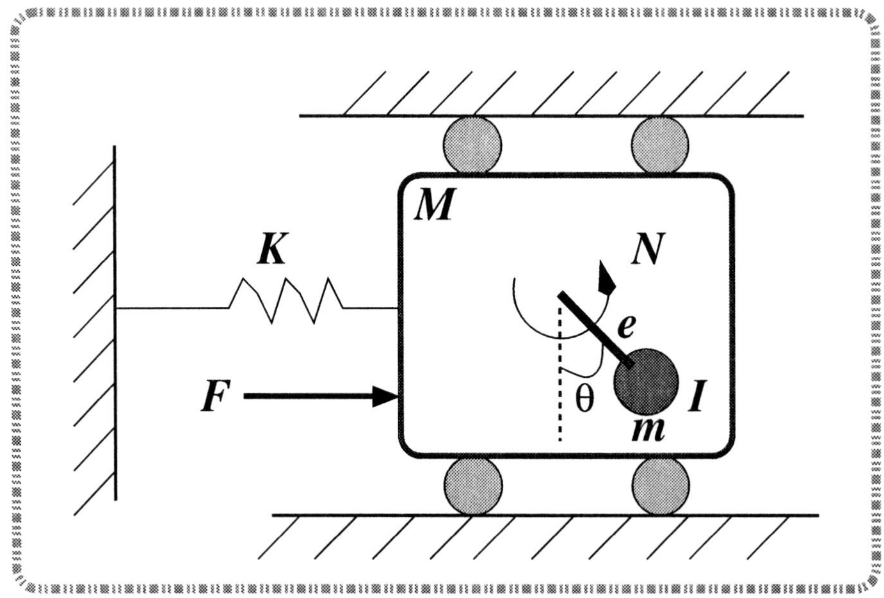

Figure 2.1. *Rotational/translational actuator.*

2.8 Examples of Nonlinear Control Systems

In this section, we introduce three well-known nonlinear systems, namely, the rotational/translational actuator (RTAC) system, the inverted pendulum on a cart system, and the ball and beam system. It is well known that the asymptotic tracking and/or disturbance rejection problem associated with these systems present challenges to conventional input–output linearization–based method since, as will be seen shortly, all these three systems are nonminimum phase. Nevertheless, we will further show in later chapters that the output regulation theory introduced in this book can practically solve the asymptotic tracking and/or disturbance rejection problem associated with these systems.

The RTAC [2], [3]. The RTAC, depicted in Figure 2.1, consists of a cart of mass M connected to a fixed wall by a linear spring of stiffness k. The cart is constrained to have one-dimensional travel. The proof-mass actuator attached to the cart has mass m and moment of inertia I about its center of mass, which is located at a distance e from the point about which the proof-mass rotates. Its motion occurs in a horizontal plane so that no gravitational forces need to be considered. The motion of RTAC is described as follows:

$$\begin{aligned} \ddot{\xi} + \xi &= \epsilon(\dot{\theta}^2 \sin\theta - \ddot{\theta}\cos\theta) + F, \\ \ddot{\theta} &= -\epsilon\ddot{\xi}\cos\theta + u, \end{aligned} \tag{2.103}$$

where ξ is the one-dimensional displacement of the cart, θ the angular position of the proof body, F the disturbance, and u the control input. The coupling between the translational and rotational motion is captured by the parameter ϵ, which is defined by

$$\epsilon = \frac{me}{\sqrt{(I + me^2)(M + m)}},$$

where $0 < e < 1$ is the eccentricity of the proof body.

Letting $x = \mathrm{col}(x_1\ x_2\ x_3\ x_4) = \mathrm{col}(\xi\ \dot{\xi}\ \theta\ \dot{\theta})$ and $y = \xi$ yields the following state-space representation of (2.103):

$$\dot{x} = f(x) + g_1(x)u + g_2(x)F, \qquad y = x_1,$$

where

$$f = \begin{bmatrix} x_2 \\ \frac{-x_1+\epsilon x_4^2 \sin x_3}{1-\epsilon^2\cos^2 x_3} \\ x_4 \\ \frac{\epsilon \cos x_3 (x_1 - \epsilon x_4^2 \sin x_3)}{1-\epsilon^2\cos^2 x_3} \end{bmatrix}, \quad g_1 = \begin{bmatrix} 0 \\ \frac{-\epsilon \cos x_3}{1-\epsilon^2\cos^2 x_3} \\ 0 \\ \frac{1}{1-\epsilon^2\cos^2 x_3} \end{bmatrix}, \quad g_2 = \begin{bmatrix} 0 \\ \frac{1}{1-\epsilon^2\cos^2 x_3} \\ 0 \\ \frac{-\epsilon \cos x_3}{1-\epsilon^2\cos^2 x_3} \end{bmatrix}, \tag{2.104}$$

where $1 - \epsilon^2 \cos^2 x_3 \neq 0$ for all x_3 and $\epsilon < 1$.

When the disturbance F is zero, the RTAC system takes the standard form of (2.7). Let us consider the problem of finding the normal form and the zero dynamics of the RTAC system with $F = 0$. To this end, note that the relative degree of the system at the origin is 2. Define the coordinates transformation $z = T(x)$ as follows:

$$\begin{aligned} z_1 &= h(x) = x_1, \\ z_2 &= L_f h(x) = x_2, \\ z_3 &= x_3, \\ z_4 &= x_2 + \epsilon x_4 \cos x_3, \end{aligned}$$

whose inverse transformation is given by

$$\begin{aligned} x_1 &= z_1, \\ x_2 &= z_2, \\ x_3 &= z_3, \\ x_4 &= \frac{z_4 - z_2}{\epsilon \cos z_3}. \end{aligned}$$

Under the new coordinates, the system can be described by its normal form as follows:

$$\begin{aligned} \dot{z}_1 &= z_2, \\ \dot{z}_2 &= \frac{-z_1 + \epsilon(\frac{z_4 - z_2}{\epsilon \cos z_3})^2 \sin z_3}{1 - \epsilon^2 \cos^2 z_3} + \frac{-\epsilon \cos z_3}{1 - \epsilon^2 \cos^2 z_3} u, \\ \dot{z}_3 &= \dot{x}_3 = x_4 = \frac{z_4 - z_2}{\epsilon \cos z_3}, && (2.105) \\ \dot{z}_4 &= \dot{x}_2 + \dot{x}_4 \epsilon \cos x_3 - x_4 \dot{x}_3 \epsilon \sin x_3 = -z_1, && (2.106) \\ y &= z_1. \end{aligned}$$

The zero dynamics of the RTAC system can be identified from (2.105) and (2.106), which can be put into the form

$$\dot{\eta} = \gamma(\zeta, \eta),$$

where $\zeta = \mathrm{col}(z_1, z_2)$ and $\eta = \mathrm{col}(z_3, z_4)$. The zero dynamics of the system is defined by $\dot{\eta} = \gamma(0, \eta)$ or, what is the same,

$$\dot{z}_3 = \frac{z_4}{\epsilon \cos z_3}, \qquad \dot{z}_4 = 0. \tag{2.107}$$

The Jacobian matrix of the zero dynamics at (0, 0) is

$$J = \begin{bmatrix} 0 & 1/\epsilon \\ 0 & 0 \end{bmatrix}.$$

Since both of the eigenvalues of J are at the origin, we cannot determine the stability of the equilibrium point of the zero dynamics of the system based on the matrix J. Nevertheless, it can be verified that the solution of this equation is given by $\sin(z_3(t)) = \frac{z_4(0)}{\epsilon} t + \sin(z_3(0))$ and $z_4(t) = z_4(0)$. Clearly, the equilibrium point of the zero dynamic is unstable. Therefore, the system is nonminimum phase.

The zero dynamics of the RTAC system with $F = 0$ can also be identified using the algorithm described in Remark 2.46. As a matter of fact, simple calculation gives

$$D(x) = \frac{-\epsilon \cos x_3}{1 - \epsilon^2 \cos^2 x_3}, \qquad E(x) = \frac{-x_1 + \epsilon x_4^2 \sin x_3}{1 - \epsilon^2 \cos^2 x_3},$$

$$u_e(x) = \frac{-E(x)}{D(x)} = \frac{-x_1 + \epsilon x_4^2 \sin x_3}{\epsilon \cos x_3}, \qquad H(x) = \begin{bmatrix} x_1 \\ x_2 \end{bmatrix}.$$

Thus, applying the algorithm described in Remark 2.46 gives the partition $x = \mathrm{col}(x^1, x^2)$, with $x^1 = \mathrm{col}(x_1, x_2)$ and $x^2 = \mathrm{col}(x_3, x_4)$, and the following mapping:

$$x^1 = \sigma(x^2) = \begin{bmatrix} 0 \\ 0 \end{bmatrix}$$

as well as the zero dynamics of the RTAC system as follows:

$$\dot{x}_3 = x_4, \qquad \dot{x}_4 = x_4^2 \tan x_3. \tag{2.108}$$

It can be easily verified that the two representations (2.107) and (2.108) of the zero dynamics are locally diffeomorphic to each other under the coordinate transformation $z_3 = x_3$ and $z_4 = \epsilon x_4 \cos x_3$.

Inverted Pendulum on a Cart [31]. Shown in Figure 2.2 is a system known as the inverted pendulum on a cart. The pendulum is freely hinged to the cart, which is free to move on a

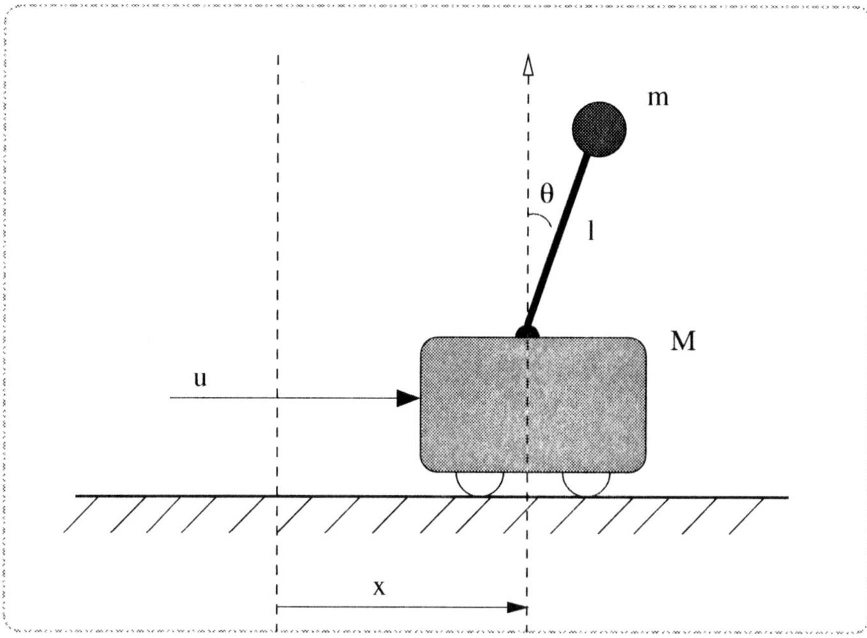

Figure 2.2. *Inverted pendulum on a cart.*

horizontal plane. The control available is a force applicable to the cart. The motion of the system can be described by

$$(M+m)\ddot{x} + ml(\ddot{\theta}\cos\theta - \dot{\theta}^2 \sin\theta) + b\dot{x} = u,$$
$$m(\ddot{x}\cos\theta + l\ddot{\theta} - g\sin\theta) = 0,$$

where M is the mass of the cart, m the mass of the block on the pendulum, l the length of the pendulum, g the acceleration due to gravity, b the coefficient of viscous friction for motion of the cart, θ the angle the pendulum makes with vertical, x the position of the cart, and u the applied force. With the choice of the state variables $x_1 = x, \quad x_2 = \dot{x}, \quad x_3 = \theta, \quad x_4 = \dot{\theta}$, the state-space equations of the system are

$$\begin{aligned}
\dot{x}_1 &= x_2, \\
\dot{x}_2 &= \frac{1}{M + m(\sin x_3)^2}\Big(u + mlx_4^2 \sin x_3 - bx_2 - mg\cos x_3 \sin x_3\Big), \\
\dot{x}_3 &= x_4, \\
\dot{x}_4 &= \frac{1}{l(M + m(\sin x_3)^2)}\Big((M+m)g\sin x_3 - u\cos x_3 \\
&\qquad + bx_2\cos x_3 - mlx_4^2 \sin x_3 \cos x_3\Big), \\
y &= x_1.
\end{aligned} \tag{2.109}$$

We can put the system (2.109) into the following standard form:

$$\dot{x} = f(x) + g(x)u, \qquad y = h(x), \tag{2.110}$$

where

$$x = \begin{bmatrix} x_1 \\ x_2 \\ x_3 \\ x_4 \end{bmatrix}, \qquad g(x) = \begin{bmatrix} 0 \\ \frac{1}{M+m(\sin x_3)^2} \\ 0 \\ \frac{-\cos x_3}{l(M+m(\sin x_3)^2)} \end{bmatrix},$$

$$f(x) = \begin{bmatrix} x_2 \\ \frac{1}{M+m(\sin x_3)^2}(mlx_4^2 \sin x_3 - bx_2 - mg\cos x_3 \sin x_3) \\ x_4 \\ \frac{1}{l(M+m(\sin x_3)^2)}((M+m)g\sin x_3 + bx_2\cos x_3 - mlx_4^2 \sin x_3 \cos x_3) \end{bmatrix},$$

and $h(x) = x_1$.

We can now see that the relative degree of (2.109) is 2, and simple calculation gives

$$\begin{aligned} D(x) &= \frac{1}{M+m(\sin x_3)^2}, \qquad H(x) = \begin{bmatrix} x_1 \\ x_2 \end{bmatrix}, \\ E(x) &= \frac{1}{M+m(\sin x_3)^2}\Big(mlx_4^2 \sin x_3 - bx_2 - mg\cos x_3 \sin x_3\Big), \\ u_e(x) &= \frac{-E(x)}{D(x)} = -(mlx_4^2 \sin x_3 - bx_2 - mg\cos x_3 \sin x_3). \end{aligned}$$

Thus, applying the algorithm described in Remark 2.46 gives the partition $x = \text{col}(x^1, x^2)$, with $x^1 = \text{col}(x_1, x_2)$ and $x^2 = \text{col}(x_3, x_4)$, and the following mapping:

$$x^1 = \sigma(x^2) = \begin{bmatrix} 0 \\ 0 \end{bmatrix}$$

as well as the zero dynamics of the system (2.109):

$$\dot{x}_3 = x_4, \qquad \dot{x}_4 = \frac{g}{l}\sin x_3. \tag{2.111}$$

Simple calculation shows that the Jacobian matrix of the zero dynamics has two eigenvalues at the origin given by $\pm\sqrt{g/l}$. Thus the system is nonminimum phase.

Ball and Beam System. Shown in Figure 2.3 is the ball and beam system. The motion equation of the system can be derived as follows:

$$\begin{aligned} 0 &= \left(\frac{J_b}{R^2} + M\right)\ddot{r} + MG\sin\theta - Mr\dot{\theta}^2, \\ \tau &= (Mr^2 + J + J_b)\ddot{\theta} + 2Mr\dot{r}\dot{\theta} + MGr\cos\theta, \end{aligned} \tag{2.112}$$

where θ and r are the beam angle and the ball position, respectively; τ is the torque applied to the beam; J is the moment of inertia of the beam; M and J_b are the mass and moment of inertia of the ball, respectively; R is the radius of the ball; and G is the acceleration of gravity.

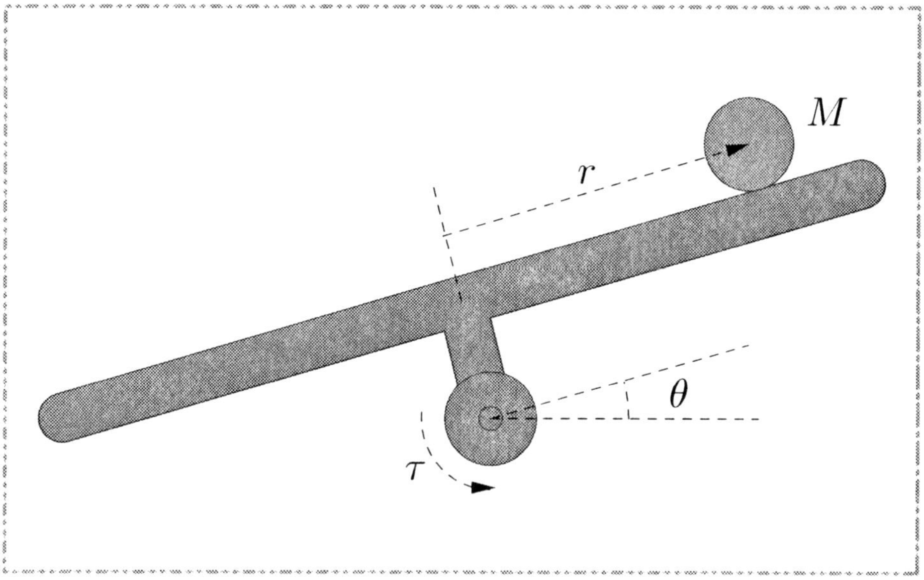

Figure 2.3. *Ball and beam system.*

Letting $x = \mathrm{col}(x_1, x_2, x_3, x_4) = \mathrm{col}(r, \dot{r}, \theta, \dot{\theta})$ and $y = r$ yields the following state-space equations:

$$
\begin{aligned}
\dot{x}_1(t) &= x_2(t),\\
\dot{x}_2(t) &= Hx_1(t)x_4^2(t) - HG\sin x_3(t),\\
\dot{x}_3(t) &= x_4(t),\\
\dot{x}_4(t) &= -\frac{2Mx_1(t)x_2(t)x_4(t) + MGx_1(t)\cos x_3(t)}{Mx_1^2(t) + J + J_b} + \frac{\tau}{Mx_1^2(t) + J + J_b},\\
y(t) &= x_1(t),
\end{aligned}
\tag{2.113}
$$

where $H = \frac{M}{J_b/R^2 + M}$.

An input transformation of the form

$$u = -\frac{2Mx_1x_2x_4 + MGx_1\cos x_3}{Mx_1^2 + J + J_b} + \frac{\tau}{Mx_1^2 + J + J_b}$$

will further simplify the system into the following:

$$
\begin{aligned}
\dot{x}_1(t) &= x_2(t),\\
\dot{x}_2(t) &= Hx_1(t)x_4^2(t) - HG\sin x_3(t),\\
\dot{x}_3(t) &= x_4(t),\\
\dot{x}_4(t) &= u(t),\\
y(t) &= x_1(t),
\end{aligned}
$$

which is in the standard form of (2.7) with

$$f(x) = \begin{bmatrix} x_2 \\ Hx_1x_4^2 - HG\sin x_3 \\ x_4 \\ 0 \end{bmatrix}, \quad g(x) = \begin{bmatrix} 0 \\ 0 \\ 0 \\ 1 \end{bmatrix}, \quad h(x) = x_1.$$

It can be verified that

$$L_f h(x) = x_2, \quad L_f^2 h(x) = Hx_1x_4^2 - HG\sin x_3, \quad L_f^3 h(x) = Hx_1x_4^2 - HGx_4\cos x_3$$

and

$$L_g h(x) = L_g L_f h(x) = 0, \quad L_g L_f^2 h(x) = 2Hx_1x_4.$$

Since there exists no open neighborhood of $x^0 = 0$ in which $L_g L_f^2 h(x) = 2Hx_1x_4$ is identically zero, the relative degree of the ball and beam system is not well defined at $x^0 = 0$.

Chapter 3

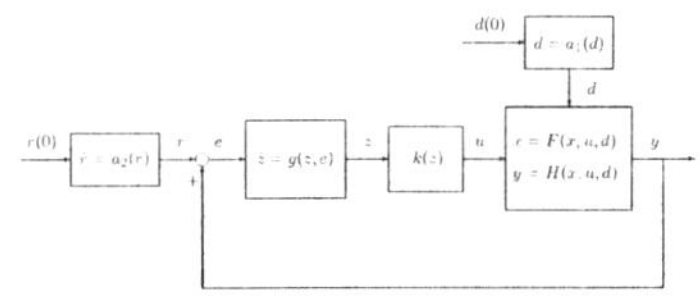

Nonlinear Output Regulation

3.1 Introduction

Beginning with this chapter, we turn to the nonlinear output regulation problem, a nonlinear analog of the linear output regulation problem studied in Chapter 1. The typical scenario studied by the nonlinear output regulation problem is shown in Figure 3.1, where we have a nonlinear plant described by

$$\begin{aligned} \dot{x}(t) &= F(x(t), u(t), d(t)), \quad x(0) = x_0, \\ y(t) &= H(x(t), u(t), d(t)), \quad t \geq 0, \end{aligned} \tag{3.1}$$

where $x(t)$ is the plant state, $u(t)$ the plant input, $y(t)$ the plant output, and $d(t)$ the disturbance signal generated by an exogenous system described by

$$\dot{d}(t) = a_1(d(t)), \quad d(0) = d_0. \tag{3.2}$$

In addition, there is a reference input also generated by an exogenous system

$$\dot{r}(t) = a_2(r(t)), \quad r(0) = r_0. \tag{3.3}$$

The tracking error is defined by

$$e(t) = y(t) - r(t). \tag{3.4}$$

To handle the nonlinear system described in (3.1), we need to go beyond the class of linear control laws described in Chapter 1 and resort to the class of nonlinear feedback control laws. A typical nonlinear feedback control law takes the following form:

$$\begin{aligned} u(t) &= k(z(t)), \\ \dot{z}(t) &= g(z(t), e(t)), \end{aligned} \tag{3.5}$$

where k and g are some nonlinear functions. This control law can be viewed as a nonlinear analog of the linear dynamic output feedback control law (1.49) described in Chapter 1.

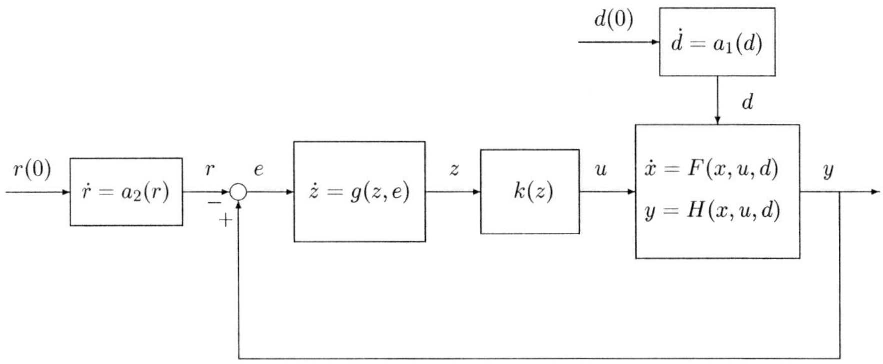

Figure 3.1. *Nonlinear output regulation problem.*

The objective of the control law is that the closed-loop system be stable in the sense to be described later and that the output be able to track the reference input asymptotically in the following sense:

$$\lim_{t\to\infty} (y(t) - r(t)) = 0.$$

The control systems as described in Section 2.8 are all nonlinear. To achieve better system performance, it is desirable to design the control system based on the nonlinear model, thus leading to the nonlinear output regulation problem.

As in the linear case, we can combine the reference input $r(t)$ and disturbance $d(t)$ into a single exogenous signal vector $v = \text{col}(r, d)$, thereby leading to a more compact notation,

$$\dot{v}(t) = a(v(t)), \quad v(0) = v_0. \tag{3.6}$$

As a result, the plant with the tracking error $e(t)$ as the output takes the following form:

$$\begin{aligned} \dot{x}(t) &= f(x(t), u(t), v(t)) \stackrel{def}{=} F(x(t), u(t), d(t)), \\ e(t) &= h(x(t), u(t), v(t)) \stackrel{def}{=} H(x(t), u(t), d(t)) - r(t). \end{aligned} \tag{3.7}$$

Thus, we can focus on the problem of driving the output e of the system of the form (3.7) to zero asymptotically. It should be noted that the plant (3.7) can be viewed as a nonautonomous nonlinear system with x as the state, u as the input, and e as the output. On the other hand, we can put the plant (3.7) and the exosystem (3.6) together as follows:

$$\begin{aligned} \dot{x}(t) &= f(x(t), u(t), v(t)), \\ \dot{v}(t) &= a(v(t)), \\ e(t) &= h(x(t), u(t), v(t)). \end{aligned} \tag{3.8}$$

Then the system (3.8), which is called a composite system, can be viewed as an autonomous nonlinear system with $\text{col}(x, v)$ as the state, u as the input, and e as the output.

Since the plant inevitably contains uncertainties, it is desirable to further require the controller to be able to maintain the property of asymptotic tracking and disturbance rejection in the closed-loop system regardless of model uncertainties. The problem of designing such controllers for nonlinear systems is called *the robust nonlinear output regulation problem*, which will be studied in Chapters 5 to 7. In this chapter, we will focus only on the case where no uncertainty is present. The results are basically extensions of those of Section 1.2 to the nonlinear setting.

In the reminder of this chapter, we first give a precise description of the nonlinear output regulation problem in Section 3.2. In Section 3.3, we study the solvability of the nonlinear output regulation problem. In analogy to the linear case, we give the characterization of the solvability conditions for the problem in terms of a set of constrained nonlinear partial differential equations, which are an extension of the regulator equations given in Chapter 1 and are called the nonlinear regulator equations. In Section 3.4, we study the solvability of the nonlinear regulator equations, through the zero dynamics algorithm, for the class of nonlinear systems whose zero dynamics have a hyperbolic equilibrium. In Section 3.5, we study the output regulation problem of nonlinear systems whose zero dynamics is not hyperbolic. Finally, we study the problem of asymptotic disturbance rejection for the RTAC system in Section 3.6.

3.2 Problem Description

We consider a nonlinear plant described by

$$\begin{aligned} \dot{x}(t) &= f(x(t), u(t), v(t)), \quad x(0) = x_0, \\ e(t) &= h(x(t), u(t), v(t)), \quad t \geq 0, \end{aligned} \tag{3.9}$$

where $x(t)$ is the n-dimensional plant state, $u(t)$ the m-dimensional plant input, $e(t)$ the p-dimensional plant output representing tracking error, and $v(t)$ the q-dimensional disturbance signal which can represent either disturbance signal or the reference input or both. It is assumed that $v(t)$ is generated by a q-dimensional autonomous differential equation

$$\dot{v}(t) = a(v(t)), \quad v(0) = v_0, \quad t \geq 0. \tag{3.10}$$

We will consider two classes of control laws as follows.

1. Static State Feedback:

$$u(t) = k(x(t), v(t)), \tag{3.11}$$

where the function $k(\cdot, \cdot)$ satisfies $k(0, 0) = 0$.

2. Dynamic Measurement Output Feedback:

$$\begin{aligned} u(t) &= k(z(t)), \\ \dot{z}(t) &= g(z(t), y_m(t)), \end{aligned} \tag{3.12}$$

where $z(t)$ is the compensator state of dimension n_z to be specified later, $y_m(t) = h_m(x(t), u(t), v(t))$ is the measurement output of dimension p_m for some integer p_m, and the functions $k(\cdot)$, $h_m(\cdot, \cdot, \cdot)$, and $g(\cdot, \cdot)$ satisfy $k(0) = 0$, $h_m(0, 0, 0) = 0$, and $g(0, 0) = 0$.

The two control laws (3.11) and (3.12) are obviously nonlinear analogs of the linear static state feedback control law (1.10) and the linear dynamic measurement output feedback control law (1.11) described in Chapter 1. It is noted that the dynamic measurement output feedback control law (3.12) is more general than the dynamic error output feedback control as described in (3.5) because it always includes the error output feedback control as a special case by letting $h_m(x, u, v) = h(x, u, v)$. In Section 3.6, we will see that the output regulation problem for the RTAC system is solvable by a measurement output feedback control but not any error output feedback control.

Our requirements will be imposed on the closed-loop composite system, that is, the system consisting of the plant (3.9), the exosystem (3.10), and the controller (3.11) or (3.12) as follows:

$$\begin{aligned} \dot{x}_c(t) &= f_c(x_c(t), v(t)), \quad x_c(0) = x_{c0}, \\ \dot{v}(t) &= a(v(t)), \\ e(t) &= h_c(x_c(t), v(t)), \quad t \geq 0, \end{aligned} \tag{3.13}$$

where, under the static state feedback control, $x_c = x$ and

$$\begin{aligned} f_c(x_c, v) &= f(x, k(x, v), v), \\ h_c(x_c, v) &= h(x, k(x, v), v), \end{aligned} \tag{3.14}$$

and under the dynamic measurement output feedback control, $x_c = \text{col}(x, z)$ and

$$\begin{aligned} h_c(x_c, v) &= h(x, k(z), v)), \\ f_c(x_c, v) &= \begin{bmatrix} f(x, k(z), v) \\ g(z, h_m(x, k(z), v)) \end{bmatrix}. \end{aligned} \tag{3.15}$$

For simplicity, all the functions involved in this setup are assumed to be sufficiently smooth and defined globally on the appropriate Euclidean spaces, with the value zero at the respective origins. Our results will be stated locally in terms of an open neighborhood V of the origin in $\mathcal{R}^q$, and we implicitly permit V to be made smaller to accommodate subsequent local arguments.

Nonlinear Output Regulation Problem (NORP): Design a controller of the form (3.11) or (3.12) such that the closed-loop system has the following two properties.

Property 3.1. For all sufficiently small x_{c0} and v_0, the trajectories $\text{col}(x_c(t), v(t))$ of the closed-loop composite system (3.13) exist and are bounded for all $t \geq 0$, and

Property 3.2. For all sufficiently small x_{c0} and v_0, the trajectory $\text{col}(x_c(t), v(t))$ of the closed-loop composite system (3.13) satisfies

$$\lim_{t\to\infty} e(t) = \lim_{t\to\infty} h_c(x_c(t), v(t)) = 0. \tag{3.16}$$

Remark 3.1. By Definition 2.2, Property 3.1 is guaranteed if the equilibrium point of the closed-loop composite system (3.13) at $\text{col}(x_c, v) = \text{col}(0, 0)$ is stable in the sense of Lyapunov. Moreover, by Theorem 2.27 and Assumption 3.1, to be introduced later, the equilibrium point of the closed-loop composite system (3.13) at $\text{col}(x_c, v) = \text{col}(0, 0)$ is stable in the sense of Lyapunov if the closed-loop composite system has the following property.

Property 3.3. All the eigenvalues of the matrix

$$\frac{\partial f_c}{\partial x_c}(0, 0) \tag{3.17}$$

have negative real parts.

As it is quite straightforward to achieve Property 3.3 by using a linear feedback control under Assumptions 3.2 and/or 3.3 to be given below, we often impose Property 3.3 instead of Property 3.1 on the closed-loop system. We will say that a controller of the form (3.11) or (3.12) solves *the output regulation problem with exponential stability* if it makes the closed-loop composite system (3.13) satisfy Properties 3.2 and 3.3. ∎

The output regulation problem that has just been described is of *local* nature in the sense that the desirable properties imposed on the closed-loop system hold only for sufficiently small initial states of the closed-loop composite system (3.13). Thus the above problem can be more precisely called the *local nonlinear output regulation problem.* Later, we will further study the global nonlinear output regulation problem in the sense to be described in Chapter 7.

If there exists a controller such that the closed-loop system satisfies Properties 3.1 and 3.2, we say that the (local) nonlinear output regulation problem is solvable, and the controller is called a nonlinear *servoregulator*. In particular, the controller in the form of (3.11) is called a *state feedback servoregulator*, and the controller in the form of (3.12) is called a *measurement output feedback servoregulator*. Alternatively, we say that the controller achieves asymptotic tracking and disturbance rejection in the plant.

Various assumptions needed for the solvability of the above problem are listed below.

Assumption 3.1. The equilibrium of exosystem (3.10) at $v = 0$ is Lyapunov stable, and all the eigenvalues of $\frac{\partial a}{\partial v}(0)$ have zero real parts.

Assumption 3.1′. The equilibrium of the exosystem (3.10) at $v = 0$ is Lyapunov stable, and there is an open neighborhood of $v = 0$ in which every point is Poisson stable in the sense to be described in Remark 3.2.

Assumption 3.2. The pair

$$\left(\frac{\partial f}{\partial x}(0, 0, 0),\ \ \frac{\partial f}{\partial u}(0, 0, 0)\right)$$

is stabilizable.

Assumption 3.3. The pair

$$\left(\begin{bmatrix} \frac{\partial h_m}{\partial x}(0, 0, 0) & \frac{\partial h_m}{\partial v}(0, 0, 0) \end{bmatrix},\ \begin{bmatrix} \frac{\partial f}{\partial x}(0, 0, 0) & \frac{\partial f}{\partial v}(0, 0, 0) \\ 0 & \frac{\partial a}{\partial v}(0) \end{bmatrix}\right)$$

is detectable.

Remark 3.2. A point $v^0 \in \mathcal{R}^q$ is said to be Poisson stable if the solution $v(t, v^0)$ exists for all $t \in \mathcal{R}$ and for each open neighborhood V^0 of v^0 and if for any real number $T > 0$, there exists a time $t_1 > T$ such that $v(t_1, v^0) \in V^0$ and a time $t_2 < -T$ such that $v(t_2, v^0) \in V^0$. ■

Remark 3.3. Assumption 3.1 is more restrictive than its linear counterpart Assumption 1.1. For example, it does not accommodate the ramp function. This is because we require that all trajectories of the closed-loop composite system (3.13) starting from sufficiently small initial states be bounded. Thus, we have to exclude any unbounded signals such as the ramp signal. Assumption 3.1′ is a somewhat strengthened version of Assumption 3.1. It always implies Assumption 3.1. Assumption 3.1′ is only used for establishing the necessary condition for the solvability of the output regulation problem and is not essential for our development. Assumption 3.2 guarantees that the plant can be locally stabilized by a state feedback control, and Assumption 3.2 together with Assumption 3.3 guarantees that the plant can be locally stabilized by a measurement output feedback control based on an estimation of the composite state $\text{col}(x, v)$. It is noted that the error output e is always measurable, but the measurement output y_m does not have to be the error output e. Thus, in some cases, for example, the RTAC system to be studied in Section 3.6, the output regulation problem may be solvable by the measurement output feedback but not the error output feedback control. ■

Example 3.4 (RTAC). Consider the RTAC system described in Section 2.8. Our objective is to design a state or measurement output feedback controller such that, despite the presence of a sinusoidal disturbance of the form $F(t) = A_m \sin \omega t$, the closed-loop system is asymptotically stable, and the position of the cart can asymptotically approach the origin. For this purpose, let us introduce the following exosystem:

$$\dot{v} = A_1 v, \quad t \geq 0, \quad v(0) = v_0, \tag{3.18}$$

with

$$v = \begin{bmatrix} v_1 \\ v_2 \end{bmatrix}, \quad A_1 = \begin{bmatrix} 0 & \omega \\ -\omega & 0 \end{bmatrix}, \quad v_0 = \begin{bmatrix} 0 \\ A_m \end{bmatrix}. \tag{3.19}$$

Let $h(x, v) = x_1$. Then the disturbance rejection problem can be formulated as an output regulation problem of the following composite system:

$$\begin{aligned} \dot{x} &= f(x) + g_1(x)u + g_2(x)v_1, \\ \dot{v} &= A_1 v, \quad v(0) = v_0, \\ e &= h(x, v). \end{aligned} \tag{3.20}$$

Assuming that the position of the cart x_1 and the angular position of the proof-mass x_3 are measurable, then as will be shown in Section 3.6, the above output regulation problem is solvable by a dynamic measurement output feedback control with $h_m(x, u, v) = \text{col}(x_1, x_3)$. ■

Example 3.5 (Asymptotic Tracking of Inverted Pendulum on a Cart). Consider the problem of designing a state or output feedback controller for the inverted pendulum on a cart system described in Section 2.8 such that the position of the cart can asymptotically track a sinusoidal input $y_d(t) = A_m \sin \omega t$. For this purpose, we need to design a feedback controller to locally stabilize the closed-loop system and to achieve

$$\lim_{t\to\infty} (y(t) - y_d(t)) = 0. \tag{3.21}$$

To this end, again we can introduce the same exosystem as described in (3.18) and (3.19). Then, clearly, $y_d(t) = v_1(t)$. Let $h(x, v) = x_1 - v_1$. Then, the above asymptotic tracking problem can be formulated as the output regulation problem of the following composite system:

$$\begin{aligned} \dot{x} &= f(x) + g(x)u, \\ \dot{v} &= A_1 v, \quad v(0) = v_0, \\ e &= h(x, v). \end{aligned} \tag{3.22}$$

We will show in Chapter 4 that the output regulation problem for this system is solvable by either state feedback control or error output feedback control. ∎

3.3 Solvability of the Nonlinear Output Regulation Problem

The idea of synthesizing a controller to solve the nonlinear output regulation problem is similar to what has been used to solve the linear output regulation problem, that is, using a feedback control to achieve Property 3.3 and a feedforward control to achieve Property 3.2. Since Property 3.3 is a property of the linearization of the plant, it can be achieved by the same control techniques as used in Chapter 1 based on Lyapunov's linearization method. However, in the present case, the feedforward control is much more difficult to find since, as will be seen shortly, it is determined by a set of nonlinear partial differential and algebraic equations, which is a nonlinear analog of the regulator equations encountered in Chapter 1. In this section, we will focus on relating the solvability of the nonlinear output regulation problem to that of the nonlinear regulator equations. Solvability of the nonlinear regulator equations will be given only for the special case where the exogenous signals are constant. The more general case will be studied in the next section.

We first establish a result parallel to Lemma 1.4.

Lemma 3.6. *Under Assumption 3.1′, suppose the closed-loop composite system (3.13) resulting from the controller (3.11) or (3.12) has Property 3.3. Then, it also has Property 3.2 if and only if there exists a sufficiently smooth function* $\mathbf{x}_c(v)$ *with* $\mathbf{x}_c(0) = 0$ *that satisfies, for* $v \in V$*, where* V *is an open neighborhood of* $0 \in \mathcal{R}^q$*, the following partial differential equations:*

$$\frac{\partial \mathbf{x}_c}{\partial v} a(v) = f_c(\mathbf{x}_c(v), v), \tag{3.23}$$

$$0 = h_c(\mathbf{x}_c(v), v). \tag{3.24}$$

Proof. First note that Assumption 3.1′ implies Assumption 3.1; thus the exosystem has a stable equilibrium point at the origin and all the eigenvalues of its Jacobian matrix have zero real parts. Since the closed-loop composite system (3.13) has Property 3.3, by Theorem 2.25, there exists a center manifold for the closed-loop composite system (3.13). That is, there exists a sufficiently smooth function $\mathbf{x}_c(v)$ with $\mathbf{x}_c(0) = 0$ that satisfies (3.23) for $v \in V$. Moreover, by Theorem 2.27, the equilibrium point of the closed-loop system (3.13) at the origin is Lyapunov stable. Thus, the solution of the closed-loop composite system (3.13) starting from any sufficiently small initial state exists for all $t \geq 0$.

(If part): Since the function $\mathbf{x}_c(v)$ with $\mathbf{x}_c(0) = 0$ that satisfies (3.23) for $v \in V$ defines a center manifold $x_c = \mathbf{x}_c(v)$ for the closed-loop composite system (3.13), by Theorem 2.28, there exist positive constants δ and λ such that for all sufficiently small $x_c(0)$ and $v(0)$, the trajectories $\text{col}(x_c(t),\ v(t))$ of the closed-loop composite system (3.13) satisfy

$$||x_c(t) - \mathbf{x_c}(v(t))|| \leq \delta e^{-\lambda t}||x_c(0) - \mathbf{x_c}(v(0))||, \quad t \geq 0. \tag{3.25}$$

Furthermore, there exists a compact set S_c in $\mathcal{R}^{n+n_z+q}$ such that, for $t \geq 0$, $\text{col}(x_c(t),\ v(t)) \in S_c$, $\text{col}(\mathbf{x}_c(v(t)),\ v(t)) \in S_c$. Also, there exists a finite constant L such that

$$\left\| \frac{\partial h_c}{\partial x_c}(x_c,\ v) \right\| < L \tag{3.26}$$

for $(x_c,\ v) \in S_c$. Thus, if the function $\mathbf{x}_c(v)$ also satisfies (3.24), then we have

$$\begin{aligned} \lim_{t\to\infty} ||e(t)|| &= \lim_{t\to\infty} ||h_c(x_c(t),\ v(t))|| \\ &= \lim_{t\to\infty} ||h_c(x_c(t),\ v(t)) - h_c(\mathbf{x_c}(v(t)),\ v(t))|| \\ &\leq \lim_{t\to\infty} L||x_c(t) - \mathbf{x_c}(v(t))|| = 0; \end{aligned} \tag{3.27}$$

that is, the closed-loop system also has Property 3.2.

(Only if part): Assume the closed-loop system has both Property 3.2 and Property 3.3, yet (3.24) is not true. Then there exists a sufficiently small $v_0 \in V$ such that the solution of the closed-loop system (3.13) satisfying $\text{col}(x_c(0),\ v(0)) = \text{col}(\mathbf{x}_c(v_0),\ v_0)$, denoted by $\text{col}(x_c(t,\mathbf{x}_c(v_0)),\ v(t,\ v_0))$, exists for all $t \geq 0$ and satisfies

$$\lim_{t\to\infty} ||h_c(x_c(t,\mathbf{x}_c(v_0)),\ v(t,\ v_0))|| = 0, \tag{3.28}$$

yet

$$||h_c(\mathbf{x}_c(v_0),\ v_0)|| > 0.$$

Thus there exists an open neighborhood $V_0 \subset V$ of v_0 and some real number $R > 0$ such that

$$||h_c(\mathbf{x}_c(v),\ v)|| > R$$

for all $v \in V_0$. Clearly, $x_c(t,\mathbf{x}_c(v_0)) = \mathbf{x}_c(v(t,\ v_0))$, since $x_c(0,\mathbf{x}_c(v_0)) = \mathbf{x}_c(v_0) = \mathbf{x}_c(v(0,\ v_0))$, and (3.23) implies

$$\frac{d\mathbf{x}_c}{dt} = f_c(\mathbf{x}_c(v(t,\ v_0)),\ v(t,\ v_0)), \quad t \geq 0.$$

But, since the exosystem satisfies Assumption 3.1′, we can assume that v_0 is small enough so that it is Poisson stable. Therefore, given any $T > 0$, there exists $t_1 > T$ such that $v(t_1, v_0) \in V_0$. Thus,

$$||h_c(x_c(t_1, \mathbf{x}_c(v_0)), v(t_1, v_0))|| = ||h_c(\mathbf{x}_c(v(t_1, v_0)), v(t_1, v_0))|| > R,$$

which contradicts (3.28). □

In what follows, we call the manifold $x_c = \mathbf{x}_c(v)$, where $\mathbf{x}_c(v)$ satisfies (3.23) and (3.24), a *zero error center manifold* for (3.13).

Remark 3.7. A systemic interpretation to Lemma 3.6 can be given as follows. First consider the special case where the exogenous signals are constant. Then, (3.23) and (3.24) reduce to the following algebraic equations:

$$\begin{aligned} 0 &= f_c(\mathbf{x}_c(v), v), \\ 0 &= h_c(\mathbf{x}_c(v), v), \end{aligned} \tag{3.29}$$

since $a(v) = 0$ in this case. Thus, the solution $\mathbf{x}_c(v)$ of (3.29) defines an equilibrium manifold $M_c = \{(x_c, v) \in \mathcal{R}^{n+n_z} \times V \mid x_c = \mathbf{x}_c(v)\ \}$ of the closed-loop composite system on which the output is identically zero. For the general case, the existence of the sufficiently smooth function $\mathbf{x}_c(v)$ satisfying (3.23) simply says that the manifold M_c is a *stable center manifold* of the closed-loop composite system (3.13). Thus the trajectory $\text{col}(x(t), v(t))$ of the closed-loop composite system (3.13) starting from any sufficiently small initial state $\text{col}(x(0), v(0))$ will approach this manifold asymptotically. The fact that $\mathbf{x}_c(v)$ also satisfies (3.24) means that the center manifold M_c is contained in the kernel of the output mapping $h_c(x_c, v)$. Thus, as the trajectory approaches the center manifold, the output e will approach zero asymptotically. Lemma 3.6 has also led to an equivalent characterization of Property 3.2 in terms of a set of partial differential and algebraic equations resulting from the center manifold theory. Thus the asymptotic property of the system can be addressed using the center manifold theory. Also, we emphasize that Assumption 3.1′ is only used for establishing the necessary condition. It suffices to use Assumption 3.1 to establish the sufficient condition. ■

Next we will establish the solvability of the state feedback output regulation problem in terms of the given plant.

Theorem 3.8. *Under Assumptions 3.1′ and 3.2, the nonlinear output regulation problem with exponential stability is solvable by a static state feedback control of the form (3.11) if and only if there exist two sufficiently smooth functions* $\mathbf{x}(v)$ *and* $\mathbf{u}(v)$ *defined for* $v \in V$ *satisfying* $\mathbf{x}(0) = 0$ *and* $\mathbf{u}(0) = 0$ *such that*

$$\begin{aligned} \frac{\partial \mathbf{x}}{\partial v} a(v) &= f(\mathbf{x}(v), \mathbf{u}(v), v), \\ 0 &= h(\mathbf{x}(v), \mathbf{u}(v), v). \end{aligned} \tag{3.30}$$

Proof. Assume a controller of the form $u = k(x, v)$ solves the nonlinear output regulation problem. Then, by Lemma 3.6, there exists a sufficiently smooth function $\mathbf{x}_c(v)$ that satisfies

(3.23) and (3.24) for $v \in V$. Let $\mathbf{x}(v) = \mathbf{x}_c(v)$ and $\mathbf{u}(v) = k(\mathbf{x}(v), v)$. Then, $\mathbf{x}(v)$ and $\mathbf{u}(v)$ satisfy (3.30). On the other hand, assume $\mathbf{x}(v)$ and $\mathbf{u}(v)$ satisfy (3.30) for $v \in V$. Let $K_x \in \mathcal{R}^{m\times n}$ be any constant matrix such that the eigenvalues of the following matrix:

$$\frac{\partial f}{\partial x}(0, 0, 0) + \frac{\partial f}{\partial u}(0, 0, 0)K_x \tag{3.31}$$

have negative real parts. Due to Assumption 3.2, K_x always exists. Let

$$k(x, v) = \mathbf{u}(v) + K_x(x - \mathbf{x}(v)). \tag{3.32}$$

Then, the closed-loop system (3.13) under $k(x, v)$ satisfies Property 3.3. Moreover, letting $\mathbf{x}_c(v) = \mathbf{x}(v)$ leads to

$$\begin{aligned} f_c(\mathbf{x}_c(v), v) &= f(\mathbf{x}_c(v), k(\mathbf{x}_c(v), v), v) = f(\mathbf{x}(v), \mathbf{u}(v), v) = \frac{\partial \mathbf{x}}{\partial v}a(v) = \frac{\partial \mathbf{x}_c}{\partial v}a(v), \\ h_c(\mathbf{x}_c(v), v) &= h(\mathbf{x}_c(v), k(\mathbf{x}_c(v), v), v) = h(\mathbf{x}(v), \mathbf{u}(v), v) = 0, \end{aligned}$$

as $\mathbf{x}(v)$ and $\mathbf{u}(v)$ satisfy the regulator equations (3.30). By Lemma 3.6, the controller solves the nonlinear output regulation problem. □

Remark 3.9. Equations (3.30) are clearly a nonlinear analogue of the linear regulator equations (1.21) encountered in Chapter 1. In fact, suppose equations (3.9) and (3.10) are linear, that is,

$$\begin{aligned} f(x, u, v) &= Ax + Bu + Ev, \\ h(x, u, v) &= Cx + Du + Fv, \\ a(v) &= A_1 v, \end{aligned}$$

where A, B, E, C, D, F, and A_1 are constant matrices of appropriate dimensions. Let $\mathbf{x}(v) = Xv$ and $\mathbf{u}(v) = Uv$ for some matrices X and U. Then equations (3.30) become

$$\begin{aligned} XA_1v &= AXv + BUv + Ev, \\ 0 &= CXv + DUv + Fv. \end{aligned} \tag{3.33}$$

Since equations (3.33) hold for all $v \in V$, they are equivalent to the following:

$$\begin{aligned} XA_1 &= AX + BU + E, \\ 0 &= CX + DU + F, \end{aligned} \tag{3.34}$$

which are exactly the linear regulator equations (1.21). ∎

Remark 3.10. We can also give a systemic interpretation to Theorem 3.8. First consider the special case where the exogenous signals are constant. The nonlinear regulator equations (3.30) are reduced to the following algebraic equations:

$$\begin{aligned} 0 &= f(\mathbf{x}(v), \mathbf{u}(v), v), \\ 0 &= h(\mathbf{x}(v), \mathbf{u}(v), v). \end{aligned} \tag{3.35}$$

The solution of (3.35) gives the desired control $\mathbf{u}(v)$ under which the plant has an equilibrium state $\mathbf{x}(v)$ at which the output is identically zero. For the general case, the solvability of the regulator equations (3.30) simply means that the composite system (3.8) has an output zeroing manifold characterized by

$$M = \{(x, v) \in \mathcal{R}^n \times V \mid x = \mathbf{x}(v)\}. \tag{3.36}$$

In fact, the first equation of (3.30) means that M is a control invariant manifold of the composite system (3.8) rendered by the state feedback control $u = \mathbf{u}(v)$, and the second equation of (3.30) means that this manifold is contained in the kernel of the output mapping $h(x, \mathbf{u}(v), v)$. Thus, Theorem 3.8 can be interpreted as follows: if the composite system has an output zeroing manifold as defined by the solution of the regulator equations (3.30), and the plant satisfies Assumption 3.2, then there exists a state feedback control $u = k(x, v)$ such that the output zeroing manifold M is also a stable center manifold M_c of the closed-loop composite system (3.13) which is contained in the kernel of the mapping $h_c(x_c, v)$. Note that $\mathbf{x}(v)$ can be viewed as the *steady-state* state of the closed-loop system since the trajectory $x_c(t)$ of the closed-loop system starting from any sufficiently small initial state $(x_c(0), v(0))$ necessarily satisfies, by (3.25),

$$\lim_{t\to\infty} (x_c(t) - \mathbf{x}(v(t))) = \lim_{t\to\infty} (x_c(t) - \mathbf{x}_c(v(t))) = 0.$$

Correspondingly, the control input also approaches its steady state in the following sense:

$$\begin{aligned}
&\lim_{t\to\infty} (u(t) - \mathbf{u}(v(t))) \\
&\quad = \lim_{t\to\infty} (u(t) - k(\mathbf{x}(v(t)), v(t))) \\
&\quad = \lim_{t\to\infty} (u(t) - k(x_c(t), v(t))) + \lim_{t\to\infty} (k(x_c(t), v(t)) - k(\mathbf{x}(t), v(t))) \\
&\quad = \lim_{t\to\infty} (u(t) - k(x_c(t), v(t))) \\
&\quad = 0. \quad \blacksquare
\end{aligned}$$

By the same token as Remark 1.8, we will call the functions $\mathbf{u}(v)$ and $\mathbf{x}(v)$ the *zero-error constrained input* and *zero-error constrained state* for the plant and the exosystem, respectively.

In the linear case, the solvability of the regulator equations can be related to the locations of the system's transmission zeros. For the nonlinear case, a similar condition can also be established. Here we only study the special case when the exogenous signals are constant. The general case will be studied in Section 3.4.

Proposition 3.11. *Under the assumption that the exogenous signals are constant, there exist sufficiently smooth functions* $\mathbf{u}(v)$ *and* $\mathbf{x}(v)$ *satisfying equations (3.35) if*

$$\operatorname{rank} \begin{bmatrix} \frac{\partial f}{\partial x}(0,0,0) & \frac{\partial f}{\partial u}(0,0,0) \\ \frac{\partial h}{\partial x}(0,0,0) & \frac{\partial h}{\partial u}(0,0,0) \end{bmatrix} = n + p. \tag{3.37}$$

Proof. The conclusion is a straightforward application of the Implicit Function Theorem. □

Remark 3.12. Let

$$\hat{x} = x - \mathbf{x}(v), \;\; \hat{u} = u - \mathbf{u}(v). \tag{3.38}$$

Then

$$\begin{aligned} \dot{\hat{x}} &= \hat{f}(\hat{x}, \hat{u}, v) \stackrel{def}{=} f(\hat{x} + \mathbf{x}(v), \hat{u} + \mathbf{u}(v), v) - f(\mathbf{x}(v), \mathbf{u}(v), v), \\ e &= \hat{h}(\hat{x}, \hat{u}, v) \stackrel{def}{=} h(\hat{x} + \mathbf{x}(v), \hat{u} + \mathbf{u}(v), v). \end{aligned} \tag{3.39}$$

It can easily be verified that $\hat{f}$ and $\hat{h}$ satisfy

$$0 = \hat{f}(0, 0, v), \qquad 0 = \hat{h}(0, 0, v). \tag{3.40}$$

Thus, if any state feedback controller of the form $\hat{u} = k(\hat{x}, v)$ with $k(0, v) = 0$ stabilizes the equilibrium point at the origin of the system $\dot{\hat{x}} = \hat{f}(\hat{x}, \hat{u}, v)$, then the state feedback controller $u = \mathbf{u}(v) + k(x - \mathbf{x}(v), v)$ solves the output regulation problem of the original system. Therefore, the solution of the regulator equations provides a coordinate and input transformation such that the stabilization solution of the transformed system (3.39) leads to the solution of the output regulation problem of the original plant. ∎

Remark 3.13. Once the solution of equations (3.30) is available, there are a variety of ways to synthesize a state feedback servoregulator $k(x, v)$. In fact, it can be verified that any controller of the form $u = k(x, v)$ satisfying $k(\mathbf{x}(v), v) = \mathbf{u}(v)$ will make $\mathbf{x}_c(v) \stackrel{def}{=} \mathbf{x}(v)$ satisfy equations (3.23) and (3.24). If, in addition, the controller also renders all eigenvalues of the matrix (3.31) negative real parts, then the controller solves the state feedback output regulation problem. Clearly, the controller given in (3.32) satisfies the above conditions. A more general controller is given as follows:

$$k(x, v) = \mathbf{u}(v) - \hat{k}(0, v) + \hat{k}(x(t) - \mathbf{x}(v), v), \tag{3.41}$$

where $\hat{k}(x, v)$ is any state feedback control such that the closed-loop system satisfies Property 3.3. For example, let $K(v)$ be a sufficiently smooth function such that all the eigenvalues of the matrix

$$\frac{\partial f}{\partial x}(\mathbf{x}(v), \mathbf{u}(v), v) + \frac{\partial f}{\partial u}(\mathbf{x}(v), \mathbf{u}(v), v)\; K(v) \tag{3.42}$$

are fixed complex numbers with negative real parts for all v in an open neighborhood V of the origin of $\mathcal{R}^q$. Let $\hat{k}(x, v) = K(v)x$. Then (3.41) gives

$$k(x, v) = \mathbf{u}(v) + K(v)(x - \mathbf{x}(v)). \tag{3.43}$$

This controller can uniformly place the eigenvalues of the linearization of the closed-loop system to be fixed complex numbers for all $v \in V$ and is expected to be able to accommodate larger exogenous signals. ∎

Example 3.14. To illustrate the mechanism of the design process, consider the following example:

$$\begin{aligned} \begin{bmatrix} \dot{x}_1 \\ \dot{x}_2 \end{bmatrix} &= \begin{bmatrix} -x_1 - x_2 + v_1 \\ 1 - e^{-x_2} + u \end{bmatrix}, \\ e &= x_1 + v_1 - v_2, \end{aligned} \tag{3.44}$$

where the disturbance signal v_1 and reference input v_2 are generated by the following exosystem:

$$\dot{v}_1 = 0, \qquad \dot{v}_2 = 0.$$

For this simple system, the regulator equations (3.35) can easily be solved to give the following solution:

$$\begin{bmatrix} \mathbf{x}_1(v_1, v_2) \\ \mathbf{x}_2(v_1, v_2) \end{bmatrix} = \begin{bmatrix} v_2 - v_1 \\ 2v_1 - v_2 \end{bmatrix},$$
$$\mathbf{u}(v_1, v_2) = e^{v_2 - 2v_1} - 1.$$

The Jacobian linearization of this system along the output zeroing manifold is given by

$$\frac{\partial f}{\partial x}(\mathbf{x}(v), \mathbf{u}(v), v) = \begin{bmatrix} -1 & -1 \\ 0 & e^{-2v_1+v_2} \end{bmatrix}, \quad \frac{\partial f}{\partial u}(\mathbf{x}(v), \mathbf{u}(v), v) = \begin{bmatrix} 0 \\ 1 \end{bmatrix}.$$

Given a Hurwitz polynomial, for example,

$$p(\lambda) = (\lambda + 2)^2,$$

we can compute a feedback gain $K(v_1, v_2)$ such that the eigenvalues of the matrix (3.42) are given by the roots of the above polynomial for all v_1 and v_2. Doing so yields

$$K(v_1, v_2) = \begin{bmatrix} 1 & -3 - e^{(v_2 - 2v_1)} \end{bmatrix}.$$

Then a state feedback controller of the form (3.43) is given by

$$\begin{aligned} u = k(x, v) &= \mathbf{u}(v_1, v_2) + K(v_1, v_2)(x - \mathbf{x}(v_1, v_2)) \\ &= e^{(v_2 - 2v_1)} - 1 + (x_1 - v_2 + v_1) - (3 + e^{(v_2 - 2v_1)})(x_2 - 2v_1 + v_2). \end{aligned} \tag{3.45}$$

If, instead of controller (3.45), a controller of the form (3.32), that is,

$$u = k(x, v) = \mathbf{u}(v_1, v_2) + K(0, 0)(x - \mathbf{x}(v_1, v_2)), \tag{3.46}$$

is adopted, then the Jacobian matrix of the closed-loop system on the manifold $\{(x, v) \mid x = \mathbf{x}(v)\}$ is

$$\frac{\partial f}{\partial x}(\mathbf{x}(v), \mathbf{u}(v), v) + \frac{\partial f}{\partial u}(\mathbf{x}(v), \mathbf{u}(v), v)\ K(0, 0) = \begin{bmatrix} -1 & -1 \\ 1 & e^{(v_2 - 2v_1)} - 4 \end{bmatrix},$$

which is unstable for all (v_1, v_2) such that $5 - e^{(v_2 - 2v_1)} < 0$. ∎

When the plant state and/or disturbance state is not available, one can consider using the measurement output feedback controller to solve the output regulation problem. The basic idea is similar to what has been used in Chapter 1 and is described as follows. Consider a dynamic controller of the form

$$u(t) = k(z_1(t), z_2(t)), \tag{3.47}$$
$$\dot{z}(t) = g(z(t), y_m(t)), \tag{3.48}$$

where $\text{col}(z_1,\ z_2) = z$ with $z_1 \in \mathcal{R}^n$ and $z_2 \in \mathcal{R}^q$, and $g(z, y_m)$ and $k(z_1, z_2)$ are such that the solution of the closed-loop composite system composed of the composite system (3.8) and the controller (3.47) and (3.48) satisfies, for all sufficiently small initial states $\text{col}(x(0), v(0), z(0))$,

$$\lim_{t \to \infty} \left(z(t) - \begin{bmatrix} x(t) \\ v(t) \end{bmatrix} \right) = 0.$$

In other words, the dynamic system (3.48) can be considered as a (local) asymptotic observer of the composite system (3.8).

To implement the above idea, we first establish a result that translates the requirement on the closed-loop system as given by Lemma 3.6 into the requirements on the controller (3.12).

Lemma 3.15. *Under Assumption 3.1′, suppose there exists a dynamic measurement output feedback control law of the form (3.12) such that the closed-loop system (3.13) has Property 3.3. Then the following are equivalent:*

(i) *The nonlinear output regulation problem is solvable by the dynamic measurement output feedback controller (3.12).*

(ii) *There exists a sufficiently smooth function $\mathbf{x}_c(v)$ with $\mathbf{x}_c(0) = 0$ such that*

$$\begin{aligned} \frac{\partial \mathbf{x}_c}{\partial v} a(v) &= f_c(\mathbf{x}_c(v), v), \\ 0 &= h_c(\mathbf{x}_c(v), v). \end{aligned} \tag{3.49}$$

(iii) *There exist sufficiently smooth functions $(\mathbf{x}(v), \mathbf{u}(v), \mathbf{z}(v))$ with $(\mathbf{x}(0), \mathbf{u}(0), \mathbf{z}(0)) = (0, 0, 0)$ such that $\mathbf{x}(v)$ and $\mathbf{u}(v)$ are the solution of the nonlinear regulator equations (3.30) and $\mathbf{z}(v)$ is the solution of the nonlinear partial differential equation*

$$\frac{\partial \mathbf{z}}{\partial v} a(v) = g(\mathbf{z}(v), h_m(\mathbf{x}(v), \mathbf{u}(v), v)), \tag{3.50}$$

which satisfies

$$\mathbf{u}(v) = k(\mathbf{z}(v)). \tag{3.51}$$

Proof. (i) $\leftrightarrow$ (ii). This has actually been done by the proof of Lemma 3.6.

(ii) $\leftrightarrow$ (iii). Assume (ii) holds. Partition $\mathbf{x}_c(v)$ as

$$\mathbf{x}_c(v) = \begin{bmatrix} \mathbf{x}(v) \\ \mathbf{z}(v) \end{bmatrix}, \tag{3.52}$$

where $\mathbf{x}(v) \in \mathcal{R}^n$ and $\mathbf{z}(v) \in \mathcal{R}^{n_z}$. Since $(f_c(x_c, v), h_c(x_c, v))$ is given by (3.15), expanding (3.49) gives

$$\begin{aligned} \frac{\partial \mathbf{x}}{\partial v} a(v) &= f(\mathbf{x}(v), k(\mathbf{z}(v)), v), \\ \frac{\partial \mathbf{z}}{\partial v} a(v) &= g(\mathbf{z}(v), h_m(\mathbf{x}(v), k(\mathbf{z}(v)), v)), \\ 0 &= h(\mathbf{x}(v), k(\mathbf{z}(v)), v). \end{aligned} \tag{3.53}$$

Letting $\mathbf{u}(v) = k(\mathbf{z}(v))$ gives (3.51), and using (3.51) in the second equation of (3.53) gives (3.50). Finally, using (3.51) in the first and the third equations of (3.53) shows that $\mathbf{x}(v)$ and $\mathbf{u}(v)$ satisfy the regulator equations (3.30). On the other hand, assume (iii) holds with $(\mathbf{x}(v), \mathbf{u}(v))$ being the solution of the regulator equations (3.30). Let $\mathbf{z}(v)$ be the solution of (3.50) that satisfies (3.51). We need to show that $(\mathbf{x}(v), \mathbf{z}(v))$ satisfies (3.53). To this end, using (3.51) in (3.50) gives the second equation of (3.53), and using (3.51) in (3.30) shows that $(\mathbf{x}(v), \mathbf{z}(v))$ satisfy the first and third equations of (3.53). Let $\mathbf{x}_c(v)$ be given by (3.52). Then clearly (3.53) implies that $\mathbf{x}_c(v)$ satisfies (3.49). □

Theorem 3.16. *Under Assumptions 3.1′, 3.2, and 3.3, the nonlinear output regulation problem with exponential stability is solvable by a dynamic measurement output feedback control law of the form (3.12) if and only if there exist two sufficiently smooth functions* $\mathbf{x}(v)$ *and* $\mathbf{u}(v)$ *with* $\mathbf{x}(0) = 0$ *and* $\mathbf{u}(0) = 0$ *that satisfy the nonlinear regulator equations (3.30).*

Proof. The necessity part is actually implied by the equivalence of parts (i) and (iii) of Lemma 3.15. To show the sufficiency part, note that, by Theorem 3.8, under Assumptions 3.1 and 3.2 and the assumption that there exist sufficiently smooth functions $\mathbf{x}(v)$ and $\mathbf{u}(v)$ with $\mathbf{x}(0) = 0$ and $\mathbf{u}(0) = 0$ that satisfy regulator equations (3.30), there exists a static state feedback control law of the form $u = k(x, v)$ satisfying $\mathbf{u}(v) = k(\mathbf{x}(v), v)$ that solves the state feedback nonlinear output regulation problem. By Assumption 3.3, there exist constant matrices L_1 and L_2 such that all the eigenvalues of the matrix

$$A_L \stackrel{def}{=} \begin{bmatrix} \frac{\partial f}{\partial x}(0,0,0) & \frac{\partial f}{\partial v}(0,0,0) \\ 0 & \frac{\partial a}{\partial v}(0) \end{bmatrix} - \begin{bmatrix} L_1 \\ L_2 \end{bmatrix} \begin{bmatrix} \frac{\partial h_m}{\partial x}(0,0,0) & \frac{\partial h_m}{\partial v}(0,0,0) \end{bmatrix}$$

have negative real parts. Let $z = \operatorname{col}(z_1, z_2)$ with $z_1 \in \mathcal{R}^n$ and $z_2 \in \mathcal{R}^q$, and

$$\begin{aligned} u &= k(z) = k(z_1, z_2), \\ \dot{z} &= g(z, y_m) = \begin{bmatrix} f(z_1, k(z_1, z_2), z_2) + L_1(y_m - h_m(z_1, k(z_1, z_2), z_2)) \\ a(z_2) + L_2(y_m - h_m(z_1, k(z_1, z_2), z_2)) \end{bmatrix}. \end{aligned} \tag{3.54}$$

This controller yields a closed-loop system with

$$x_c = \operatorname{col}(x, z_1, z_2), \quad h_c(x_c, v) = h(x, k(z_1, z_2), v),$$

and

$$f_c(x_c, v) = \begin{bmatrix} f(x, k(z_1, z_2), v) \\ f(z_1, k(z_1, z_2), z_2) + L_1[h_m(x, k(z_1, z_2), v) - h_m(z_1, k(z_1, z_2), z_2)] \\ a(z_2) + L_2[h_m(x, k(z_1, z_2), v) - h_m(z_1, k(z_1, z_2), z_2)] \end{bmatrix}. \tag{3.55}$$

We first show that the closed-loop system has Property 3.3. For convenience of the notation, let

$$\begin{aligned} A &= \frac{\partial f}{\partial x}(0,0,0), & B &= \frac{\partial f}{\partial u}(0,0,0), & E &= \frac{\partial f}{\partial v}(0,0,0), \\ C_m &= \frac{\partial h_m}{\partial x}(0,0,0), & F_m &= \frac{\partial h_m}{\partial v}(0,0,0), & A_1 &= \frac{\partial a}{\partial v}(0), \\ K_x &= \frac{\partial k}{\partial x}(0,0), & K_v &= \frac{\partial k}{\partial v}(0,0), & A_c &= \frac{\partial f_c}{\partial x_c}(0,0). \end{aligned}$$

Then, a simple calculation gives

$$A_c = \begin{bmatrix} A & BK_x & BK_v \\ 0 & A+BK_x & E+BK_v \\ 0 & 0 & A_1 \end{bmatrix} + \begin{bmatrix} 0 \\ L_1 \\ L_2 \end{bmatrix} \begin{bmatrix} C_m & -C_m & -F_m \end{bmatrix}. \quad (3.56)$$

As in the proof of Theorem 1.14, in (3.56), subtracting the first row from the second row and adding the second column to the first column shows that A_c is equivalent to

$$\begin{bmatrix} A+BK_x & BK_x & BK_v \\ 0 & A & E \\ 0 & 0 & A_1 \end{bmatrix} + \begin{bmatrix} 0 \\ L_1 \\ L_2 \end{bmatrix} \begin{bmatrix} 0 & -C_m & -F_m \end{bmatrix}$$
$$= \begin{bmatrix} A+BK_x & BK_x & BK_v \\ 0 & A-L_1C_m & E-L_1F_m \\ 0 & -L_2C_m & A_1-L_2F_m \end{bmatrix}. \quad (3.57)$$

Thus $\sigma(A_c) = \sigma(A+BK_x) \cup \sigma(A_L)$.

Next we show that there exists a sufficiently smooth function $\mathbf{z}(v)$ with $\mathbf{z}(0) = 0$ that satisfies equations (3.50) and (3.51). Indeed, let $(\mathbf{x}(v), \mathbf{u}(v))$ be the solution of the regulator equations (3.30), let $\mathbf{z}_1(v) = \mathbf{x}(v)$ and $\mathbf{z}_2(v) = v$, and let

$$\mathbf{z}(v) = \begin{bmatrix} \mathbf{z}_1(v) \\ \mathbf{z}_2(v) \end{bmatrix} = \begin{bmatrix} \mathbf{x}(v) \\ v \end{bmatrix}.$$

Then,

$$k(\mathbf{z}(v)) = k(\mathbf{z}_1(v), \mathbf{z}_2(v)) = k(\mathbf{x}(v), v) = \mathbf{u}(v) \quad (3.58)$$

and

$$\frac{\partial \mathbf{z}}{\partial v} a(v) = \begin{bmatrix} \frac{\partial \mathbf{x}}{\partial v} a(v) \\ \frac{\partial v}{\partial v} a(v) \end{bmatrix}$$
$$= \begin{bmatrix} f(\mathbf{x}(v), \mathbf{u}(v), v) \\ a(v) \end{bmatrix}$$
$$= \begin{bmatrix} f(\mathbf{x}(v), k(\mathbf{z}(v)), v) \\ a(v) \end{bmatrix} = g(\mathbf{z}(v), h_m(\mathbf{x}(v), \mathbf{u}(v), v)). \quad \square$$

Remark 3.17. From the statement of Lemma 3.15, we can see that a measurement output feedback servoregulator of the form (3.12) can be characterized as follows:

(i) It makes the closed-loop system satisfy Property 3.3.

(ii) It is such that the following equation:

$$\frac{\partial \mathbf{z}}{\partial v} a(v) = g(\mathbf{z}(v), h_m(\mathbf{x}(v), \mathbf{u}(v), v))$$

has a local solution $\mathbf{z}(v)$ satisfying $\mathbf{z}(0) = 0$ and

$$\mathbf{u}(v) = k(\mathbf{z}(v)).$$

As a result, the controller given in (3.54) is not unique. In particular, similar to Remark 3.13, the observer gains (L_1, L_2) in (3.54) need not be constant. We can choose sufficiently smooth functions $L_1(v)$ and $L_2(v)$ such that all the eigenvalues of the matrix

$$\begin{bmatrix} \frac{\partial f}{\partial x}(\mathbf{x}(v), \mathbf{u}(v), v) & \frac{\partial f}{\partial v}(\mathbf{x}(v), \mathbf{u}(v), v) \\ 0 & \frac{\partial a}{\partial v}(v) \end{bmatrix} - \begin{bmatrix} L_1(v) \\ L_2(v) \end{bmatrix} \begin{bmatrix} \frac{\partial h_m}{\partial x}(\mathbf{x}(v), \mathbf{u}(v), v) & \frac{\partial h_m}{\partial v}(\mathbf{x}(v), \mathbf{u}(v), v) \end{bmatrix} \tag{3.59}$$

are fixed complex numbers with negative real parts for all v in an open neighborhood of $\mathcal{R}^q$. Then let

$$\begin{aligned} k(z) &= \mathbf{u}(z_2) + K(z_2)(z_1 - \mathbf{x}(z_2)), \\ g(z, y_m) &= \begin{bmatrix} f(z_1, k(z), z_2) + L_1(z_2)(y_m - h_m(z_1, k(z), z_2)) \\ a(z_2) + L_2(z_2)(y_m - h_m(z_1, k(z), z_2)) \end{bmatrix}. \end{aligned} \tag{3.60}$$

This control law is also expected to be capable of accommodating larger exogenous signals. ∎

3.4 Solvability of the Regulator Equations

As we have seen in the last section, the key condition to the solvability of the nonlinear output regulation problem is the solvability of the regulator equations. By Remark 3.10, the solvability of the regulator equations is related to the existence of a particular type of the output zeroing manifold M of the composite system (3.8) described in (3.36). This manifold must be contained in the maximal output zeroing manifold of (3.8). Thus, we will begin this section by introducing the following assumption.

Assumption 3.4. There exists a (locally) maximal output zeroing manifold M_e for composite system (3.8), which is characterized by

$$M_e = \{ \text{col}(x, v) \mid \text{col}(x, v) \in \Gamma_e \,,\ H_e(x, v) = 0 \}, \tag{3.61}$$

where Γ_e is an open neighborhood of the origin of $\mathcal{R}^{n+q}$ and $H_e(x, v) : \mathcal{R}^{n+q} \to \mathcal{R}^r$ for some integer r is a sufficiently smooth function satisfying $H_e(0, 0) = 0$ and

$$\text{rank}\ \frac{\partial H_e}{\partial x}(0, 0) = r. \tag{3.62}$$

Remark 3.18. By condition (3.62), there exist some partition $x = \text{col}(x^1, x^2)$ with $x^1 \in \mathcal{R}^r$ and $x^2 \in \mathcal{R}^{n-r}$ and a locally defined sufficiently smooth function $x^1 = \sigma(x^2, v)$ satisfying $\sigma(0, 0) = 0$ such that $H_e(\sigma(x^2, v), x^2, v) = 0$. Moreover, by the definition given in Section 2.6, the fact that M_e is an output zeroing manifold for (3.8) implies the existence of a locally defined sufficiently smooth feedback control $u_e(x, v)$ satisfying $u_e(0, 0) = 0$ such that, under the control $u = u_e(x, v)$, M_e is an invariant manifold of system (3.8), which is contained in the kernel of the mapping $h(x, u_e(x, v), v)$. More specifically, corresponding to the partition $x = \text{col}(x^1, x^2)$, we can rewrite system (3.8) as follows:

$$\begin{aligned}\dot{x}^1 &= f^1(x^1, x^2, u, v),\\ \dot{x}^2 &= f^2(x^1, x^2, u, v),\\ \dot{v} &= a(v),\\ e &= h(x^1, x^2, u, v).\end{aligned} \tag{3.63}$$

Then the fact that M_e is an output zeroing manifold for (3.8) means the existence of $\sigma(x^2, v)$ and $u_e(x, v)$ such that

$$\begin{aligned}\frac{\partial\sigma}{\partial x^2} f^2(\sigma(x^2, v), x^2, u_e(\sigma(x^2, v), x^2, v), v) + \frac{\partial\sigma}{\partial v} a(v)&\\ = f^1(\sigma(x^2, v), x^2, u_e(\sigma(x^2, v), x^2, v), v),& \qquad (3.64)\\ 0 = h(\sigma(x^2, v), x^2, u_e(\sigma(x^2, v), x^2, v), v).& \qquad (3.65)\end{aligned}$$

Furthermore, the two functions $\sigma(x^2, v)$ and $u_e(x, v)$ will induce a subsystem from system (3.63) as follows:

$$\begin{aligned}\dot{x}^2 &= \delta(x^2, v) \stackrel{def}{=} f^2(\sigma(x^2, v), x^2, u_e(\sigma(x^2, v), x^2, v), v),\\ \dot{v} &= a(v),\end{aligned} \tag{3.66}$$

which is the zero dynamics of the composite system (3.63). ∎

Proposition 3.19. *Under Assumption 3.4, there exist sufficiently smooth functions* $\mathbf{x}(v)$ *and* $\mathbf{u}(v)$ *defined for* $v \in V$ *with* $\mathbf{x}(0) = 0$ *and* $\mathbf{u}(0) = 0$ *satisfying the regulator equations if there exists a sufficiently smooth function* $\mathbf{x}^2 : V \to \mathcal{R}^{n-r}$ *with* $\mathbf{x}^2(0) = 0$ *such that*

$$\frac{\partial \mathbf{x}^2}{\partial v} a(v) = \delta(\mathbf{x}^2(v), v). \tag{3.67}$$

Proof. Assume (3.67) has a solution $\mathbf{x}^2(v)$. Let $\mathbf{x}^1(v) = \sigma(\mathbf{x}^2(v), v)$, $\mathbf{x}(v) = \text{col}(\mathbf{x}^1(v), \mathbf{x}^2(v))$, and $\mathbf{u}(v) = u_e(\mathbf{x}(v), v)$. Then combining (3.64), (3.66), and (3.67) gives

$$\begin{aligned}\frac{\partial \mathbf{x}^1}{\partial v} a(v) &= f^1(\sigma(\mathbf{x}^2(v), v), \mathbf{x}^2(v), u_e(\sigma(\mathbf{x}^2(v), v), \mathbf{x}^2(v), v), v)\\ &= f^1(\mathbf{x}^1(v), \mathbf{x}^2(v), \mathbf{u}(v), v)\\ &= f^1(\mathbf{x}(v), \mathbf{u}(v), v),\\ \frac{\partial \mathbf{x}^2}{\partial v} a(v) &= \delta(\mathbf{x}^2(v), v)\\ &= f^2(\sigma(\mathbf{x}^2(v), v), \mathbf{x}^2(v), u_e(\sigma(\mathbf{x}^2(v), v), \mathbf{x}^2(v), v), v)\\ &= f^2(\mathbf{x}^1(v), \mathbf{x}^2(v), \mathbf{u}(v), v)\\ &= f^2(\mathbf{x}(v), \mathbf{u}(v), v),\end{aligned}$$

and using (3.65) gives

$$\begin{aligned} 0 &= h(\sigma(\mathbf{x}^2(v), v), \mathbf{x}^2(v), u_e(\sigma(\mathbf{x}^2(v), v), \mathbf{x}^2(v), v), v) \\ &= h(\mathbf{x}^1(v), \mathbf{x}^2(v), \mathbf{u}(v), v) \\ &= h(\mathbf{x}(v), \mathbf{u}(v), v). \end{aligned}$$

That is, the two functions $\mathbf{x}(v)$ and $\mathbf{u}(v)$ satisfy the regulator equations associated with (3.63). □

By Theorem 2.25, if all the eigenvalues of the matrix

$$\frac{\partial \delta}{\partial x^2}(0, 0)$$

have nonzero real parts, then there exists a sufficiently smooth function $\mathbf{x}^2 : V \to \mathcal{R}^{n-r}$ with $\mathbf{x}^2(0) = 0$, which satisfies (3.67). Thus we have reached the following corollary.

Corollary 3.20. *Under Assumption 3.4, suppose all the eigenvalues of the matrix*

$$\frac{\partial \delta}{\partial x^2}(0, 0)$$

have nonzero real parts. Then there exist locally defined sufficiently smooth functions $\mathbf{x}(v)$ *and* $\mathbf{u}(v)$ *with* $\mathbf{x}(0) = 0$ *and* $\mathbf{u}(0) = 0$ *satisfying the regulator equations.*

As described in Remark 2.24, the equation of the form (3.67) is an invariant manifold equation. In what follows, we will further call (3.67) a center manifold equation if all the eigenvalues of the matrix

$$\frac{\partial \delta}{\partial x^2}(0, 0)$$

have nonzero real parts and all the eigenvalues of $\frac{\partial a}{\partial v}(0)$ have zero real parts. Note that it is the special form of the zero dynamics (3.66) of (3.8), which contains the exosystem as a subsystem, that reduces the solvability of the regulator equations into that of the invariant equation (3.67). Also note that the mere existence of an output zeroing manifold for (3.8) is not enough to make the zero dynamics (3.66) of (3.8) satisfy the desired form. The additional condition (3.62) has to be imposed on the output zeroing manifold.

From the above discussion, we need to find out the conditions under which the composite system has a maximal output zeroing manifold satisfying condition (3.62). This issue can be addressed by the concepts of the normal form and zero dynamics described in Chapter 2. For convenience of notation, we will focus on the class of nonlinear systems described as follows:

$$\begin{aligned} \dot{x} &= f(x, v) + \sum_{i=1}^{m} g_i(x, v)u_i, \\ e_i &= h_i(x, v), \quad i = 1, \ldots, p, \end{aligned} \tag{3.68}$$

where $x \in \mathcal{R}^n$, u_j, $j = 1, \ldots, m$, are m scalar plant inputs; e_j, $j = 1, \ldots, p$, are p scalar plant outputs; $f : \mathcal{R}^{n+q} \to \mathcal{R}^n$ and $g_j : \mathcal{R}^{n+q} \to \mathcal{R}^n$, $j = 1, \ldots, m$, are sufficiently smooth functions; and $h_j : \mathcal{R}^{n+q} \to \mathcal{R}^1$, $j = 1, \ldots, p$, are sufficiently smooth scalar functions. Let $g(x, v) = [g_1(x, v), \ldots, g_m(x, v)]$, and

$$u = \begin{bmatrix} u_1 \\ u_2 \\ \vdots \\ u_m \end{bmatrix}, \quad e = \begin{bmatrix} e_1 \\ e_2 \\ \vdots \\ e_p \end{bmatrix}, \quad h(x, v) = \begin{bmatrix} h_1(x, v) \\ h_2(x, v) \\ \vdots \\ h_p(x, v) \end{bmatrix}.$$

Then system (3.68) can be put into the following compact form:

$$\begin{aligned} \dot{x} &= f(x, v) + g(x, v)u, \\ e &= h(x, v). \end{aligned} \tag{3.69}$$

The composite system composed of (3.69) and the exosystem $\dot{v} = a(v)$ can be put into the standard form of the nonlinear affine system as follows:

$$\begin{aligned} \dot{x}_a &= f_a(x_a) + g_a(x_a)u, \\ e &= h(x_a), \end{aligned} \tag{3.70}$$

where $x_a = \mathrm{col}(x, v)$, $f_a(x_a) = \mathrm{col}(f(x, v), a(v))$, and $g_a(x_a) = \mathrm{col}(g(x, v), 0_{q \times m})$. Also, we can define another nonlinear affine system out of (3.69) as follows:

$$\begin{aligned} \dot{x} &= f_0(x) + g_0(x)u, \\ e &= h_0(x), \end{aligned} \tag{3.71}$$

where $f_0(x) = f(x, 0)$, $g_0(x) = g(x, 0)$, and $h_0(x) = h(x, 0)$.

The regulator equations associated with (3.69) can be written as follows:

$$\begin{aligned} \frac{\partial \mathbf{x}}{\partial v} a(v) &= f(\mathbf{x}(v), v) + g(\mathbf{x}(v), v)\mathbf{u}(v), \\ 0 &= h(\mathbf{x}(v), v). \end{aligned} \tag{3.72}$$

Remark 3.21. Assume (3.70) has a (vector) relative degree $\{r_1, r_2, \ldots, r_p\}$ at $x_a = 0$, that is, that there exist integers r_i, $i = 1, \ldots, p$, such that

(i) for each $i = 1, \ldots, p$,

$$L_{g_a} L_{f_a}^k h_i(x, v) = 0_{1 \times m}$$

for all $0 \leq k < r_i - 1$ and for all x_a in an open neighborhood of the origin of $\mathcal{R}^{n+q}$; and

(ii) the $p \times m$ matrix

$$D_a(x_a) = \begin{bmatrix} L_{g_a} L_{f_a}^{r_1 - 1} h_1(x, v) \\ L_{g_a} L_{f_a}^{r_2 - 1} h_2(x, v) \\ \vdots \\ L_{g_a} L_{f_a}^{r_p - 1} h_p(x, v) \end{bmatrix} \tag{3.73}$$

has full row rank at $x_a = 0$.

Then, from Chapter 2, the hypersurface $H_a(x, v) = 0$ defines the maximal output zeroing manifold of the composite system (3.70), where $H_a(x, v)$ is the H vector of (3.70) defined in Chapter 2 and is described as follows:

$$H_a(x,v) = \begin{bmatrix} h_1(x,v) \\ L_{f_a} h_1(x,v) \\ \vdots \\ L_{f_a}^{r_1-1} h_1(x,v) \\ \vdots \\ h_p(x,v) \\ L_{f_a} h_p(x,v) \\ \vdots \\ L_{f_a}^{r_p-1} h_p(x,v) \end{bmatrix}, \tag{3.74}$$

and the corresponding state feedback control $u_e(x, v)$ is governed by the following equation:

$$E_a(x,v) + D_a(x,v)u_e(x,v) = 0, \tag{3.75}$$

where $D_a(x, v)$ is given by (3.73) and $E_a(x, v)$ is the E vector of the system (3.70) defined in Chapter 2 and described as follows:

$$E_a(x,v) = \begin{bmatrix} L_{f_a}^{r_1} h_1(x,v) \\ L_{f_a}^{r_2} h_2(x,v) \\ \vdots \\ L_{f_a}^{r_p} h_p(x,v) \end{bmatrix}.$$

We will call the restriction of the flow of the composite system (3.70) to the manifold $H_a(x, v) = 0$ the zero dynamics of the composite system (3.70). If the vector $H_a(x, v)$ further satisfies condition (3.62), then the zero dynamics of (3.70) will admit a form of (3.66); that is, the zero dynamics of (3.70) will include the exosystem as a subsystem. ∎

In what follows, we will show that if the composite system (3.70) has a vector relative degree at the origin, then the vector $H_a(x, v)$ indeed satisfies condition (3.62).

Lemma 3.22. *For $i = 1, \ldots, p$, the functions $h_i(x, v)$, the vector field f_a, and the mapping g_a associated with (3.70) satisfy*

$$L_{f_a}^k h_i(x,v) = L_f^k h_i(x,v) + a^T(v)\lambda_k(x,v), \; k = 1, 2, \ldots, \tag{3.76}$$

$$L_{g_a} L_{f_a}^k h_i(x,v) = L_g L_f^k h_i(x,v) + a^T(v)\gamma_k(x,v), \; k = 1, 2, \ldots, \tag{3.77}$$

where, with some abuse of the notation,

$$L_f h_i \stackrel{def}{=} \frac{\partial h_i}{\partial x} f,$$

$$L_f^k h_i \stackrel{def}{=} \frac{\partial L_f^{k-1} h_i}{\partial x} f, \; k > 1,$$

$$L_g L_f^k h_i \stackrel{def}{=} \frac{\partial L_f^k h_i}{\partial x} g, \; k \geq 1,$$

and $\lambda_k : \mathcal{R}^{n+q} \to \mathcal{R}^{q\times 1}$ and $\gamma_k : \mathcal{R}^{n+q} \to \mathcal{R}^{q\times m}$ are sufficiently smooth functions.

Proof. By definition,

$$L_{f_a} h_i(x, v) = \frac{\partial h_i(x, v)}{\partial x} f(x, v) + \frac{\partial h_i(x, v)}{\partial v} a(v)$$
$$= L_f h_i(x, v) + a^T(v)\lambda_1(x, v),$$

with $\lambda_1^T(x, v) \stackrel{def}{=} (\partial h_i(x, v))/(\partial v)$, and

$$L_{g_a} L_{f_a} h_i(x, v) = \frac{\partial L_{f_a} h_i}{\partial x_a} g_a(x, v)$$
$$= \frac{\partial L_{f_a} h_i(x, v)}{\partial x} g(x, v)$$
$$= \frac{\partial L_f h_i(x, v)}{\partial x} g(x, v) + \frac{\partial (a^T(v)\lambda_1(x, v))}{\partial x} g(x, v)$$
$$= L_g L_f h_i(x, v) + a^T(v)\gamma_1(x, v),$$

where $\gamma_1(x, v)$ is some sufficiently smooth function. Thus both (3.76) and (3.77) hold for $k = 1$.

Next assume that both (3.76) and (3.77) hold for all positive integers less than or equal to some integer $k > 0$. Then

$$L_{f_a}^{k+1} h_i(x, v) = \frac{\partial L_{f_a}^k h_i(x, v)}{\partial x} f(x, v) + \frac{\partial L_{f_a}^k h_i(x, v)}{\partial v} a(v)$$
$$= \frac{\partial L_f^k h_i(x, v)}{\partial x} f(x, v) + \frac{\partial (a^T(v)\lambda_k(x, v))}{\partial x} f(x, v) + \frac{\partial L_{f_a}^k h_i(x, v)}{\partial v} a(v)$$
$$= L_f^{k+1} h_i(x, v) + a^T(v)\lambda_{k+1}(x, v),$$

where $\lambda_{k+1}(x, v)$ is some sufficiently smooth function. Also,

$$L_{g_a} L_{f_a}^{k+1} h_i(x, v) = \frac{\partial L_{f_a}^{k+1} h_i(x, v)}{\partial x_a} g_a(x, v)$$
$$= \frac{\partial L_{f_a}^{k+1} h_i(x, v)}{\partial x} g(x, v)$$
$$= \frac{\partial (L_f^{k+1} h_i(x, v) + a^T(v)\lambda_{k+1}(x, v))}{\partial x} g(x, v)$$
$$= L_g L_f^{k+1} h_i(x, v) + a^T(v)\gamma_{k+1}(x, v)$$

with $\gamma_{k+1}(x, v)$ some sufficiently smooth function. □

Corollary 3.23. *If system (3.70) has a relative degree $\{r_1, r_2, \ldots, r_p\}$ at $x_a = 0$, then system (3.71) also has a relative degree $\{r_1, r_2, \ldots, r_p\}$ at $x = 0$.*

Proof. Due to (3.76) and (3.77), for $i = 1, \ldots, p$, and $k = 1, 2, \ldots,$

$$L_{f_a}^k h_i(x, 0) = L_f^k h_i(x, 0),$$
$$L_{g_a} L_{f_a}^k h_i(x, 0) = L_g L_f^k h_i(x, 0).$$

Using induction on k, it can be easily verified that

$$L_f^k h_i(x, 0) = L_{f_0}^k h_{0i}(x),$$
$$L_g L_f^k h_i(x, 0) = L_{g_0} L_{f_0}^k h_{0i}(x),$$

where $h_{0i}(x)$ is the ith component of $h_0(x)$. Thus, we have

$$L_{g_a} L_{f_a}^k h_i(x, 0) = L_{g_0} L_{f_0}^k h_{0i}(x).$$

As a result, if $L_{g_a} L_{f_a}^k h_i(x, v) = 0$ in an open neighborhood of $x_a = 0$, then $L_{g_0} L_{f_0}^k h_{0i}(x) = 0$ in an open neighborhood of $x = 0$. Moreover, let $D_0(x)$ be the decoupling matrix of (3.71). Then

$$D_a(x, 0) = D_0(x),$$

and therefore rank $D_a(0, 0) =$ rank $D_0(0)$. □

Due to (3.76), let $H_0(x)$ and $E_0(x)$ be the H and E vectors of (3.71), respectively. Then

$$H_a(x, 0) = H_0(x), \tag{3.78}$$
$$E_a(x, 0) = E_0(x). \tag{3.79}$$

Moreover, by Remark 2.48 of Chapter 2, the fact that (3.71) has a relative degree $\{r_1, r_2, \ldots, r_p\}$ at $x = 0$ implies

$$\text{rank } \frac{\partial H_0}{\partial x}(0) = r_1 + r_2 + \cdots + r_p.$$

Thus (3.78) implies

$$\text{rank } \frac{\partial H_a}{\partial x}(0, 0) = r_1 + r_2 + \cdots + r_p. \tag{3.80}$$

Thus, we have reached the following result.

Proposition 3.24. *Assume (3.70) has a relative degree $\{r_1, r_2, \ldots, r_p\}$ at $x_a = 0$ with $r_1 + r_2 + \cdots + r_p = r$. Then, the vector $H_a(x, v)$ satisfies condition (3.62).*

By Propositions 3.19 and 3.24, if the system (3.70) has a relative degree at the origin, then it will induce a subsystem of the form (3.66) such that the solvability of the regulator equations is reduced to the solvability of the center manifold equation (3.67). If the equilibrium point of the system $\dot{x}^2 = \delta(x^2, 0)$ is hyperbolic, then the center manifold equation is always solvable. As discussed in Remark 2.50, when $p = m$, the subsystem (3.66) is uniquely determined within the coordinate transformations. However, when $p < m$, the subsystem (3.66) is not uniquely determined. Thus, it is of interest to further characterize the normal form of the system (3.70). For this purpose, let $r = r_1 + r_2 + \cdots + r_p$. Then, there are r components of x denoted by $x^1 = \mathrm{col}(x_{j_1}, \ldots, x_{j_r})$ such that

$$\mathrm{rank}\ \frac{\partial H_a}{\partial x^1}(0, 0) = r. \tag{3.81}$$

Let $x^2 = \mathrm{col}(x_{j_{r+1}}, \ldots, x_{j_n})$. By the Implicit Function Theorem, there exists a sufficiently smooth function $\alpha : \mathcal{R}^{n+q} \to \mathcal{R}^r$ satisfying $\alpha(0) = 0$ such that, for sufficiently small $\xi \in \mathcal{R}^r$,

$$\xi = H_a(x, v)|_{x^1=\alpha(\xi, x^2, v)}.$$

Lemma 3.25. *Assume (3.70) has a relative degree $\{r_1, r_2, \ldots, r_p\}$ at $x_a = 0$ with $r_1 + r_2 + \cdots + r_p = r$. Then, (3.70) is locally diffeomorphic to the following system:*

$$\begin{aligned}
\dot{z}^i &= A_i z^i + B_i \left(L_{f_a}^{r_i} h_i(x, v) + L_{g_a} L_{f_a}^{r_i - 1} h_i(x, v) u \right)\Big|_{x^1=\alpha(\xi, x^2, v)}, \\
e_i &= C_i z^i, \quad i = 1, \ldots, p, \\
\dot{x}^2 &= \pi(\xi, x^2, v, u), \\
\dot{v} &= a(v),
\end{aligned} \tag{3.82}$$

where

$$z^i = \begin{bmatrix} h_i(x, v) \\ L_{f_a} h_i(x, v) \\ \vdots \\ L_{f_a}^{r_i - 1} h_i(x, v) \end{bmatrix}, \quad \xi = \begin{bmatrix} z^1 \\ z^2 \\ \vdots \\ z^p \end{bmatrix}, \quad A_i = \begin{bmatrix} 0 & 1 & 0 & \cdots & 0 \\ 0 & 0 & 1 & \cdots & 0 \\ \cdots & \cdots & \cdots & \cdots & \cdots \\ 0 & 0 & 0 & \cdots & 1 \\ 0 & 0 & 0 & \cdots & 0 \end{bmatrix},$$

$$B_i = \begin{bmatrix} 0 \\ 0 \\ \vdots \\ 0 \\ 1 \end{bmatrix}, \quad C_i = [1, \ldots, 0],$$

and

$$\pi(\xi, x^2, v, u) = \begin{bmatrix} (f_{j_{r+1}}(x, v) + g^{j_{r+1}}(x, v) u)|_{x^1=\alpha(\xi, x^2, v)} \\ \vdots \\ (f_{j_n}(x, v) + g^{j_n}(x, v) u)|_{x^1=\alpha(\xi, x^2, v)} \end{bmatrix}, \tag{3.83}$$

where g^{j_i}, $i = r + 1, \ldots, n$, is the j_ith row of g.

Proof. Let

$$z_a = \begin{bmatrix} H_a(x,v) \\ x_{j_{r+1}} \\ \vdots \\ x_{j_n} \\ v \end{bmatrix} \stackrel{def}{=} T_a(x,v).$$

Clearly T_a is a local diffeomorphism in an open neighborhood of the origin of $\mathcal{R}^{n+q}$ into its image.

Now let $z_j^i = L_{f_a}^{j-1} h_i(x,v), i = 1,\ldots,p$ and $j = 1,\ldots,r_i$, and let $z_a = \mathrm{col}(\xi, x^2, v)$. Then the components of z_a satisfy

$$\begin{aligned} \dot{z}_1^i &= z_2^i, \\ &\cdots \\ \dot{z}_{r_i-1}^i &= z_{r_i}^i, \\ \dot{z}_{r_i}^i &= \left(L_{f_a}^{r_i} h_i(x,v) + L_{g_a} L_{f_a}^{r_i-1} h_i(x,v) u\right)\Big|_{x^1=\alpha(\xi,x^2,v)}, \\ e_i &= z_1^i, \quad i = 1,\ldots,p, \\ \dot{x}_{j_i} &= \left(f_{j_i}(x,v) + g^{j_i}(x,v)u\right)\big|_{x^1=\alpha(\xi,x^2,v)}, \quad i = r+1,\ldots,n, \\ \dot{v} &= a(v). \end{aligned} \tag{3.84}$$

Clearly, (3.84) is in the form of (3.82) with $x^2 = \mathrm{col}(x_{j_{r+1}},\ldots,x_{j_n})$ and $\pi(\xi, x^2, v, u)$ being given by (3.83). □

Since D_a is of full row rank at $x_a = 0$, there exists a function $u_e : \mathcal{R}^{n+q} \to \mathcal{R}^m$ satisfying (3.75). Letting $\sigma(x^2, v) = \alpha(0, x^2, v)$ shows $H_a(x,v)|_{x^1=\sigma(x^2,v)} = 0$, and defining $\delta(x^2, v) = \pi(0, x^2, v, u_e(\sigma(x^2, v), x^2, v))$ gives the zero dynamics of (3.70) as follows:

$$\dot{x}^2 = \delta(x^2, v), \qquad \dot{v} = a(v). \tag{3.85}$$

Now applying Proposition 3.19 and Corollary 3.20 gives the main result of this section as follows.

Theorem 3.26. *Suppose the composite system (3.70) has a relative degree $\{r_1, r_2, \ldots, r_p\}$ at $(x,v) = (0,0)$ with $r_1 + r_2 + \cdots + r_p = r$. Assume, for some sufficiently smooth function $u_e(x,v)$ satisfying (3.75), that there exists a sufficiently smooth function $\mathbf{x}^2 : \mathcal{R}^q \to \mathcal{R}^{n-r}$ with $\mathbf{x}^2(0) = 0$ such that*

$$\frac{\partial \mathbf{x}^2}{\partial v} a(v) = \delta(\mathbf{x}^2(v), v). \tag{3.86}$$

Then the two functions $\mathbf{x}(v) = (\mathbf{x}^1(v), \mathbf{x}^2(v))$ and $\mathbf{u}(v) = u_e(\mathbf{x}(v), v)$, where $\mathbf{x}^1(v) = \sigma(\mathbf{x}^2(v), v)$, are the solution of the regulator equations (3.72).

Corollary 3.27. *Suppose the composite system (3.70) has a relative degree $\{r_1, r_2, \ldots, r_p\}$ at $(x,v) = (0,0)$ with $r_1 + r_2 + \cdots + r_p = r$. Then there exist locally defined sufficiently* *smooth functions* $\mathbf{x}(v)$ *and* $\mathbf{u}(v)$ *with* $\mathbf{x}(0) = 0$ *and* $\mathbf{u}(0) = 0$ *satisfying the regulator equations (3.72) if there exists some sufficiently smooth feedback control* $u = u_e(x, v)$ *satisfying* $u_e(0, 0) = 0$ *such that all the eigenvalues of the matrix*

$$\frac{\partial \delta}{\partial x^2}(0, 0) \tag{3.87}$$

have nonzero real parts.

Remark 3.28. As discussed in Remark 2.50, if $p = m$, the feedback control $u_e(x, v)$ is uniquely determined by $u_e(x, v) = -D_a^{-1}(x, v)E_a(x, v)$. Thus, the zero dynamics of (3.70) is also unique within coordinate transformations. The fact that all the eigenvalues of the matrix (3.87) evaluated at $x^2 = 0$ have nonzero real parts simply means that the plant (3.71) has a hyperbolic zero dynamics. If $p < m$, there exist a partition $u = \text{col}(u^1, u^2)$ with $u^1 \in \mathcal{R}^p$, $u^2 \in \mathcal{R}^{m-p}$ and a function $k_u : \mathcal{R}^{n+q+m-p} \to \mathcal{R}^m$ such that

$$E_a(x, v) + D_a(x, v)k_u(x, v, u^2) = 0 \tag{3.88}$$

regardless of the values of u^2. Letting $k(\xi, x^2, v, u^2) = k_u(x, v, u^2)|_{x^1=\alpha(\xi, x^2, v)}$ and substituting $u = k(\xi, x^2, v, u^2)$ into $\dot{x}^2 = \pi(\xi, x^2, v, u)$ gives

$$\dot{x}^2 = \pi(\xi, x^2, v, k(\xi, x^2, v, u^2)).$$

Thus, for any sufficiently smooth feedback control $u^2 = \theta(x^2)$ satisfying $\theta(0) = 0$, the following system:

$$\dot{x}^2 = \delta(x^2, v), \qquad \dot{v} = a(v),$$

where $\delta(x^2, v) = \pi(0, x^2, v, k(0, x^2, v, \theta(x^2)))$, is the zero dynamics of (3.70). If, for some $\theta(x^2)$, all the eigenvalues of the matrix

$$\frac{\partial \delta}{\partial x^2}(0, 0)$$

have nonzero real parts, then the regulator equations are solvable. Therefore, one can take the advantage of the $m - p$ extra control components to modify the zero dynamics of system (3.70). ∎

Remark 3.29. Though the identification of the zero dynamics of (3.70) involves a coordinate transformation, there is no need to perform the coordinate transformation in order to solve the regulator equations. Indeed, similar to the zero dynamics algorithm described in Remark 2.46, we can reduce the regulator equations to an invariant manifold equation of the form (3.86) through a simple algorithm summarized below.

(i) Solve the equation

$$H_a(x, v) = 0$$

for r components of x in terms of the remaining $n - r$ components of x and v. By property (3.81) and the Implicit Function Theorem, there exist a partition $x =$

$\text{col}(x^1, x^2)$, with $x^1 = \text{col}(x_{j_1}, \dots, x_{j_r})$ and $x^2 = \text{col}(x_{j_{r+1}}, \dots, x_{j_n})$, and a mapping $\sigma : \mathcal{R}^{(n-r+q)} \to \mathcal{R}^r$ such that

$$H_a(x, v)|_{x^1=\sigma(x^2,v)} = 0.$$

(ii) Solve $u_e(x, v)$ from the equation $E_a(x, v) + D_a(x, v)u_e(x, v) = 0$.

(iii) Solve the invariant manifold equation associated with the following system

$$\dot{x}_{j_i} = \left(f_{j_i}(x, v) + g^{j_i}(x, v)u_e(x, v)\right)\Big|_{x^1=\sigma(x^2,v)}, \quad i = r+1, \dots, n, \tag{3.89}$$

and denote the solution by $\mathbf{x}^2(v)$. Let $\mathbf{x}^1(v) = \sigma(\mathbf{x}^2(v), v)$. Then the solution of the regulator equations is given by $\mathbf{x}(v) = \text{col}(\mathbf{x}^1(v), \mathbf{x}^2(v))$ and $\mathbf{u}(v) = u_e(\mathbf{x}(v), v)$. ∎

Remark 3.30. It can be verified that, in the special case in which $f(x, v) = f(x)$, $g(x, v) = g(x)$, and $h(x, v) = h(x) - d(v)$, if (3.71) has a relative degree $\{r_1, r_2, \dots, r_p\}$ at $x = 0$, then (3.70) also has a relative degree $\{r_1, r_2, \dots, r_p\}$ at $(x, v) = (0, 0)$. Thus, a somehow simpler algorithm can be obtained. For this purpose, let $D(x)$, $E(x)$, and $H(x)$ be the decoupling matrix, E vector, and H vector of the system $\dot{x} = f(x) + g(x)u$ and $e = h(x)$, and let

$$H_d(v) = \begin{bmatrix} d_1(v) \\ L_a d_1(v) \\ \vdots \\ L_a^{r_1-1} d_1(v) \\ \vdots \\ d_p(v) \\ L_a d_p(v) \\ \vdots \\ L_a^{r_p-1} d_p(v) \end{bmatrix}, \quad E_d(v) = \begin{bmatrix} L_a^{r_1} d_1(v) \\ L_a^{r_2} d_2(v) \\ \vdots \\ L_a^{r_p} d_p(v) \end{bmatrix}.$$

Then we can simplify the algorithm described in Remark 3.29 as follows.

(i) Solve the equation

$$H(x) = H_d(v)$$

for r components of x in terms of the rest $n - r$ components of x and v. By property (3.81) and the Implicit Function Theorem, there exists a partition $x = \text{col}(x^1, x^2)$, with $x^1 = \text{col}(x_{j_1}, \dots, x_{j_r})$ and $x^2 = \text{col}(x_{j_{r+1}}, \dots, x_{j_n})$, and a mapping $\sigma : \mathcal{R}^{(n-r+q)} \to \mathcal{R}^r$ such that

$$H(x)|_{x^1=\sigma(x^2,v)} = H_d(v).$$

(ii) Solve $u_e(x, v)$ from the equation $E_d(v) = E(x) + D(x)u_e(x, v)$.

(iii) Solve the invariant manifold equation associated with the following system:

$$\dot{x}_{j_i} = \left(f_{j_i}(x) + g^{j_i}(x)u_e(x, v)\right)\Big|_{x^1=\sigma(x^2,v)}, \quad i = r+1, \dots, n, \tag{3.90}$$

and denote the solution by $\mathbf{x}^2(v)$. Let $\mathbf{x}^1(v) = \sigma(\mathbf{x}^2(v), v)$. Then the solution of the regulator equations is given by $\mathbf{x}(v) = \text{col}(\mathbf{x}_1(v), \mathbf{x}_2(v))$ and $\mathbf{u}(v) = u_e(\mathbf{x}(v), v)$. ∎

Example 3.31. Consider the following system:

$$\begin{bmatrix} \dot{x}_1 \\ \dot{x}_2 \\ \dot{x}_3 \\ \dot{x}_4 \end{bmatrix} = \begin{bmatrix} u_1 \\ x_4 + x_3 v_2 \\ x_3 + x_4 + \sin(x_1 v_2) u_2 \\ u_2 \end{bmatrix},$$
$$\begin{bmatrix} \dot{v}_1 \\ \dot{v}_2 \end{bmatrix} = \begin{bmatrix} v_2 \\ -v_1 \end{bmatrix},$$
$$\begin{bmatrix} e_1 \\ e_2 \end{bmatrix} = \begin{bmatrix} x_1 - v_1 \\ x_2 - v_2 \end{bmatrix}. \tag{3.91}$$

It is easy to verify that the system has a relative degree $\{1, 2\}$ at the origin with

$$D_a(x, v) = \begin{bmatrix} 1 & 0 \\ 0 & 1 + v_2 \sin(x_1 v_2) \end{bmatrix}, \quad H_a(x, v) = \begin{bmatrix} x_1 - v_1 \\ x_2 - v_2 \\ x_3 v_2 + x_4 + v_1 \end{bmatrix},$$

$$E_a(x, v) = \begin{bmatrix} -v_2 \\ (1 + x_3 + x_4) v_2 - x_3 v_1 \end{bmatrix}.$$

Furthermore, using the algorithm described in Remark 3.30 gives the zero dynamics of (3.71) as follows:

$$\dot{x}_3 = x_3.$$

Thus, by Corollary 3.27, the regulator equations associated with (3.91) are solvable. As a matter of fact, applying the algorithm described in Remark 3.29 gives the partition $x = \text{col}(x^1, x^2)$ with $x^1 = \text{col}(x_1, x_2, x_4)$ and $x^2 = x_3$ and the following functions:

$$x^1 = \sigma(x^2, v) = \begin{bmatrix} v_1 \\ v_2 \\ -x_3 v_2 - v_1 \end{bmatrix}, \quad u_e(x, v) = \begin{bmatrix} v_2 \\ -\dfrac{(1 + x_3 + x_4) v_2 - x_3 v_1}{1 + v_2 \sin(x_1 v_2)} \end{bmatrix},$$

$$u_e(x, v)|_{x^1 = \sigma(x^2, v)} = \begin{bmatrix} v_2 \\ -\dfrac{x_3(v_2 - v_2^2 - v_1) + v_2 - v_1 v_2}{1 + v_2 \sin(v_1 v_2)} \end{bmatrix}$$

as well as the zero dynamics of (3.91):

$$\begin{bmatrix} \dot{x}_3 \\ \dot{v}_1 \\ \dot{v}_2 \end{bmatrix} = \begin{bmatrix} x_3 + (-x_3 v_2 - v_1) - \sin(v_1 v_2) \dfrac{x_3(v_2 - v_2^2 - v_1) + v_2 - v_1 v_2}{1 + v_2 \sin(v_1 v_2)} \\ v_2 \\ -v_1 \end{bmatrix}.$$

As a result, $\mathbf{x}_3(v)$ can be obtained by solving the following center manifold equation:

$$\frac{\partial \mathbf{x}_3}{\partial v} a(v) = \mathbf{x}_3(v) + (-\mathbf{x}_3(v) v_2 - v_1) - \sin(v_1 v_2) \frac{\mathbf{x}_3(v)(v_2 - v_2^2 - v_1) + v_2 - v_1 v_2}{1 + v_2 \sin(v_1 v_2)}.$$

Therefore, the solution of the regulator equations is given by

$$\mathbf{x}(v) = \begin{bmatrix} v_1 \\ v_2 \\ \mathbf{x}_3(v) \\ -\mathbf{x}_3(v)v_2 - v_1 \end{bmatrix},$$

$$\mathbf{u}(v) = u_e(\mathbf{x}(v), v) = \begin{bmatrix} v_2 \\ -\dfrac{(1 + \mathbf{x}_3(v) - \mathbf{x}_3(v)v_2 - v_1)v_2 - \mathbf{x}_3(v)v_1}{1 + v_2 \sin(v_1 v_2)} \end{bmatrix}. \quad \blacksquare$$

Example 3.32. The RTAC system described in Section 2.8 is also in the form of (3.68). The composite system has a relative degree 2 at the origin. We have already shown in Section 2.8 that the zero dynamics of the system (3.71) is as follows:

$$\dot{x}_3 = x_4, \qquad \dot{x}_4 = x_4^2 \tan x_3.$$

The Jacobian matrix of this system at the origin has two eigenvalues at the origin. Therefore, Corollary 3.27 cannot tell whether or not the regulator equations (3.72) have a solution. Nevertheless, it is still possible to show, in the last section of this chapter, that the regulator equations of the RTAC system will admit a solution. $\blacksquare$

3.5 Output Regulation of Nonlinear Systems with Nonhyperbolic Zero Dynamics

As shown in Section 3.4, if the composite system (3.8) satisfies Assumption 3.4, then the solvability of the regulator equations associated with (3.8) can be reduced to the solvability of an invariant manifold equation of the form (3.67). In the case when the equilibrium of (3.66) is not hyperbolic, we cannot guarantee the solvability of the regulator equations, and hence we cannot guarantee the solvability of the output regulation problem. Nevertheless, under certain conditions, it is still possible to solve the output regulation problem for systems with nonhyperbolic zero dynamics. In this section, we will develop a procedure to handle this case which involves a reduction of the plant dynamics and an augmentation of the exosystem.

We assume that the system (3.8) satisfies Assumption 3.4. To save the notation, we can start from the system (3.63) and assume that the zero dynamics of (3.63), that is, $\dot{x}^2 = \delta(x^2, v)$, are described by (3.66). Now assume that the equilibrium of $\delta(x^2, 0)$ is not hyperbolic; then, without loss of generality, we can decompose $\dot{x}^2 = \delta(x^2, v)$ into the following:

$$\begin{aligned} \dot{x}_1^2 &= Ax_1^2 + g_1^2(x_1^2, x_2^2, v), \\ \dot{x}_2^2 &= Bx_2^2 + g_2^2(x_1^2, x_2^2, v), \end{aligned} \tag{3.92}$$

where $x_1^2 \in \mathcal{R}^{n_1}$, $x_2^2 \in \mathcal{R}^{n_2}$ with $n_1 + n_2 = n - r$, all the eigenvalues of the matrix A have nonzero real parts, all the eigenvalues of the matrix B have zero real parts, and g_1^2 and g_2^2

are sufficiently smooth functions satisfying

$$g_1^2(0,0,0)=0, \quad g_2^2(0,0,0)=0,$$
$$\frac{\partial g_1^2}{\partial(x_1^2,x_2^2)}(0,0,0)=0, \quad \frac{\partial g_2^2}{\partial(x_1^2,x_2^2)}(0,0,0)=0.$$

Otherwise, we can always find a coordinate transformation matrix T such that, under the new coordinate $z = Tx^2$, the system $\dot{x}^2 = \delta(x^2, v)$ can be decomposed as in (3.92).

Since A is hyperbolic, by the Center Manifold Theorem, there exists a locally defined function $\mathbf{x}_1^2(x_2^2, v)$ satisfying $\mathbf{x}_1^2(0,0) = 0$ such that

$$\begin{aligned}\frac{\partial \mathbf{x}_1^2}{\partial x_2^2}\left(Bx_2^2 + g_2^2(\mathbf{x}_1^2(x_2^2,v),x_2^2,v)\right) + \frac{\partial \mathbf{x}_1^2}{\partial v}a(v)\\ = A\mathbf{x}_1^2(x_2^2,v) + g_1^2(\mathbf{x}_1^2(x_2^2,v),x_2^2,v). \qquad (3.93)\end{aligned}$$

In terms of the partition $x = \mathrm{col}(x^1, x_1^2, x_2^2)$, we can write the composite system (3.8) as follows

$$\dot{x}^1 = f^1(x^1, x_1^2, x_2^2, u, v), \qquad (3.94)$$
$$\dot{x}_1^2 = f_1^2(x^1, x_1^2, x_2^2, u, v), \qquad (3.95)$$
$$\dot{x}_2^2 = f_2^2(x^1, x_1^2, x_2^2, u, v), \qquad (3.96)$$
$$\dot{v} = a(v), \qquad (3.97)$$
$$e = h(x^1, x_1^2, x_2^2, u, v). \qquad (3.98)$$

Note that in conjunction with (3.92), the notation used in (3.94) to (3.98) implies

$$\begin{aligned}f_1^2(\sigma(x^2,v),x_1^2,x_2^2,u_e(\sigma(x^2,v),x_1^2,x_2^2,v),v) &= Ax_1^2 + g_1^2(x_1^2,x_2^2,v),\\ f_2^2(\sigma(x^2,v),x_1^2,x_2^2,u_e(\sigma(x^2,v),x_1^2,x_2^2,v),v) &= Bx_2^2 + g_2^2(x_1^2,x_2^2,v), \qquad (3.99)\end{aligned}$$

where the functions $\sigma(x^2, v)$ and $u_e(x, v)$ are defined in (3.64) and (3.65). Now if $f_2^2(x^1, x_1^2, x_2^2, u, v)$ does not depend on $\mathrm{col}(x^1, x_1^2, u)$, then we may be able to solve the output regulation problem for the plant (3.94) to (3.98) by considering (3.94) and (3.95) as the plant and (3.96) and (3.97) as the exosystem. However, what makes this problem interesting is that it may be solved under much less restrictive conditions. Indeed, it suffices to assume the following.

Assumption 3.5. The input u does not appear in the function f_2^2; that is

$$\dot{x}_2^2 = f_2^2(x^1, x_1^2, x_2^2, v). \qquad (3.100)$$

Assumption 3.6.

$$\frac{\partial f_2^2}{\partial(x^1, x_1^2)}(0,0,0,0) = 0.$$

Remark 3.33. Assumption 3.5 is made so that the dynamics of (3.100) is not affected by any feedback control. This assumption is not as restrictive as it might appear. In fact, it is satisfied for a large class of nonlinear systems. For example, the affine SISO nonlinear system with well-defined relative degree at the origin always has a normal form described in Remark 2.42. Clearly, Assumption 3.5 is satisfied for this class of systems. Assumption 3.6 is made for invoking the Center Manifold Theorem (Theorem 2.25) later in the proof of Theorem 3.34. ∎

Theorem 3.34. *Under Assumptions 3.1 and 3.4 to 3.6, suppose that the pair*

$$\left(\begin{bmatrix} \frac{\partial f^1}{\partial x^1}(0,0,0,0,0) & \frac{\partial f^1}{\partial x_1^2}(0,0,0,0,0) \\ \frac{\partial f_1^2}{\partial x^1}(0,0,0,0,0) & \frac{\partial f_1^2}{\partial x_1^2}(0,0,0,0,0) \end{bmatrix}, \begin{bmatrix} \frac{\partial f^1}{\partial u}(0,0,0,0,0) \\ \frac{\partial f_1^2}{\partial u}(0,0,0,0,0) \end{bmatrix}\right)$$

is stabilizable and the equilibrium point at the origin of the following system:

$$\begin{aligned} \dot{\hat{v}} &= f_2^2(\sigma(\mathbf{x}_1^2(\hat{v}, v), \hat{v}, v), \mathbf{x}_1^2(\hat{v}, v), \hat{v}, v), \\ \dot{v} &= a(v) \end{aligned} \tag{3.101}$$

is stable in the sense of Lyapunov. Then there exists a state feedback control law of the form $u = k(x^1, x_1^2, x_2^2, v)$ *such that the equilibrium of the composite system (3.94) to (3.98) at* $(x, v) = (0, 0)$ *is stable in the sense of Lyapunov, and for all sufficiently small initial states* x_0 *and* v_0, *the tracking error* $e(t)$ *satisfies*

$$\lim_{t\to\infty} e(t) = 0.$$

Proof. Let

$$\mathbf{x}_r(x_2^2, v) = \begin{bmatrix} \sigma(\mathbf{x}_1^2(x_2^2, v), x_2^2, v) \\ \mathbf{x}_1^2(x_2^2, v) \end{bmatrix}, \quad \mathbf{u}_r(x_2^2, v) = u_e(\mathbf{x}_r(x_2^2, v), x_2^2, v). \tag{3.102}$$

Then, combining (3.64), (3.65), and (3.93) shows that $\mathbf{x}_r(x_2^2, v)$ and $\mathbf{u}_r(x_2^2, v)$ satisfy

$$\begin{bmatrix} \frac{\partial \mathbf{x}_r}{\partial x_2^2}(x_2^2, v) & \frac{\partial \mathbf{x}_r}{\partial v}(x_2^2, v) \end{bmatrix} \begin{bmatrix} Bx_2^2 + g_2^2(\mathbf{x}_1^2(x_2^2, v), x_2^2, v) \\ a(v) \end{bmatrix}$$
$$= \begin{bmatrix} f^1(\sigma(\mathbf{x}_1^2(x_2^2, v), x_2^2, v), \mathbf{x}_1^2(x_2^2, v), x_2^2, \mathbf{u}_r(x_2^2, v), v) \\ f_1^2(\sigma(\mathbf{x}_1^2(x_2^2, v), x_2^2, v), \mathbf{x}_1^2(x_2^2, v), x_2^2, \mathbf{u}_r(x_2^2, v), v) \end{bmatrix}, \tag{3.103}$$

$$0 = h(\sigma(\mathbf{x}_1^2(x_2^2, v), x_2^2, v), \mathbf{x}_1^2(x_2^2, v), x_2^2, \mathbf{u}_r(x_2^2, v), v). \tag{3.104}$$

Also, by the stabilizability assumption, there exists a matrix K_r such that all the eigenvalues of the matrix

$$\begin{bmatrix} \frac{\partial f^1}{\partial x^1}(0,0,0,0,0) & \frac{\partial f^1}{\partial x_1^2}(0,0,0,0,0) \\ \frac{\partial f_1^2}{\partial x^1}(0,0,0,0,0) & \frac{\partial f_1^2}{\partial x_1^2}(0,0,0,0,0) \end{bmatrix} + \begin{bmatrix} \frac{\partial f^1}{\partial u}(0,0,0,0,0) \\ \frac{\partial f_1^2}{\partial u}(0,0,0,0,0) \end{bmatrix} K_r \tag{3.105}$$

have negative real parts. Define a state feedback controller as follows:

$$u = k(x^1, x_1^2, x_2^2, v) = \mathbf{u}_r(x_2^2, v) + K_r \begin{bmatrix} x^1 - \sigma(\mathbf{x}_1^2(x_2^2, v), x_2^2, v) \\ x_1^2 - \mathbf{x}_1^2(x_2^2, v) \end{bmatrix}. \tag{3.106}$$

We now show that this controller solves the output regulation problem for the composite system (3.94) to (3.98). To this end, consider the closed-loop system composed of the composite system (3.94) to (3.98) and the controller (3.106):

$$\begin{aligned}\dot{x}^1 &= f^1(x^1, x_1^2, x_2^2, k(x^1, x_1^2, x_2^2, v), v),\\ \dot{x}_1^2 &= f_1^2(x^1, x_1^2, x_2^2, k(x^1, x_1^2, x_2^2, v), v),\\ \dot{x}_2^2 &= f_2^2(x^1, x_1^2, x_2^2, v),\\ \dot{v} &= a(v),\end{aligned} \tag{3.107}$$

which has the following properties:

(i) Due to (3.105), all the eigenvalues of the Jacobian matrix at the origin of the reduced-order closed-loop system composed of (3.94), (3.95), and (3.106) have negative real parts.

(ii) Due to Assumption 3.6 and the decomposition (3.92), we have

$$\begin{bmatrix} f_2^2(x^1, x_1^2, x_2^2, v) \\ a(v) \end{bmatrix} = \begin{bmatrix} B & 0 \\ 0 & A_1 \end{bmatrix} \begin{bmatrix} x_2^2 \\ v \end{bmatrix} + g(x^1, x_1^2, x_2^2, v), \tag{3.108}$$

where all the eigenvalues of B and A_1 have zero real parts by assumption, and the function g vanishes at $(0, 0, 0, 0)$ together with its first-order partial derivatives with respect to x.

(iii)

$$k(\sigma(\mathbf{x}_1^2(x_2^2, v), x_2^2, v), \mathbf{x}_1^2(x_2^2, v), x_2^2, v) = \mathbf{u}_r(x_2^2, v). \tag{3.109}$$

These facts, together with (3.103), show that $\mathrm{col}(x^1, x_1^2) = \mathbf{x}_r(x_2^2, v)$ is a center manifold for (3.107). Since the equilibrium of the augmented exosystem (3.101) is stable by assumption, it follows from Theorem 2.27 that the equilibrium of the closed-loop system (3.107) is also stable. Thus system (3.107) satisfies Property 3.1. Moreover, by Theorem 2.28, for sufficiently small $x(0)$ and $v(0)$, there exist real numbers $\delta > 0$ and $\lambda > 0$ such that the solution of (3.107) satisfies

$$\begin{aligned}&\left\| \begin{bmatrix} x^1(t) - \sigma(\mathbf{x}_1^2(x_2^2(t), v(t)), x_2^2(t), v(t)) \\ x_1^2(t) - \mathbf{x}_1^2(x_2^2(t), v(t)) \end{bmatrix} \right\| \\ &\le \delta e^{-\lambda t} \left\| \begin{bmatrix} x^1(0) - \sigma(\mathbf{x}_1^2(x_2^2(0), v(0)), x_2^2(0), v(0)) \\ x_1^2(0) - \mathbf{x}_1^2(x_2^2(0), v(0)) \end{bmatrix} \right\|.\end{aligned} \tag{3.110}$$

We now show that (3.107) also satisfies Property 3.2. In fact, from (3.104) and (3.109), we have

$$\begin{aligned}e &= h(x^1, x_1^2, x_2^2, k(x^1, x_1^2, x_2^2, v), v)\\ &= h(x^1, x_1^2, x_2^2, k(x^1, x_1^2, x_2^2, v), v)\\ &\quad - h(\sigma(\mathbf{x}_1^2(x_2^2, v), x_2^2, v), \mathbf{x}_1^2(x_2^2, v), x_2^2, k(\sigma(\mathbf{x}_1^2(x_2^2, v), x_2^2, v), \mathbf{x}_1^2(x_2^2, v), x_2^2, v), v).\end{aligned}$$

It follows from the continuous differentiability of h and k and (3.110) that

$$\lim_{t\to\infty} e(t) = 0. \quad \square$$

Remark 3.35. A reduced-order plant can be defined out of the original plant as follows:

$$\begin{aligned}\dot{x}^1 &= f^1(x^1, x_1^2, \hat{v}, u, v),\\ \dot{x}_1^2 &= f_1^2(x^1, x_1^2, \hat{v}, u, v),\\ e &= h(x^1, x_1^2, \hat{v}, u, v),\end{aligned} \tag{3.111}$$

where $\mathrm{col}(\hat{v}, v)$ is generated by system (3.101), which can be viewed as an augmented exosystem. Then clearly the two functions $\mathbf{x}_r(\hat{v}, v)$ and $\mathbf{u}_r(\hat{v}, v)$ are the solution of the regulator equations associated with the reduced-order plant (3.111) and the augmented exosystem (3.101). Thus, basically, Theorem 3.34 says that if the state feedback control law $u = k(x^1, x_1^2, \hat{v}, v)$ is the solution of the state feedback output regulation problem with exponential stability for the reduced-order plant (3.111) and the augmented exosystem (3.101), then the state feedback control law $u = k(x^1, x_1^2, x_2^2, v)$ is the solution of the state feedback output regulation problem of the composite system composed of the original plant and original exosystem. ∎

Remark 3.36. By Theorem 2.9, if the equilibrium point $\hat{v} = 0$ of the system

$$\dot{\hat{v}} = f_2^2(\sigma(\mathbf{x}_1^2(\hat{v}, 0), \mathbf{x}_1^2(\hat{v}, 0), \hat{v}, 0)$$

is asymptotically stable, then the equilibrium point of the origin of (3.101) is also Lyapunov stable since the exosystem satisfies Assumption 3.1. ∎

Example 3.37. Consider the nonlinear system

$$\begin{aligned}\dot{x}_1 &= x_2,\\ \dot{x}_2 &= a(x_1, x_2, x_3, x_4) + b(x_1, x_2, x_3, x_4)u,\\ \dot{x}_3 &= x_1 + x_3 + x_2x_4,\\ \dot{x}_4 &= -(x_4)^3 + x_1x_2 + x_1x_3,\\ y &= x_1 - v_1,\\ \dot{v}_1 &= v_2,\\ \dot{v}_2 &= -v_1,\end{aligned} \tag{3.112}$$

where $a(\cdot, \cdot, \cdot, \cdot)$ and $b(\cdot, \cdot, \cdot, \cdot)$ are sufficiently smooth scalar functions, $a(0, 0, 0, 0) = 0$, and $b(0, 0, 0, 0) \neq 0$. The system is in the form (3.70). Using the approach given in Section 3.4, we can obtain the zero dynamics and the associated control as follows:

$$\begin{aligned}\dot{x}_3 &= v_1 + x_3 + v_2x_4,\\ \dot{x}_4 &= -(x_4)^3 + v_1v_2 + v_1x_3,\\ \dot{v}_1 &= v_2,\\ \dot{v}_2 &= -v_1,\end{aligned} \tag{3.113}$$

and

$$u_e(x_1, x_2, x_3, x_4, v_1, v_2) = -\frac{v_1 + a(x_1, x_2, x_3, x_4)}{b(x_1, x_2, x_3, x_4)}. \tag{3.114}$$

Clearly the equilibrium of the subsystem $\dot{x}_3 = x_3,\ \dot{x}_4 = -(x_4)^3$ is not hyperbolic. Nevertheless, the subsystem

$$\begin{aligned}\dot{x}_3 &= v_1 + x_3 + v_2 x_4,\\ \dot{x}_4 &= -(x_4)^3 + v_1 v_2 + v_1 x_3\end{aligned} \tag{3.115}$$

admits the form given by (3.92). Thus (3.113) has a center manifold denoted by $x_3 = \mathbf{x}_3(x_4, v_1, v_2)$. Now let the reduced-order plant be

$$\begin{aligned}\dot{x}_1 &= x_2,\\ \dot{x}_2 &= a(x_1, x_2, x_3, \hat{v}) + b(x_1, x_2, x_3, \hat{v})u,\\ \dot{x}_3 &= x_1 + x_3 + x_2 \hat{v},\end{aligned} \tag{3.116}$$

and the augmented exosystem be

$$\begin{aligned}\dot{\hat{v}} &= -\hat{v}^3 + v_1 v_2 + v_1 \mathbf{x}_3(\hat{v}, v_1, v_2),\\ \dot{v}_1 &= v_2,\\ \dot{v}_2 &= -v_1.\end{aligned} \tag{3.117}$$

It can be verified that the linearization of the reduced plant is controllable. Moreover, since the equilibrium of $\dot{\hat{v}} = -(\hat{v})^3$ is asymptotically stable and the equilibrium of $\dot{v}_1 = v_2,\ \dot{v}_2 = -v_1$ is Lyapunov stable, by Theorem 2.9, the equilibrium of (3.117) is also Lyapunov stable. Therefore, Theorem 3.34 concludes that, for system (3.112), the output regulation with Lyapunov stability can be achieved using the state feedback control (3.106). ∎

3.6 Disturbance Rejection of the RTAC System

Now we turn our attention to the disturbance attenuation problem of the RTAC system formulated in Section 3.2. Let us first consider the solvability of the regulator equations associated with the RTAC system. As pointed out in Example 3.32, Corollary 3.27 cannot tell whether or not the regulator equations (3.72) have a solution, since the zero dynamics of the system with the disturbance being set to zero is not hyperbolic. Nevertheless, we will show that the regulator equations of the RTAC system admit a solution.

For this purpose, consider the composite system consisting of the RTAC system and the exosystem as follows:

$$\begin{bmatrix}\dot{x}_1\\ \dot{x}_2\\ \dot{x}_3\\ \dot{x}_4\\ \dot{v}_1\\ \dot{v}_2\end{bmatrix} = \begin{bmatrix} x_2\\ \frac{-x_1+\epsilon x_4^2 \sin x_3 + v_1 - \epsilon(\cos x_3)u}{1-\epsilon^2\cos^2 x_3}\\ x_4\\ \frac{\epsilon\cos x_3(x_1 - \epsilon x_4^2 \sin x_3) - \epsilon(\cos x_3)v_1 + u}{1-\epsilon^2\cos^2 x_3}\\ \omega v_2\\ -\omega v_1\end{bmatrix},$$

$$e = x_1. \tag{3.118}$$

Differentiating the error output e twice gives

$$\begin{aligned}\dot{e} &= \dot{x}_1 = x_2,\\ \ddot{e} &= \dot{x}_2 = \frac{-x_1 + \epsilon x_4^2 \sin x_3 + v_1 - \epsilon(\cos x_3)u}{1 - \epsilon^2 \cos^2 x_3}.\end{aligned}$$

Thus the composite system has a well-defined relative degree 2 at the origin with

$$\begin{aligned}D_a(x, v) &= \frac{-\epsilon \cos x_3}{1 - \epsilon^2 \cos^2 x_3},\\ E_a(x, v) &= \frac{-x_1 + \epsilon x_4^2 \sin x_3 + v_1}{1 - \epsilon^2 \cos^2 x_3},\\ H_a(x, v) &= \begin{bmatrix} x_1 \\ x_2 \end{bmatrix}.\end{aligned}$$

Applying the algorithm described in Remark 3.29 gives the partition $x = \mathrm{col}(x^1, x^2)$ with $x^1 = \mathrm{col}(x_1, x_2)$ and $x^2 = \mathrm{col}(x_3, x_4)$ and the following functions:

$$\begin{aligned}x^1 = \sigma(x^2, v) &= \begin{bmatrix} 0 \\ 0 \end{bmatrix},\\ u_e(x, v) &= -\frac{E_a(x, v)}{D_a(x, v)} = -\frac{-x_1 + \epsilon x_4^2 \sin x_3 + v_1}{-\epsilon \cos x_3},\\ u_e(x, v)|_{x^1=\sigma(x^2,v)} &= -\frac{\epsilon x_4^2 \sin x_3 + v_1}{-\epsilon \cos x_3},\end{aligned}$$

as well as the zero dynamics of (3.118)

$$\begin{aligned}\dot{x}_3 &= x_4,\\ \dot{x}_4 &= x_4^2 \tan x_3 + \frac{v_1}{\epsilon \cos x_3},\\ \dot{v}_1 &= \omega v_2,\\ \dot{v}_2 &= -\omega v_1.\end{aligned} \tag{3.119}$$

Therefore, the solution of the regulator equations is given by

$$\mathbf{x}(v) = \begin{bmatrix} 0 \\ 0 \\ \mathbf{x}_3(v) \\ \mathbf{x}_4(v) \end{bmatrix}, \quad \mathbf{u}(v) = u_e(\mathbf{x}(v), v) = \mathbf{x}_4^2(v) \tan \mathbf{x}_3(v) + \frac{v_1}{\epsilon \cos \mathbf{x}_3(v)}, \tag{3.120}$$

with $\mathbf{x}_3(v)$ and $\mathbf{x}_4(v)$ satisfying

$$\begin{aligned}\frac{\partial \mathbf{x}_3}{\partial v} A_1 v &= \mathbf{x}_4(v),\\ \frac{\partial \mathbf{x}_4}{\partial v} A_1 v &= \alpha(\mathbf{x}_3(v), \mathbf{x}_4(v), v) = \mathbf{x}_4^2(v) \tan \mathbf{x}_3(v) + \frac{v_1}{\epsilon \cos \mathbf{x}_3(v)},\end{aligned} \tag{3.121}$$

where $\alpha(x_3, x_4, v) = x_4^2 \tan x_3 + \frac{v_1}{\epsilon \cos x_3}$. Equations (3.121) can be viewed as the invariant manifold equation associated with the zero dynamics (3.119).

It is usually impossible to obtain an analytic solution for a nonlinear partial differential equation of the form (3.121). However, by taking advantage of the special structure of (3.121), it is possible to solve (3.121) as follows. First note that equations (3.121) hold if and only if, for all sufficiently small trajectories $v(t)$ of the exosystem,

$$\frac{d\mathbf{x}_3(v)}{dt} = \mathbf{x}_4(v), \tag{3.122}$$

$$\frac{d\mathbf{x}_4(v)}{dt} = \mathbf{x}_4^2(v)\frac{\sin \mathbf{x}_3(v)}{\cos \mathbf{x}_3(v)} + \frac{v_1}{\epsilon \cos \mathbf{x}_3(v)}. \tag{3.123}$$

Equation (3.123) can be written as

$$\cos \mathbf{x}_3(v)\frac{d\mathbf{x}_4(v)}{dt} - \mathbf{x}_4^2(v) \sin \mathbf{x}_3(v) = -\frac{1}{\epsilon\omega}\frac{dv_2}{dt}. \tag{3.124}$$

Using the identity

$$\frac{d((\cos x_3)x_4)}{dt} = -(\sin x_3)x_4\frac{dx_3}{dt} + (\cos x_3)\frac{dx_4}{dt} = -(\sin x_3)x_4^2 + (\cos x_3)\frac{dx_4}{dt}$$

in (3.124) gives

$$\frac{d((\cos \mathbf{x}_3(v))\mathbf{x}_4(v))}{dt} = -\frac{1}{\epsilon\omega}\frac{dv_2}{dt},$$

which further yields, upon noting $\mathbf{x}_4(0) = 0$,

$$\mathbf{x}_4(v) = \frac{-v_2}{\epsilon\omega \cos \mathbf{x}_3(v)} = \frac{-1}{\epsilon\omega^2 \cos \mathbf{x}_3(v)}\frac{dv_1}{dt}. \tag{3.125}$$

Combining (3.122) and (3.125) gives

$$\frac{d \sin \mathbf{x}_3(v)}{dt} = -\frac{1}{\epsilon\omega^2}\frac{dv_1}{dt},$$

which further yields, upon noting $\mathbf{x}_3(0) = 0$,

$$\sin \mathbf{x}_3(v) = -\frac{v_1}{\epsilon\omega^2}, \tag{3.126}$$

or equivalently,

$$\mathbf{x}_3(v) = \arcsin \frac{-v_1}{\epsilon\omega^2}. \tag{3.127}$$

Substituting (3.127) into (3.125) gives

$$\mathbf{x}_4(v) = \frac{-v_2}{\epsilon\omega}\frac{1}{\sqrt{1 - \left(\frac{v_1}{\epsilon\omega^2}\right)^2}}, \tag{3.128}$$

where $-\epsilon\omega^2 < v_1 < \epsilon\omega^2$.

Once we obtain the solution of the regulator equations, we can obtain a state feedback controller as follows:

$$u = \mathbf{u}(v) + K_x(x - \mathbf{x}(v)),$$

where K_x is such that $\frac{\partial f}{\partial x}(0) + g_1(0)K_x$ is Hurwitz. A simple calculation gives

$$\frac{\partial f}{\partial x}(0) = \begin{bmatrix} 0 & 1 & 0 & 0 \\ \frac{-1}{1-\epsilon^2} & 0 & 0 & 0 \\ 0 & 0 & 0 & 1 \\ \frac{\epsilon}{1-\epsilon^2} & 0 & 0 & 0 \end{bmatrix}, \quad g_1(0) = \begin{bmatrix} 0 \\ \frac{-\epsilon}{1-\epsilon^2} \\ 0 \\ \frac{1}{1-\epsilon^2} \end{bmatrix}.$$

Clearly, the pair $\left(\frac{\partial f}{\partial x}(0), g_1(0)\right)$ is controllable for all $\epsilon > 0$. However, it can be verified that the pair $\left(\frac{\partial h}{\partial v}(0, 0, 0), \ \frac{\partial f}{\partial x}(0, 0, 0)\right)$ is not detectable. Thus the problem cannot be solved by an error output feedback controller.

Nevertheless, since the angular position of the proof-mass actuator x_3 is also measurable, we can define the measurement output as $y_m = h_m(x, u, v) = \text{col}(x_1, x_3)$. Let

$$C_m = \begin{bmatrix} \frac{\partial h_m}{\partial x}(0, 0, 0) & \frac{\partial h_m}{\partial v}(0, 0, 0) \end{bmatrix} = \begin{bmatrix} 1 & 0 & 0 & 0 & 0 & 0 \\ 0 & 0 & 1 & 0 & 0 & 0 \end{bmatrix}$$

and

$$B_v = \left.\frac{\partial g_2(x)v_1}{\partial v}\right|_{x=0, v=0} = \begin{bmatrix} 0 & 0 \\ \frac{1}{1-\epsilon^2} & 0 \\ 0 & 0 \\ \frac{-\epsilon}{1-\epsilon^2} & 0 \end{bmatrix}.$$

Then it can be verified that the following pair:

$$\left(C_m, \ \begin{bmatrix} \frac{\partial f}{\partial x}(0) & B_v \\ 0 & A_1 \end{bmatrix}\right)$$

is detectable. Thus the problem can be solvable by the dynamic measurement output feedback control.

Let $L = \text{col}(L_1, L_2)$ with $L_1 \in \mathcal{R}^{4\times 2}$ and $L_2 \in \mathcal{R}^{2\times 2}$ be such that

$$\begin{bmatrix} \frac{\partial f}{\partial x}(0) & B_v \\ 0 & A_1 \end{bmatrix} - \begin{bmatrix} L_1 \\ L_2 \end{bmatrix} C_m \tag{3.129}$$

is Hurwitz, and $z = \text{col}(z_1, z_2)$ with $z_1 \in \mathcal{R}^4$ and $z_2 \in \mathcal{R}^2$. Then a dynamic measurement output feedback controller that solves the output regulation problem for the RTAC system can be given as follows:

$$\begin{aligned} u &= k(z_1, z_2) = \mathbf{u}(z_2) + K_x(z_1 - \mathbf{x}(z_2)), \\ \dot{z} &= g(z, y_m) \\ &= \begin{bmatrix} f(z_1) + g_1(z_1)k(z_1, z_2) + g_2(z_1)[1, \ 0]z_2 + L_1(y_m - h_m(z_1, k(z_1, z_2), z_2)) \\ A_1 z_2 + L_2(y_m - h_m(z_1, k(z_1, z_2), z_2)) \end{bmatrix}. \end{aligned}$$

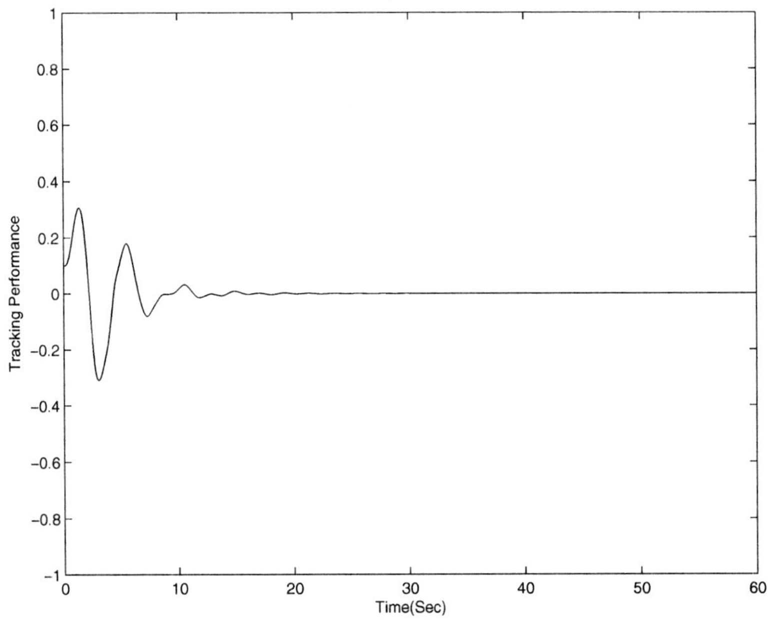

Figure 3.2. *The profile of the displacement* x_1 *with* $\epsilon = 0.2$, $\omega = 3$, *and* $A_m = 0.5$.

To evaluate the performance of this controller by computer simulation, let us give the specific gains K_x and L for the case where $\epsilon = 0.20$ and $\omega = 3$. First, letting $K_x = [-16.52\ -83.52\ -15.4\ -20.7]$ places the eigenvalues of $\frac{\partial f}{\partial x}(0) + g_1(0)K_x$ at $[(-0.848 \pm 2.52j),\ (-1.25 \pm 0.828j)]$. The above eigenvalues are based on the **ITAE** (**i**ntegral of the **t**ime multiplied by the **a**bsolute value of the **e**rror) prototype design with cutoff frequency equal to 1 (described in Appendix B).

Next, letting the eigenvalues of (3.129) be given by

$$\begin{bmatrix} -0.1871 \pm j3.0918 & -0.7065 \pm j1.1866 & -1.3627 & -12.6325 \end{bmatrix}$$

gives

$$L = \begin{bmatrix} 3.4152 & -3.0473 \\ 1.9628 & 5.5501 \\ -3.4819 & 11.6188 \\ -4.5875 & 1.6509 \\ -3.3591 & -1.0914 \\ -1.0312 & -1.7223 \end{bmatrix}.$$

Simulation has been run for the initial state $x(0) = \mathrm{col}(0.1, 0, 0, 0)$, $z(0) = 0$, and various values of the amplitude A_m. With $\omega = 3$, Figure 3.2 shows the profile of the displacement x_1 of the closed-loop system, Figure 3.3 shows the profile of the other three state variables, x_2, x_3, x_4, and Figure 3.4 shows the profile of the control input $u(t)$.

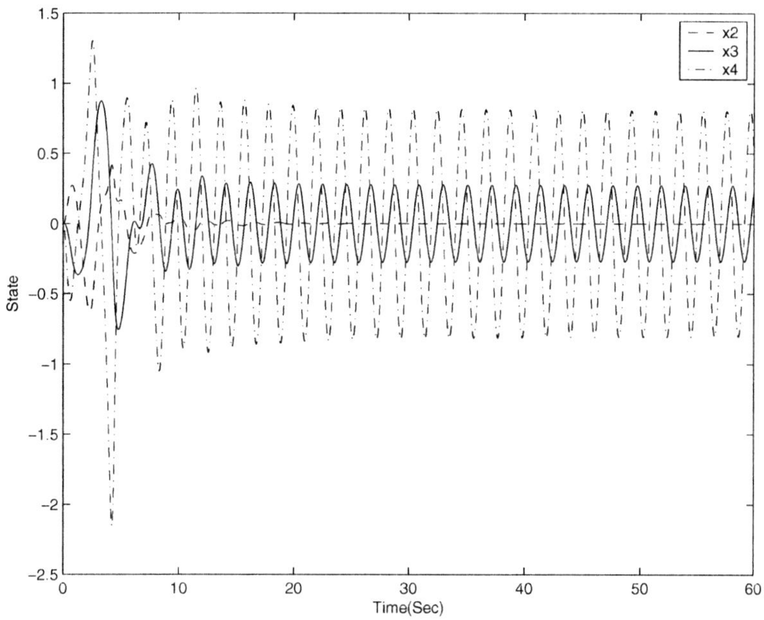

Figure 3.3. *The profiles of the state variables* (x_2, x_3, x_4) *with* $\epsilon = 0.2$, $\omega = 3$, *and* $A_m = 0.5$.

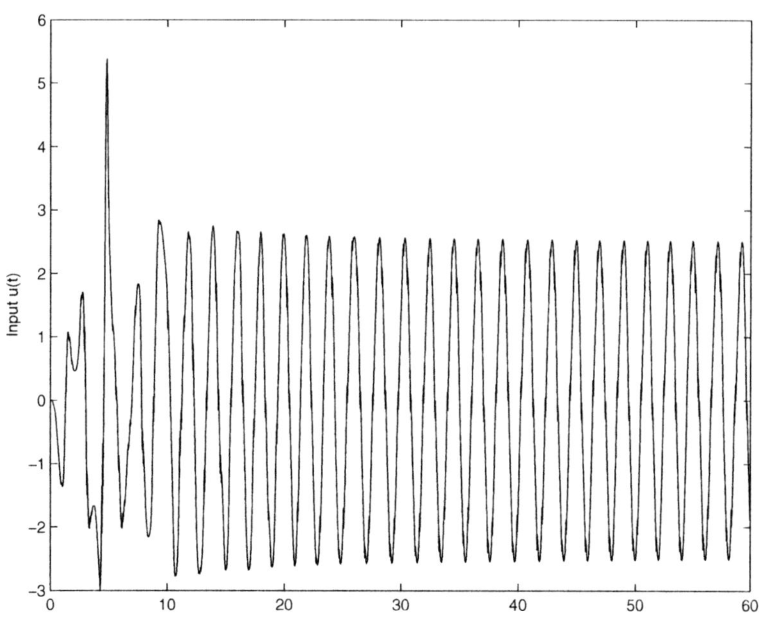

Figure 3.4. *The profile of the control input u with* $\epsilon = 0.2$, $\omega = 3$, *and* $A_m = 0.5$.

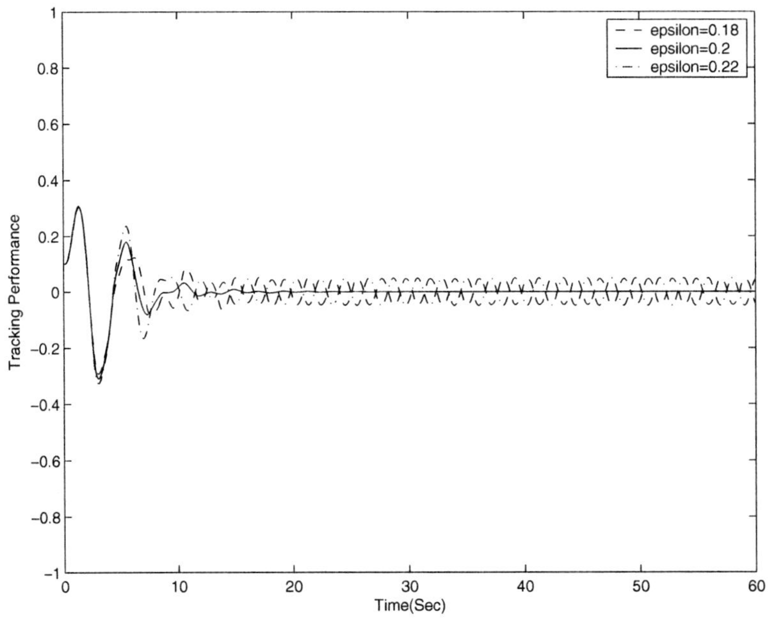

Figure 3.5. *The profiles of the displacement x_1 when ϵ undergoes perturbation.*

It is known that the feedforward part of the controller depends on the solution of the regulator equations, and thus demands precise knowledge of the plant. It is interesting to know what will happen if some parameters of the plant undergo some perturbations. Figure 3.5 shows the profiles of the displacement x_1 of the closed-loop system under the same controller with the parameter ϵ being equal to 0.18, 0.20, and 0.22, respectively. It can be seen that when the parameter ϵ deviates from its nominal value 0.20, the displacement x_1 displays a sizable nondecaying oscillation. Thus we have seen that the performance of this controller is not robust with respect to parameter variations. It is desirable to have a regulator that can maintain its performance in the presence of small parameter variations. Such a regulator is called a robust regulator and will be introduced in Chapter 5. A robust regulator for the same RTAC system will be designed in Chapter 6.

Chapter 4

Approximation Method for the Nonlinear Output Regulation

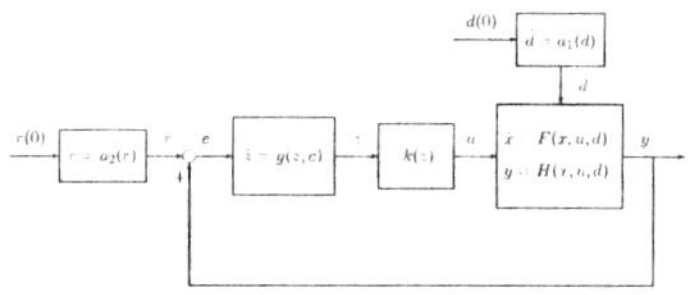

As we have seen in Chapter 3, the construction of the control laws for solving the output regulation problem relies on the solution of the nonlinear regulator equations (3.30), which are repeated below for convenience:

$$\begin{aligned}\frac{\partial \mathbf{x}(v)}{\partial v}a(v) &= f(\mathbf{x}(v), \mathbf{u}(v), v),\\ 0 &= h(\mathbf{x}(v), \mathbf{u}(v), v).\end{aligned} \tag{4.1}$$

Since (4.1) are a set of nonlinear partial differential and algebraic equations, it is rarely possible to find the closed-form solution for them. Therefore, it is desirable to have a numerical approach that can solve (4.1) approximately. This chapter will present an approximation approach to the solution of the nonlinear output regulation problem that is based on the approximate solution of (4.1) in terms of power series. The chapter is organized as follows. Section 4.1 introduces the kth-order nonlinear output regulation problem and gives its solvability conditions by both state feedback and measurement output feedback controls. Section 4.2 presents an approximate solution of the regulator equations in terms of power series. Section 4.3 further gives an approximation solution of the center manifold equations in terms of the power series. Finally, the approximation approach developed in this chapter is applied, in Section 4.4, to design a state feedback control law to approximately solve the asymptotic tracking problem of the inverted pendulum on a cart system.

4.1 kth-Order Approximate Solution of Nonlinear Output Regulation Problem

In this chapter, we will study the same class of nonlinear plants, exosystems, and control laws as those described in Chapter 3. All assumptions introduced in Chapter 3 will be adopted. We will first introduce another property for the closed-loop system described by (3.13) as follows.

Definition 4.1. *Let V be an open neighborhood of the origin of $\mathcal{R}^q$. A function $o_s^k : V \to \mathcal{R}^s$ is said to be zero up to kth order if it is sufficiently smooth and vanishes at the origin together* *with all partial derivatives of order less than or equal to k. The notation $o^k(v)$ will be used to denote a* generic *function of v which is zero up to kth order regardless of the dimension of its range space.*

kth-Order Nonlinear Output Regulation Problem (KNORP): Design a control law of the form (3.11) or (3.12) such that the closed-loop composite system (3.13) has Property 3.3 as well as the following property.

Property 4.1. For all sufficiently small x_{c0} and v_0, the trajectories $\text{col}(x_c(t), v(t))$ of the closed-loop composite system (3.13) satisfy

$$\lim_{t\to\infty}(e(t) - o^k(v(t))) = \lim_{t\to\infty}(h_c(x_c(t), v(t)) - o^k(v(t))) = 0. \tag{4.2}$$

If the closed-loop composite system has Properties 3.3 and 4.1, then we say that the steady-state tracking error of the closed-loop system is zero up to kth order. In what follows, a controller that solves the kth-order nonlinear output regulation problem will be called *kth-order servoregulator*. In particular, (3.11) and (3.12) are called, respectively, the *kth-order state feedback servoregulator* and the *kth-order measurement output feedback servoregulator*.

To study the solvability of the kth-order nonlinear output regulation problem, we first establish an equivalent characterization of Property 4.1 for the closed-loop composite system.

Lemma 4.2. *Under Assumption 3.1′, suppose the closed-loop composite system (3.13) has Property 3.3. Then the following are equivalent:*

(i) *The closed-loop composite system (3.13) has Property 4.1.*

(ii) *There exists a sufficiently smooth function $\mathbf{x}_c(v)$ with $\mathbf{x}_c(0) = 0$ that satisfies, for $v \in V$, the following partial differential and algebraic equations:*

$$\begin{aligned}\frac{\partial \mathbf{x}_c(v)}{\partial v}a(v) &= f_c(\mathbf{x}_c(v), v),\\ o^k(v) &= h_c(\mathbf{x}_c(v), v).\end{aligned} \tag{4.3}$$

(iii) *There exists a sufficiently smooth function $\mathbf{x}_c^{(k)}(v)$ with $\mathbf{x}_c^{(k)}(0) = 0$ that satisfies, for $v \in V$, the following partial differential and algebraic equations:*

$$\begin{aligned}\frac{\partial \mathbf{x}_c^{(k)}(v)}{\partial v}a(v) &= f_c(\mathbf{x}_c^{(k)}(v), v) + o^k(v),\\ o^k(v) &= h_c(\mathbf{x}_c^{(k)}(v), v).\end{aligned} \tag{4.4}$$

Proof. (i) $\leftrightarrow$ (ii). Define another system as follows:

$$\begin{aligned}&\dot{x}_c(t) = f_c(x_c(t), v(t)), \quad x_c(0) = x_{c0},\\ &\dot{v}(t) = a(v(t)), \quad v(0) = v_0,\\ &e^k(t) = h_c^k(x_c(t), v(t)), \quad t \geq 0,\end{aligned} \tag{4.5}$$

where

$$h_c^k(x_c, v) = h_c(x_c, v) - o^k(v). \tag{4.6}$$

Clearly, the system (3.13) has Property 4.1 if and only if (4.5) has Property 3.2. By Lemma 3.6 of Chapter 3, if (4.5) has Property 3.3, then it also has Property 3.2 if and only if there exists a sufficiently smooth function $\mathbf{x}_c(v)$ with $\mathbf{x}_c(0) = 0$ that satisfies, for $v \in V$,

$$\begin{aligned} \frac{\partial \mathbf{x}_c(v)}{\partial v} a(v) &= f_c(\mathbf{x}_c(v), v), \\ 0 &= h_c^k(\mathbf{x}_c(v), v), \end{aligned} \tag{4.7}$$

or, equivalently, the function $\mathbf{x}_c(v)$ satisfies (4.3).

(ii) $\leftrightarrow$ (iii). (ii) trivially implies (iii) by letting $\mathbf{x}_c^{(k)}(v) = \mathbf{x}_c(v)$. To show that (iii) also implies (ii), let $\mathbf{x}_c^{(k)}(v)$ satisfy, for $v \in V$, (4.4). Since (3.13) has Property 3.3, by Theorem 2.26, there exists a sufficiently smooth function $\mathbf{x}_c(v)$ with $\mathbf{x}_c(0) = 0$ that satisfies the first equation of (4.3). Moreover,

$$\mathbf{x}_c(v) = \mathbf{x}_c^{(k)}(v) + o^k(v). \tag{4.8}$$

We need to show that $\mathbf{x}_c(v)$ also satisfies the second equation of (4.3). Indeed,

$$h_c(\mathbf{x}_c(v), v) = h_c(\mathbf{x}_c^{(k)}(v) + o^k(v), v) = h_c(\mathbf{x}_c^{(k)}(v), v) + o^k(v) = o^k(v). \quad \square \tag{4.9}$$

Lemma 4.2 leads to the following characterization of the control law that solves the kth-order nonlinear output regulation problem.

Theorem 4.3. *Under Assumptions 3.1′ and 3.2, the kth-order nonlinear output regulation problem is solvable by a static state feedback controller*

$$u = k(x, v) \tag{4.10}$$

if and only if there exist two sufficiently smooth functions $\mathbf{x}^{(k)}(v)$ *and* $\mathbf{u}^{(k)}(v)$ *satisfying* $\mathbf{x}^{(k)}(0) = 0$ *and* $\mathbf{u}^{(k)}(0) = 0$ *such that*

$$\begin{aligned} \frac{\partial \mathbf{x}^{(k)}(v)}{\partial v} a(v) &= f(\mathbf{x}^{(k)}(v), \mathbf{u}^{(k)}(v), v) + o^k(v), \\ o^k(v) &= h(\mathbf{x}^{(k)}(v), \mathbf{u}^{(k)}(v), v). \end{aligned} \tag{4.11}$$

Proof. Assume that the controller (4.10) solves the kth-order nonlinear output regulation problem. Then, by Lemma 4.2, there exists a sufficiently smooth function $\mathbf{x}_c^{(k)}(v)$ that satisfies (4.4) for $v \in V$. Let $\mathbf{x}^{(k)}(v) = \mathbf{x}_c^{(k)}(v)$ and $\mathbf{u}^{(k)}(v) = k(\mathbf{x}^{(k)}(v), v)$. Then, clearly, $\mathbf{x}^{(k)}(v)$ and $\mathbf{u}^{(k)}(v)$ satisfy (4.11). On the other hand, let $\mathbf{x}^{(k)}(v)$ and $\mathbf{u}^{(k)}(v)$ satisfy (4.11). Using the same argument as used in the proof of Theorem 3.8, there exists a state feedback controller $k(x, v)$ with $k(0, 0) = 0$ such that the closed-loop system has Property 3.3. Furthermore, if $k(x, v)$ satisfies

$$k(\mathbf{x}^{(k)}(v), v) = \mathbf{u}^{(k)}(v),$$

for example,

$$k(x, v) = \mathbf{u}^{(k)}(v) + K_x(x - \mathbf{x}^{(k)}(v)), \tag{4.12}$$

where K_x is some constant feedback gain, then, clearly, this controller is such that the closed-loop system

$$\dot{x} = f(x, k(x, v), v), \quad e = h(x, k(x, v), v)$$

still has Property 3.3. Moreover, letting $\mathbf{x}_c^{(k)}(v) = \mathbf{x}^{(k)}(v)$ leads to

$$\begin{aligned} f_c(\mathbf{x}_c^{(k)}(v), v) &= f(\mathbf{x}^{(k)}(v), k(\mathbf{x}^{(k)}(v), v), v) \\ &= f(\mathbf{x}^{(k)}(v), \mathbf{u}^{(k)}(v), v) = \frac{\partial \mathbf{x}_c^{(k)}(v)}{\partial v} a(v) - o^k(v), \\ h_c(\mathbf{x}_c^{(k)}(v), v) &= h(\mathbf{x}^{(k)}(v), k(\mathbf{x}^{(k)}(v), v), v) \\ &= h(\mathbf{x}^{(k)}(v), \mathbf{u}^{(k)}(v), v) = o^k(v). \end{aligned}$$

Thus, by Lemma 4.2, the controller solves the kth-order nonlinear output regulation problem. □

Analogous to Lemma 3.15, we can also establish the following result on the solvability of the kth-order nonlinear output regulation problem via a measurement output feedback controller of the form

$$u = k(z), \quad \dot{z} = g(z, y_m). \tag{4.13}$$

Lemma 4.4. *Under Assumption 3.1′, assume that there exists a measurement output feedback control law of the form (4.13) such that the closed-loop composite system (3.13) has Property 3.3. Then the following are equivalent:*

(i) *The kth-order nonlinear output regulation problem is solvable by the measurement output feedback controller (4.13).*

(ii) *There exists a sufficiently smooth function $\mathbf{x}_c^{(k)}(v)$ with $\mathbf{x}_c^{(k)}(0) = 0$ such that*

$$\begin{aligned} \frac{\partial \mathbf{x}_c^{(k)}(v)}{\partial v} a(v) &= f_c(\mathbf{x}_c^{(k)}(v), v) + o^k(v), \\ o^k(v) &= h_c(\mathbf{x}_c^{(k)}(v), v). \end{aligned} \tag{4.14}$$

(iii) *There exist sufficiently smooth functions $(\mathbf{x}^{(k)}(v), \mathbf{u}^{(k)}(v), \mathbf{z}^{(k)}(v))$ with $(\mathbf{x}^{(k)}(0), \mathbf{u}^{(k)}(0), \mathbf{z}^{(k)}(0)) = (0, 0, 0)$ such that $\mathbf{x}^{(k)}(v)$ and $\mathbf{u}^{(k)}(v)$ satisfy equations (4.11) and $\mathbf{z}^{(k)}(v)$ satisfies*

$$\frac{\partial \mathbf{z}^{(k)}(v)}{\partial v} a(v) = g(\mathbf{z}^{(k)}(v), h_m(\mathbf{x}^{(k)}(v), \mathbf{u}^{(k)}(v), v)) + o^k(v). \tag{4.15}$$

Moreover,

$$\mathbf{u}^{(k)}(v) = k(\mathbf{z}^{(k)}(v)). \tag{4.16}$$

Proof. (i) $\leftrightarrow$ (ii). The proof is similar to that of Lemma 4.2 and is thus omitted.

(ii) $\leftrightarrow$ (iii). Assume (ii) holds. Partition $\mathbf{x}_c^{(k)}(v)$ as

$$\mathbf{x}_c^{(k)}(v) = \begin{bmatrix} \mathbf{x}^{(k)}(v) \\ \mathbf{z}^{(k)}(v) \end{bmatrix}, \tag{4.17}$$

where $\mathbf{x}^{(k)}(v) \in \mathcal{R}^n$ and $\mathbf{z}^{(k)}(v) \in \mathcal{R}^{n_z}$. Since $(f_c(x_c, v), h_c(x_c, v))$ is given by (3.15), expanding (4.14) gives

$$\begin{aligned} \frac{\partial \mathbf{x}^{(k)}(v)}{\partial v} a(v) &= f(\mathbf{x}^{(k)}(v), k(\mathbf{z}^{(k)}(v)), v) + o^k(v), \\ \frac{\partial \mathbf{z}^{(k)}(v)}{\partial v} a(v) &= g(\mathbf{z}^{(k)}(v), h_m(\mathbf{x}^{(k)}(v), k(\mathbf{z}^{(k)}(v)), v)) + o^k(v), \\ o^k(v) &= h(\mathbf{x}^{(k)}(v), k(\mathbf{z}^{(k)}(v)), v). \end{aligned} \tag{4.18}$$

Letting $\mathbf{u}^{(k)}(v) = k(\mathbf{z}^{(k)}(v))$ gives (4.16), and using (4.16), in the second equation of (4.18) gives (4.15). Finally, using (4.16) in the first and third equations of (4.18) shows that $\mathbf{x}^{(k)}(v)$ and $\mathbf{u}^{(k)}(v)$ satisfy (4.11). On the other hand, assume (iii) holds. Let $(\mathbf{x}^{(k)}(v), \mathbf{u}^{(k)}(v))$ be the solution of (4.11). Let $\mathbf{z}^{(k)}(v)$ satisfy (4.15) and (4.16). We will show that $\mathbf{x}^{(k)}(v)$ and $\mathbf{z}^{(k)}(v)$ satisfy (4.18). To this end, using (4.16) in (4.15) gives the second equation of (4.18), and using (4.16) in (4.11) shows that $(\mathbf{x}^{(k)}(v), \mathbf{z}^{(k)}(v))$ satisfies the first and third equations of (4.18). Thus, letting $\mathbf{x}_c^{(k)}(v)$ be given by (4.17) shows that $\mathbf{x}_c^{(k)}(v)$ satisfies (4.14). $\square$

Theorem 4.5. *Under Assumptions 3.1 to 3.3, suppose there exist two sufficiently smooth functions $\mathbf{x}^{(k)}(v)$ and $\mathbf{u}^{(k)}(v)$ with $\mathbf{x}^{(k)}(0) = 0$ and $\mathbf{u}^{(k)}(0) = 0$ that satisfy (4.11). Then, there exists a measurement output feedback control law that solves the kth-order nonlinear output regulation problem.*

Proof. Under the assumptions of Theorem 4.5, there exists a state feedback control law of the form $k(x, v)$ that solves the kth-order nonlinear output regulation problem. By Assumption 3.3, there exist constant matrices L_1 and L_2 such that all the eigenvalues of the matrix

$$\begin{bmatrix} \frac{\partial f}{\partial x}(0,0,0) & \frac{\partial f}{\partial v}(0,0,0) \\ 0 & \frac{\partial a}{\partial v}(0) \end{bmatrix} - \begin{bmatrix} L_1 \\ L_2 \end{bmatrix} \begin{bmatrix} \frac{\partial h_m}{\partial x}(0,0,0) & \frac{\partial h_m}{\partial v}(0,0,0) \end{bmatrix}$$

have negative real parts. Now let

$$\begin{aligned} k(z) &= k(z_1, z_2), \\ g(z, y_m) &= \begin{bmatrix} f(z_1, k(z_1, z_2), z_2) + L_1(y_m - h_m(z_1, k(z_1, z_2), z_2)) \\ a(z_2) + L_2(y_m - h_m(z_1, k(z_1, z_2), z_2)) \end{bmatrix}. \end{aligned} \tag{4.19}$$

Then, in a fashion similar to the proof of Theorem 3.16, it can be verified that the closed-loop system under this controller has Properties 3.3 and 4.1. Details are left to the reader. $\square$

4.2 Power Series Approach to Solving Regulator Equations

By Theorems 4.3 and 4.5, the key to the solvability of the kth-order nonlinear output regulation problem is to find the solution of the nonlinear regulator equations (4.1) up to kth order.

In this section, we will consider a power series approximation approach to solving (4.1). Our consideration will involve power series representations for the unknown functions $\mathbf{x}(v)$ and $\mathbf{u}(v)$, and this entails the following notation. For any matrix K, we will use the Kronecker product notation

$$K^{(0)} = 1,\ K^{(1)} = K,\ K^{(i)} = \underbrace{K \otimes K \otimes \cdots \otimes K}_{i\ factors},\ i = 2,\ 3, \ldots. \tag{4.20}$$

Then we can write the problem description in terms of the series expansions

$$\begin{aligned} f(x, u, v) &= \sum_{l \geq 1} \sum_{\substack{i+j+k=l \\ i,j,k \geq 0}} F_{ijk} x^{(i)} \otimes u^{(j)} \otimes v^{(k)}, \\ h(x, u, v) &= \sum_{l \geq 1} \sum_{\substack{i+j+k=l \\ i,j,k \geq 0}} H_{ijk} x^{(i)} \otimes u^{(j)} \otimes v^{(k)}, \\ a(v) &= \sum_{i \geq 1} A_i v^{(i)}. \end{aligned} \tag{4.21}$$

To obtain unique representations for the coefficients in series expansions of the unknown functions $\mathbf{x}(v)$ and $\mathbf{u}(v)$, the following notation will be used. For the $q \times 1$ vector $v = \mathrm{col}(v_1, \ldots, v_q)$, let $v^{[l]}$ denote the vector

$$v^{[l]} = [v_1^l,\ v_1^{l-1} v_2, \ldots, v_1^{l-1} v_q,\ v_1^{l-2} v_2^2,\ v_1^{l-2} v_2 v_3, \ldots, v_1^{l-2} v_2 v_q, \ldots, v_q^l]^T. \tag{4.22}$$

Then the Taylor series of the functions $\mathbf{x}(v)$ and $\mathbf{u}(v)$ can be uniquely expressed as follows:

$$\mathbf{x}(v) = \sum_{l \geq 1} X_l v^{[l]},\quad \mathbf{u}(v) = \sum_{l \geq 1} U_l v^{[l]}, \tag{4.23}$$

where X_l and U_l are constant coefficient matrices. We need to find these matrices such that equations (4.1) are satisfied formally. Note that the dimensions of $v^{[l]}$ and $v^{(l)}$ are, respectively,

$$\left[\begin{array}{c} q + l - 1 \\ l \end{array} \right] \times 1,\ q^l \times 1, \tag{4.24}$$

and that there exist matrices M_l and N_l of appropriate dimensions such that

$$v^{[l]} = M_l v^{(l)},\ v^{(l)} = N_l v^{[l]}. \tag{4.25}$$

For example, with $q = 2$, $v^{(2)}$ and $v^{[2]}$ are given by, respectively,

$$v^{(2)} = \begin{bmatrix} v_1^2 \\ v_1 v_2 \\ v_2 v_1 \\ v_2^2 \end{bmatrix},\quad v^{[2]} = \begin{bmatrix} v_1^2 \\ v_1 v_2 \\ v_2^2 \end{bmatrix},$$

and M_2 and N_2 are given by, respectively,

$$M_2 = \begin{bmatrix} 1 & 0 & 0 & 0 \\ 0 & 1 & 0 & 0 \\ 0 & 0 & 0 & 1 \end{bmatrix}, \quad N_2 = \begin{bmatrix} 1 & 0 & 0 \\ 0 & 1 & 0 \\ 0 & 1 & 0 \\ 0 & 0 & 1 \end{bmatrix}.$$

Although M_l is not unique, it is easy to check that $M_l N_l$ is an identity matrix regardless of the specific form of M_l.

Our purpose is to derive explicit equations that generate all matrices X_l and U_l, $l = 1, 2, \ldots$. To this end, we first list some useful identities involving the Kronecker product as follows.

Lemma 4.6.

(i) *For* $l \geq 1$,

$$\frac{\partial v^{(l)}}{\partial v} = \sum_{i=1}^{l} v^{(i-1)} \otimes I_q \otimes v^{(l-i)}. \tag{4.26}$$

(ii) *For any integers* $i, j, k \geq 0$, *and any matrix* T *of dimension* q *by* q^k,

$$(v^{(i)} \otimes I_q \otimes v^{(j)})Tv^{(k)} = (I_q^{(i)} \otimes T \otimes I_q^{(j)})v^{(i+j+k)}. \tag{4.27}$$

(iii) *For* $k, l \geq 1$, *and any matrix* T *of dimension* q *by* q^k,

$$\frac{\partial v^{[l]}}{\partial v} T v^{(k)} = M_l \left[\sum_{i=1}^{l} I_q^{(i-1)} \otimes T \otimes I_q^{(l-i)} \right] N_{l+k-1} v^{[l+k-1]}. \tag{4.28}$$

Proof. Equation (4.26) follows straightforwardly from the definition of the Kronecker product. Equation (4.27) can be proved as follows:

$$\begin{aligned}
(v^{(i)} \otimes I_q \otimes v^{(j)})Tv^{(k)} &= (v^{(i)} \otimes (I_q \otimes v^{(j)}))(1 \otimes (Tv^{(k)} \otimes 1)) \\
&= v^{(i)} \otimes (I_q \otimes v^{(j)})(Tv^{(k)} \otimes 1) \\
&= v^{(i)} \otimes (Tv^{(k)} \otimes v^{(j)}) \\
&= (I_q^{(i)} v^{(i)}) \otimes (Tv^{(k)}) \otimes v^{(j)} \\
&= (I_q^{(i)} \otimes T)v^{(i+k)} \otimes v^{(j)} \\
&= (I_q^{(i)} \otimes T)v^{(i+k)} \otimes (I_q^{(j)} v^{(j)}) \\
&= (I_q^{(i)} \otimes T \otimes I_q^{(j)})v^{(i+j+k)}.
\end{aligned}$$

Note that in deriving equation (4.27), we have repeatedly utilized the identity

$$(A \otimes B)(C \otimes D) = (AC) \otimes (BD),$$

which can be found in Appendix A.

To show (4.28) using (4.25), (4.26), (4.27), and (4.25) sequentially gives

$$\begin{aligned}
\frac{\partial v^{[l]}}{\partial v} T v^{(k)} &= M_l \frac{\partial v^{(l)}}{\partial v} T v^{(k)} \\
&= M_l \left[\sum_{i=1}^{l} v^{(i-1)} \otimes I_q \otimes v^{(l-i)} \right] T v^{(k)} \\
&= M_l \left[\sum_{i=1}^{l} (I_q^{(i-1)} \otimes T \otimes I_q^{(l-i)}) \right] v^{(l+k-1)} \\
&= M_l \left[\sum_{i=1}^{l} (I_q^{(i-1)} \otimes T \otimes I_q^{(l-i)}) \right] N_{l+k-1} v^{[l+k-1]}. \quad \square
\end{aligned}$$

Substituting equations (4.21) and (4.23) into equations (4.1), expanding equations (4.1) into the Taylor series, and identifying the coefficients of $v^{[l]}$, $l = 1, 2, \ldots$, yields the following result.

Lemma 4.7. *The power series (4.23) formally satisfies equations (4.1) if and only if the following linear equations are satisfied for $l = 1, 2, \ldots$:*

$$\begin{aligned}
&X_l M_l \left[\sum_{i=1}^{l} I_q^{(i-1)} \otimes A_1 \otimes I_q^{(l-i)} \right] N_l = \frac{\partial f}{\partial x}(0,0,0) X_l + \frac{\partial f}{\partial u}(0,0,0) U_l + E_l, \\
&\frac{\partial h}{\partial x}(0,0,0) X_l + \frac{\partial h}{\partial u}(0,0,0) U_l + F_l = 0,
\end{aligned} \tag{4.29}$$

where

$$E_1 = F_{001} = \frac{\partial f}{\partial v}(0,0,0), \quad F_1 = H_{001} = \frac{\partial h}{\partial v}(0,0,0), \tag{4.30}$$

and, for $l = 2, 3, \ldots$,

$$E_l = \left(\sum_{n=2}^{l} \sum_{\substack{i+j+k=n \\ i,j,k \ge 0}} F_{ijk} G_{l-n}^{ij} - \sum_{k=1}^{l-1} X_k M_k \left[\sum_{i=1}^{k} I_q^{(i-1)} \otimes A_{l-k+1} \otimes I_q^{(k-i)} \right] \right) N_l, \tag{4.31}$$

$$F_l = \left(\left[\sum_{n=2}^{l} \sum_{\substack{i+j+k=n \\ i,j,k \ge 0}} H_{ijk} G_{l-n}^{ij} \right] \right) N_l, \tag{4.32}$$

$$G_m^{ij} = \begin{cases} 0, & i = j = 0, m > 0, \\ 1, & i = j = 0, m = 0, \\ \delta_{i,i+m}, & j = 0, i = 1, 2, \ldots, \\ \lambda_{j,j+m}, & i = 0, j = 1, 2, \ldots, \\ \sum_{k=0}^{m} \delta_{i,i+k} \otimes \lambda_{j,j+m-k}, & i, j = 1, 2, \ldots, \end{cases} \tag{4.33}$$

$$\delta_{i,j} = \sum_{\substack{j_1+j_2+\cdots+j_i=j \\ j_1,j_2,\ldots,j_i\geq 1}} X_{j_1}M_{j_1} \otimes X_{j_2}M_{j_2} \otimes \cdots \otimes X_{j_i}M_{j_i}, \quad i = 1, 2, \ldots, j \geq i, \tag{4.34}$$

$$\lambda_{i,j} = \sum_{\substack{j_1+j_2+\cdots+j_i=j \\ j_1,j_2,\ldots,j_i\geq 1}} U_{j_1}M_{j_1} \otimes U_{j_2}M_{j_2} \otimes \cdots \otimes U_{j_i}M_{j_i}, \quad i = 1, 2, \ldots, j \geq i. \tag{4.35}$$

Proof. Substituting equations (4.21) and (4.23) into equations (4.1) yields the following equations:

$$\left[\frac{\partial}{\partial v}\sum_{i\geq 1} X_i v^{[i]}\right]\sum_{j\geq 1} A_j v^{(j)} = \sum_{l\geq 1}\sum_{\substack{i+j+k=l \\ i,j,k\geq 0}} F_{ijk}\mathbf{x}^{(i)}(v) \otimes \mathbf{u}^{(j)}(v) \otimes v^{(k)}, \tag{4.36}$$

$$0 = \sum_{l\geq 1}\sum_{\substack{i+j+k=l \\ i,j,k\geq 0}} H_{ijk}\mathbf{x}^{(i)}(v) \otimes \mathbf{u}^{(j)}(v) \otimes v^{(k)}. \tag{4.37}$$

The left-hand side of (4.36) can be written as

$$\begin{aligned}\left[\frac{\partial}{\partial v}\sum_{i\geq 1} X_i v^{[i]}\right]\sum_{j\geq 1} A_j v^{(j)} &= \left[\sum_{i\geq 1} X_i \frac{\partial v^{[i]}}{\partial v}\right]\sum_{j\geq 1} A_j v^{(j)} \\ &= \sum_{l\geq 1}\left[\sum_{k=1}^{l} X_k \frac{\partial v^{[k]}}{\partial v} A_{l-k+1} v^{(l-k+1)}\right].\end{aligned} \tag{4.38}$$

Thus using (4.28) in (4.38) gives

$$\begin{aligned}&\left[\frac{\partial}{\partial v}\sum_{i\geq 1} X_i v^{[i]}\right]\sum_{j\geq 1} A_j v^{(j)} \\ &\quad = \sum_{l\geq 1}\left[\sum_{k=1}^{l} X_k M_k \left(\sum_{i=1}^{k} I_q^{(i-1)} \otimes A_{l-k+1} \otimes I_q^{(k-i)}\right)\right] N_l v^{[l]}.\end{aligned} \tag{4.39}$$

Also, we can write

$$\mathbf{x}^{(i)}(v) = \left(\sum_{k\geq 1} X_k M_k v^{(k)}\right)^{(i)} = \sum_{l\geq i} \delta_{i,l} v^{(l)} \tag{4.40}$$

and

$$\mathbf{u}^{(i)}(v) = \left(\sum_{k\geq 1} U_k M_k v^{(k)}\right)^{(i)} = \sum_{l\geq i} \lambda_{i,l} v^{(l)}, \tag{4.41}$$

where $\delta_{i,l}$ and $\lambda_{i,l}$ are given by equations (4.34) and (4.35). Then

$$\begin{aligned}\mathbf{x}^{(i)}(v)\otimes\mathbf{u}^{(j)}(v)\otimes v^{(k)} &= \left[\sum_{l\geq i}\delta_{i,l}v^{(l)}\right]\otimes\left[\sum_{m\geq j}\lambda_{j,m}v^{(m)}\right]\otimes v^{(k)}\\ &= \sum_{m\geq 0}G_m^{ij}v^{(i+j+k+m)},\end{aligned} \tag{4.42}$$

where G_m^{ij} is given by equation (4.33). This permits the right-hand sides of equations (4.36) and (4.37) to be written as

$$\sum_{l\geq 1}\left[F_{100}G_{l-1}^{10}+F_{010}G_{l-1}^{01}+F_{001}G_{l-1}^{00}+\sum_{n=2}^{l}\sum_{\substack{i+j+k=n\\ i,j,k\geq 0}}F_{ijk}G_{l-n}^{ij}\right]N_l v^{[l]} \tag{4.43}$$

and

$$\sum_{l\geq 1}\left[H_{100}G_{l-1}^{10}+H_{010}G_{l-1}^{01}+H_{001}G_{l-1}^{00}+\sum_{n=2}^{l}\sum_{\substack{i+j+k=n\\ i,j,k\geq 0}}H_{ijk}G_{l-n}^{ij}\right]N_l v^{[l]}, \tag{4.44}$$

respectively. Using (4.39), (4.43), and (4.44) in (4.36) and (4.37) and equating the coefficients of $v^{[l]}$ on both sides of the rewritten (4.36) and (4.37) gives, for $l\geq 1$,

$$\begin{aligned}&\left[\sum_{k=1}^{l}X_kM_k\left(\sum_{i=1}^{k}I_q^{(i-1)}\otimes A_{l-k+1}\otimes I_q^{(k-i)}\right)\right]N_l\\ &\quad=\left[F_{100}G_{l-1}^{10}+F_{010}G_{l-1}^{01}+F_{001}G_{l-1}^{00}+\sum_{n=2}^{l}\sum_{\substack{i+j+k=n\\ i,j,k\geq 0}}F_{ijk}G_{l-n}^{ij}\right]N_l,\\ &0=\left[H_{100}G_{l-1}^{10}+H_{010}G_{l-1}^{01}+H_{001}G_{l-1}^{00}+\sum_{n=2}^{l}\sum_{\substack{i+j+k=n\\ i,j,k\geq 0}}H_{ijk}G_{l-n}^{ij}\right]N_l.\end{aligned}$$

Finally, using $G_{l-1}^{10}=\delta_{1,l}=X_lM_l$, $G_{l-1}^{01}=\lambda_{1,l}=U_lM_l$, $G_0^{00}=1$, $G_{l-1}^{00}=0, l>1$, along with the fact that M_lN_l is an identity matrix, completes the proof. □

Note that E_l and F_l depend only on $X_1,\ldots,X_{l-1}$ and $U_1,\ldots,U_{l-1}$, so that equations (4.29) provide a sequence of linear matrix equations. The following result establishes the solvability condition for these equations.

Lemma 4.8. *There exists a solution (unique if $p = m$) of equations (4.29) for any E_l and F_l, $l = 1, 2, \ldots$, if and only if the plant satisfies the following assumption.*

Assumption 4.1.

$$rank \begin{bmatrix} \frac{\partial f}{\partial x}(0,0,0) - \lambda I & \frac{\partial f}{\partial u}(0,0,0) \\ \frac{\partial h}{\partial x}(0,0,0) & \frac{\partial h}{\partial u}(0,0,0) \end{bmatrix} = n + p \tag{4.45}$$

for all $\lambda \in \Lambda_l$, where

$$\Lambda_l = \{ \lambda \mid \lambda = l_1\lambda_1 + \cdots + l_q\lambda_l, l_1 + \cdots + l_q = l,\ l_1, \ldots, l_q = 0, 1, \ldots, l \}, \tag{4.46}$$

with $\lambda_1, \ldots, \lambda_q$ being the eigenvalues of the matrix $\frac{\partial a}{\partial v}(0)$.

Proof. For a given l, equations (4.29) actually take the same form as the linear regulator equations (1.21). Thus, by Theorem 1.9, equations (4.29) have a solution for any E_l and F_l if and only if equation (4.45) holds for all λ in the spectrum of

$$A^{[l]} \stackrel{def}{=} M_l \left[\sum_{i=1}^{l} I_q^{(i-1)} \otimes A_1 \otimes I_q^{(l-i)} \right] N_l. \tag{4.47}$$

We now show that the eigenvalues of $A^{[l]}$ are precisely those given by (4.46). To this end, using (4.28) with $T = A_1$ and $k = 1$ gives

$$\frac{\partial v^{[l]}}{\partial v} A_1 v = A^{[l]} v^{[l]}. \tag{4.48}$$

Note that the components of $v^{[l]}$ consist of all products of the variables $v_1, \ldots, v_q$ taken l at a time. Therefore, if we define P^l as the vector space of all homogeneous polynomials in $v_1, \ldots, v_q$ of degree l, then the components of $v^{[l]}$ give a basis of P^l. Now define a linear mapping $L_{A_1 v} : P^l \to P^l$ such that, for each $\phi \in P^l$,

$$L_{A_1 v}(\phi) = \frac{\partial \phi}{\partial v} A_1 v. \tag{4.49}$$

Then, using (4.48) shows

$$\begin{aligned} & L_{A_1 v}\left[v_1^l,\ v_1^{l-1}v_2, \ldots, v_1^{l-1}v_q,\ v_1^{l-2}v_2^2,\ v_1^{l-2}v_2v_3, \ldots, v_1^{l-2}v_2v_q, \ldots, v_q^l\right] \\ & = \left[v_1^l,\ v_1^{l-1}v_2, \ldots, v_1^{l-1}v_q,\ v_1^{l-2}v_2^2,\ v_1^{l-2}v_2v_3, \ldots, v_1^{l-2}v_2v_q, \ldots, v_q^l\right](A^{[l]})^T. \end{aligned}$$

Thus, $(A^{[l]})^T$ is the matrix of the linear mapping under the ordered basis

$$\{v_1^l,\ v_1^{l-1}v_2, \ldots, v_1^{l-1}v_q,\ v_1^{l-2}v_2^2,\ v_1^{l-2}v_2v_3, \ldots, v_1^{l-2}v_2v_q, \ldots, v_q^l\}. \tag{4.50}$$

Thus, the spectrum of $A^{[l]}$ is the same as that of the linear mapping (4.49).

Now let the Jordan canonical form of A_1 be

$$A_1 = \begin{bmatrix} J_1 & 0 & \cdots & 0 \\ 0 & J_2 & \cdots & 0 \\ \cdots & \cdots & \cdots & \cdots \\ 0 & 0 & \cdots & J_k \end{bmatrix}_{q\times q}, \tag{4.51}$$

where

$$J_i = \begin{bmatrix} \lambda_i & 1 & 0 & \cdots & 0 \\ 0 & \lambda_i & 1 & \cdots & 0 \\ \cdots & \cdots & \cdots & \cdots & \cdots \\ 0 & 0 & 0 & \cdots & \lambda_i \end{bmatrix}_{n_i\times n_i}$$

is a $n_i \times n_i$ Jordan block with eigenvalue λ_i. Suppose the generalized row eigenvectors of A_1 are

$$\zeta_{11}, \zeta_{12}, \ldots, \zeta_{1n_1}, \zeta_{21}, \ldots, \zeta_{k1}, \ldots, \zeta_{kn_k}, \tag{4.52}$$

satisfying

$$\zeta_{ij} A_1 = \begin{cases} \lambda_i \zeta_{ij}, & j = n_i, \\ \lambda_i \zeta_{ij} + \zeta_{i(j+1)}, & 1 \le j < n_i. \end{cases} \tag{4.53}$$

Clearly,

$$\left\{ (\zeta_{11}v)^{\alpha_{11}} (\zeta_{12}v)^{\alpha_{12}} \cdots (\zeta_{1n_1}v)^{\alpha_{1n_1}} \cdots (\zeta_{k1}v)^{\alpha_{k1}} \cdots (\zeta_{kn_k}v)^{\alpha_{kn_k}} \,\middle|\, \sum_{i=1}^{k} \sum_{j=1}^{n_i} \alpha_{ij} = l \right\} \tag{4.54}$$

also constitutes a basis for P^l. Furthermore,

$$L_{A_1 v}((\zeta_{ij}v)^s) = \begin{cases} s\lambda_i (\zeta_{ij}v)^s, & j = n_i, \\ s\lambda_i (\zeta_{ij}v)^s + s(\zeta_{ij}v)^{s-1}\zeta_{i(j+1)}v, & j < n_i. \end{cases} \tag{4.55}$$

Now define an order on (4.54) in the following "lexicographic" way:

$$(\zeta_{11}v)^{\alpha_{11}} \cdots (\zeta_{kn_k}v)^{\alpha_{kn_k}} > (\zeta_{11}v)^{\beta_{11}} \cdots (\zeta_{kn_k}v)^{\beta_{kn_k}} \tag{4.56}$$

if and only if there exist positive integers i_0 and $j_0 \le n_{i_0}$ such that

$$\alpha_{i_0 j_0} < \beta_{i_0 j_0} \quad \text{and} \quad \alpha_{ij} = \beta_{ij}$$

if $i < i_0$, $j \le n_i$ or $i = i_0$, $j < j_0$. Then (4.54) constitutes an ordered basis of P^l. Using (4.55) gives

$$L_{A_1 v}((\zeta_{11}v)^{\alpha_{11}} \cdots (\zeta_{kn_k}v)^{\alpha_{kn_k}}) = \left(\sum_{i=1}^{k} \sum_{j=1}^{n_i} \alpha_{ij}\lambda_i \right) (\zeta_{11}v)^{\alpha_{11}} \cdots (\zeta_{kn_k}v)^{\alpha_{kn_k}}$$
$$+ \text{ terms greater than } (\zeta_{11}v)^{\alpha_{11}} \cdots (\zeta_{kn_k}v)^{\alpha_{kn_k}}.$$

Thus, the matrix of the linear mapping $L_{A_1 v}$ on P^l under the ordered basis (4.54) with the order (4.56) is upper triangular with the diagonal elements being

$$\lambda = \sum_{j=1}^{n_1} \alpha_{1j}\lambda_1 + \sum_{j=1}^{n_2} \alpha_{2j}\lambda_2 + \cdots + \sum_{j=1}^{n_k} \alpha_{kj}\lambda_k.$$

Therefore, the eigenvalues of $L_{A_1 v}$ on P^l are exactly given by equation (4.46). □

Remark 4.9. In the case when the solution of equations (4.29) is such that (4.23) has a positive convergent radius, then (4.23) is an exact solution of equations (4.1) in power series form. In particular, if the solution of equation (4.1) is a polynomial in $v^{[l]}$, then Lemma 4.7 gives an approach to exactly solving equations (4.1). Note that equation (4.45) represents the constraints on the transmission zeros of the Jacobian linearization of the plant which can be viewed as an extension of the transmission zeros condition for the linear output regulation problem as described in Remark 1.11. ∎

Remark 4.10. Assume that the transmission zeros condition in equation (4.45) holds up to some positive integer k. Let

$$\mathbf{x}^{(k)}(v) = \sum_{l=1}^{k} X_l v^{[l]}, \quad \mathbf{u}^{(k)}(v) = \sum_{l=1}^{k} U_l v^{[l]}. \tag{4.57}$$

Then, it is not difficult to see from the proof of Lemma 4.7 that $\mathbf{x}^{(k)}(v)$ and $\mathbf{u}^{(k)}(v)$ are such that

$$\begin{aligned} \frac{\partial \mathbf{x}^{(k)}(v)}{\partial v} a(v) &= f(\mathbf{x}^{(k)}(v), \mathbf{u}^{(k)}(v), v) + o^k(v), \\ h(\mathbf{x}^{(k)}(v), \mathbf{u}^{(k)}(v), v) &= o^k(v). \end{aligned} \tag{4.58}$$

In conjunction with Theorems 4.3 and 4.5, this observation immediately leads to the following sufficient conditions for the solvability of the kth-order nonlinear output regulation problem. ∎

Theorem 4.11.

(i) *Under Assumptions 3.1, 3.2, and 4.1, the kth-order nonlinear output regulation problem is solvable by the state feedback control law of the form (3.11).*

(ii) *Under the additional Assumption 3.3, the kth-order nonlinear output regulation problem is solvable by the measurement output feedback control law of the form (3.12).*

4.3 Power Series Approach to Solving Invariant Manifold Equation

As we have seen in Section 3.4, when the composite system (3.8) satisfies Assumption 3.4, we can reduce the solvability of the regulator equations into the solvability of an invariant manifold equation of the form (3.67), which is associated with the zero dynamics (3.66) of the composite system (3.8). Since the dimension of the invariant manifold equation is smaller than that of the regulator equations, it is more convenient to solve the invariant manifold equation. To put our technical development in a more general context, in this section, we will consider a general nonlinear system of the form

$$\dot{x} = F(x, v), \tag{4.59}$$

where $x \in \mathcal{R}^n$, $v \in \mathcal{R}^q$, and $F : \mathcal{R}^{n+q} \to \mathcal{R}^n$ is a sufficiently smooth function satisfying $F(0, 0) = 0$. Associated with (4.59) and the exosystem (3.10) is a partial differential equation of the form

$$\frac{\partial \mathbf{x}(v)}{\partial v} a(v) = F(\mathbf{x}(v), v). \tag{4.60}$$

This equation can be viewed as a special case of the regulator equations when $p = m = 0$. Recall from Chapter 2 that an equation of the form (4.60) is called an invariant manifold equation. In particular, when the equilibrium point of $\dot{x} = F(x, 0)$ at $x = 0$ is hyperbolic and all the eigenvalues of $\frac{\partial a(0)}{\partial v}$ have zero real parts, (4.60) is called a center manifold equation.

Similar to the last section, we will seek series of the form

$$\mathbf{x}(v) = \sum_{l \geq 1} X_l v^{[l]} \tag{4.61}$$

such that (4.60) is satisfied. For this purpose, we can again write $F(x, v)$ and $a(v)$ in terms of Taylor series as follows:

$$\begin{aligned} F(x, v) &= \sum_{l \geq 1} \sum_{\substack{i+k=l \\ i,k \geq 0}} F_{ik} x^{(i)} \otimes v^{(k)}, \\ a(v) &= \sum_{i \geq 1} A_i v^{(i)}. \end{aligned} \tag{4.62}$$

Analogous to Lemmas 4.7 and 4.8, we can obtain the following two lemmas. The proofs of these two lemmas are omitted since they can be directly deduced from the proofs of Lemma 4.7 and Lemma 4.8, respectively.

Lemma 4.12. *The power series (4.61) formally satisfies equation (4.60) if and only if the following linear equation is satisfied for $l = 1, 2, \ldots$:*

$$X_l M_l \left[\sum_{i=1}^{l} I_q^{(i-1)} \otimes A_1 \otimes I_q^{(l-i)} \right] N_l = \frac{\partial F}{\partial x}(0, 0) X_l + E_l, \tag{4.63}$$

where

$$E_1 = F_{01} = \frac{\partial F}{\partial v}(0, 0) \tag{4.64}$$

and, for $l = 2, 3, \ldots,$

$$E_l = \left(\sum_{n=2}^{l} \sum_{\substack{i+k=n \\ i,k \geq 0}} F_{ik}\delta_{i,i+l-n} - \sum_{k=1}^{l-1} X_k M_k \left[\sum_{i=1}^{k} I_q^{(i-1)} \otimes A_{l-k+1} \otimes I_q^{(k-i)} \right] \right) N_l, \tag{4.65}$$

$$\delta_{i,j} = \begin{cases} 0, & i = 0,\ j > 0, \\ 1, & i = 0,\ m = 0, \\ \sum_{\substack{j_1+\cdots+j_i=j \\ j_1,\ldots,j_i \geq 1}} X_{j_1} M_{j_1} \otimes X_{j_2} M_{j_2} \otimes \cdots \otimes X_{j_i} M_{j_i}, & i = 1, \ldots, j \geq i. \end{cases} \tag{4.66}$$

Lemma 4.13. *There exists a unique solution of equation (4.63) for any* E_l, $l = 1, 2, \ldots,$ *if and only if none of the eigenvalues of the matrix* $\frac{\partial F}{\partial x}(0, 0)$ *coincides with any* $\lambda \in \Lambda_l$, *where*

$$\Lambda_l = \{ \lambda \mid \lambda = l_1\lambda_1 + \cdots + l_q\lambda_l, l_1 + \cdots + l_q = l,\ l_1, \ldots, l_q = 0, 1, \ldots, l \}$$

and where $\lambda_1, \ldots, \lambda_q$ *are eigenvalues of the matrix* $\frac{\partial a}{\partial v}(0)$.

Remark 4.14. In the case that the solution of equation (4.60) is such that (4.61) has a positive convergent radius, then (4.61) is an exact solution of equation (4.60) in power series form. In particular, if the solution of equation (4.60) is a polynomial in $v^{[l]}$, then Lemma 4.12 gives an approach to solving equation (4.60) exactly. ∎

Remark 4.15. When the system (4.59) represents the zero dynamics of the system $\dot{x} = f(x, u, 0)$, $e = h(x, u, 0)$, the eigenvalues of the matrix $\frac{\partial F}{\partial x}(0, 0)$ are precisely the transmission zeros of the linearization of the system $\dot{x} = f(x, u, 0)$, $e = h(x, u, 0)$. Thus the condition of Lemma 4.13 is consistent with the transmission zero condition given in Assumption 4.1. Note that this condition is much less stringent than the hyperbolicity assumption of the matrix $\frac{\partial F(0,0)}{\partial x}$ because it only prohibits the eigenvalues of the matrix $\frac{\partial F(0,0)}{\partial x}$ from belonging to a countable set. Moreover, the solvability of (4.60) in power series does not have to rely on the assumption that the eigenvalues of $\frac{\partial a(0)}{\partial v}$ are on the imaginary axis. In the next section, we will see that the invariant manifold equation associated with the inverted pendulum on a cart system admits a formal power series solution. ∎

4.4 Asymptotic Tracking of the Inverted Pendulum on a Cart

We now return to the problem of the asymptotic tracking of the inverted pendulum on a cart formulated in Section 3.2. Let us first note that the Jacobian linearization of the inverted pendulum on a cart system at the origin is as follows:

$$\frac{\partial f(0)}{\partial x} = \begin{bmatrix} 0 & 1 & 0 & 0 \\ 0 & -\frac{b}{M} & -\frac{mg}{M} & 0 \\ 0 & 0 & 0 & 1 \\ 0 & \frac{b}{lM} & \frac{(M+m)}{lM}g & 0 \end{bmatrix}, \quad g(0) = \begin{bmatrix} 0 \\ \frac{1}{M} \\ 0 \\ -\frac{1}{lM} \end{bmatrix},$$

which is controllable. Thus the system satisfies Assumption 3.2.

Recall from Section 2.8 that the relative degree of (2.109) is 2, and the zero dynamics of (2.109) is given by

$$\dot{x}_3 = x_4, \qquad \dot{x}_4 = \frac{g}{l} \sin x_3, \tag{4.67}$$

which has a hyperbolic equilibrium as the eigenvalues of the Jacobian matrix of (4.67) at the origin are given by $\pm\sqrt{g/l}$. Thus the solution of the regulator equations associated with the inverted pendulum system exists. As a matter of fact, it can be further verified that, for system (2.109),

$$\begin{aligned} D(x) &= \frac{1}{M + m(\sin x_3)^2}, \\ E(x) &= \frac{1}{M + m(\sin x_3)^2}(mlx_4^2 \sin x_3 - bx_2 - mg \cos x_3 \sin x_3), \\ H(x) &= \begin{bmatrix} x_1 \\ x_2 \end{bmatrix}, \quad H_d(v) = \begin{bmatrix} v_1 \\ \omega v_2 \end{bmatrix}, \quad E_d(v) = -\omega^2 v_1. \end{aligned}$$

Thus, applying the algorithm described in Remark 3.30 gives the partition $x = \text{col}(x^1, x^2)$ with $x^1 = \text{col}(x_1, x_2)$ and $x^2 = \text{col}(x_3, x_4)$ and the following functions:

$$\begin{aligned} x^1 = \sigma(x^2, v) &= \begin{bmatrix} v_1 \\ \omega v_2 \end{bmatrix}, \\ u_e(x, v) &= -(M + m(\sin x_3)^2)\omega^2 v_1 - (mlx_4^2 \sin x_3 - bx_2 - mg \cos x_3 \sin x_3), \end{aligned}$$

as well as the zero dynamics of the composite system (3.22):

$$\begin{aligned} \dot{x}_3 &= x_4, \\ \dot{x}_4 &= \frac{\omega^2}{l} v_1 \cos x_3 + \frac{g}{l} \sin x_3, \\ \dot{v}_1 &= \omega v_2, \\ \dot{v}_2 &= -\omega v_1. \end{aligned} \tag{4.68}$$

We can put (4.68) in the following form:

$$\dot{x}^2 = \delta(x^2, v), \qquad \dot{v} = A_1 v, \tag{4.69}$$

with

$$x^2 = \begin{bmatrix} x_3 \\ x_4 \end{bmatrix}, \quad \delta(x^2, v) = \begin{bmatrix} x_4 \\ \frac{\omega^2}{l} v_1 \cos x_3 + \frac{g}{l} \sin x_3 \end{bmatrix}.$$

The center manifold equation associated with (4.69) is given by the following partial differential equation:

$$\frac{\partial \mathbf{x}^2(v)}{\partial v} A_1 v = \delta(\mathbf{x}^2(v), v). \tag{4.70}$$

Since $\frac{\partial\delta(0,0)}{\partial x^2}$ does not have eigenvalues on the imaginary axis, the Center Manifold Theorem guarantees the existence of the solution of (4.70). Let

$$\mathbf{x}^2(v) = \begin{bmatrix} \mathbf{x}_3(v) \\ \mathbf{x}_4(v) \end{bmatrix}$$

be the solution of (4.70). Then the solution of the regulator equations associated with (3.22) is

$$\mathbf{x}(v) = \begin{bmatrix} v_1 \\ \omega v_2 \\ \mathbf{x}_3(v) \\ \mathbf{x}_4(v) \end{bmatrix}, \tag{4.71}$$

$$\begin{aligned}\mathbf{u}(v) = &-(M + m\sin^2\mathbf{x}_3(v))\omega^2 v_1 - ml\mathbf{x}_4^2(v)\sin\mathbf{x}_3(v) + b\omega v_2 \\ &+ mg\cos\mathbf{x}_3(v)\sin\mathbf{x}_3(v).\end{aligned} \tag{4.72}$$

It should be noted that even though the solvability of the equation (4.70) is guaranteed by Theorem 2.25, due to the nonlinear nature of equation (4.68) we are not able to find an explicit solution for the center manifold equation (4.70). Nevertheless, we will show that the Taylor series solution of the center manifold equation studied in Section 4.3 will give an approximate solution to (4.70).

To this end, expand equation (4.70) as follows:

$$\begin{aligned}\frac{\partial\mathbf{x}_3(v)}{\partial v_1}\omega v_2 - \frac{\partial\mathbf{x}_3(v)}{\partial v_2}\omega v_1 &= \mathbf{x}_4(v), \\ \frac{\partial\mathbf{x}_4(v)}{\partial v_1}\omega v_2 - \frac{\partial\mathbf{x}_4(v)}{\partial v_2}\omega v_1 &= \frac{\omega^2}{l}v_1\cos\mathbf{x}_3(v) + \frac{g}{l}\sin\mathbf{x}_3(v).\end{aligned} \tag{4.73}$$

We have already known that the equilibrium point of the zero dynamics of a plant with $v = 0$ is hyperbolic. Therefore, by Lemma 4.13, equation (4.73) admits a power series solution of the form (4.61). Now let us proceed to find an approximate solution to equation (4.73), and then an approximate controller based on the approximate solution to equation (4.73). Though we can use the general method given in Section 4.3 to obtain an approximate solution to equation (4.73) with the help of a computer program, it is possible to obtain a lower order approximate solution to equation (4.73) using hand calculation. For this purpose, assume that the power series expansion of $\mathbf{x}_3(v)$ and $\mathbf{x}_4(v)$ takes the following form:

$$\begin{aligned}\mathbf{x}_3(v) = &\, a_{10}v_1 + a_{01}v_2 + a_{20}v_1^2 + a_{11}v_1v_2 + a_{02}v_2^2 \\ &+ a_{30}v_1^3 + a_{21}v_1^2v_2 + a_{12}v_1v_2^2 + a_{03}v_2^3 + \cdots,\end{aligned} \tag{4.74}$$

$$\begin{aligned}\mathbf{x}_4(v) = &\, b_{10}v_1 + b_{01}v_2 + b_{20}v_1^2 + b_{11}v_1v_2 + b_{02}v_2^2 \\ &+ b_{30}v_1^3 + b_{21}v_1^2v_2 + b_{12}v_1v_2^2 + b_{03}v_2^3 + \cdots.\end{aligned} \tag{4.75}$$

Then, substituting (4.74) and (4.75) into (4.73) and identifying the coefficients gives a third-order approximation of $\mathbf{x}_3(v)$ and $\mathbf{x}_4(v)$ as follows:

$$\mathbf{x}_3^{(3)}(v) = a_{10}v_1 + a_{12}v_1v_2^2 + a_{30}v_1^3, \tag{4.76}$$

$$\mathbf{x}_4^{(3)}(v) = \omega a_{10}v_2 + \omega a_{12}v_2^3 + (3a_{30} - 2a_{12})v_1^2v_2\omega, \tag{4.77}$$

where

$$
\begin{aligned}
a_{10} &= \frac{-\alpha}{\beta + \omega^2}, \\
a_{12} &= \frac{a_{10}^2 \omega^2 (3\alpha + \beta a_{10})}{(\beta + \omega^2)(\beta + 9\omega^2)}, \\
a_{30} &= \frac{a_{10}^2 (\beta + 7\omega^2)(3\alpha + \beta a_{10})}{6(\beta + \omega^2)(\beta + 9\omega^2)}, \\
\alpha &= \frac{\omega^2}{l}, \\
\beta &= \frac{g}{l}.
\end{aligned}
$$

Using the expressions (4.71) and (4.72), we can obtain a third-order approximation of the solution of the regulator equations associated with (4.68) as follows:

$$
\mathbf{x}^{(3)}(v) = \begin{bmatrix} v_1 \\ \omega v_2 \\ \mathbf{x}_3^{(3)}(v) \\ \mathbf{x}_4^{(3)}(v) \end{bmatrix},
$$

$$
\begin{aligned}
\mathbf{u}^{(3)}(v) &= -(M + m(a_{10}v_1)^2)\omega^2 v_1 - ml(\omega a_{10} v_2)^2 (a_{10} v_1) \\
&\quad + b\omega v_2 + mg\left(\mathbf{x}_3^{(3)}(v) - \frac{2}{3}(a_{10}v_1)^3\right).
\end{aligned}
$$

Based on $\mathbf{x}^{(3)}(v)$ and $\mathbf{u}^{(3)}(v)$, an approximate controller is given as follows:

$$
u = \mathbf{u}^{(3)}(v) + K_x(x - \mathbf{x}^{(3)}(v)),
$$

where K_x is such that the matrix $\frac{\partial f(0)}{\partial x} + g(0)K_x$ is Hurwitz. Let $b = 12.98$ kg/sec, $M = 1.378$ kg, $l = 0.325$m, $g = 9.8$m/sec^2, $m = 0.051$ kg, and let the eigenvalues of the matrix $\frac{\partial f(0)}{\partial x} + g(0)K_x$ be $[(-0.848 \pm 2.52j),\ (-1.25 \pm 0.828j)]$. Then $K_x = [0.0457 \quad 13.16 \quad 16.7 \quad 1.85]$. The above eigenvalues are based on the ITAE prototype design with cutoff frequency equal to 1.

Frequency	Nonlinear controller	Linear controller
$\omega = 1.0$	0.00076	0.040
$\omega = 1.5$	0.0045	0.065
$\omega = 2.0$	0.0210	0.0825

Table 4.1. *Maximal steady-state tracking error with* $A_m = 1$.

The performance of the controller has been evaluated by computer simulation with various values of the frequency ω and fixed amplitude $A_m = 1$. Table 4.1 lists the maximal steady-state tracking errors of the closed-loop system for several different frequencies with $A_m = 1$. For comparison, we also give the maximal steady-state tracking errors resulting from a linear controller of the following form:

$$
u = \mathbf{u}^{(1)}(v) + K_x(x - \mathbf{x}^{(1)}(v)),
$$

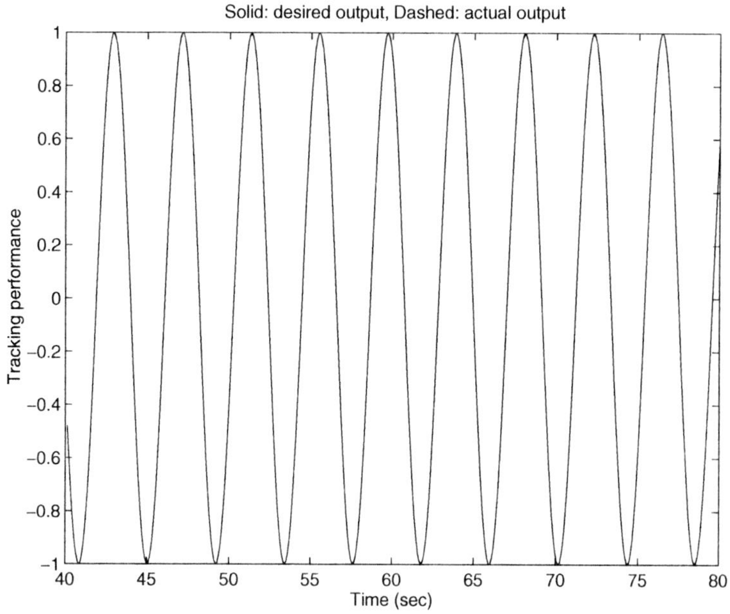

Figure 4.1. *The profile of the tracking performance of the closed-loop system under the nonlinear controller with $\omega = 1.5$ and $A_m = 1$.*

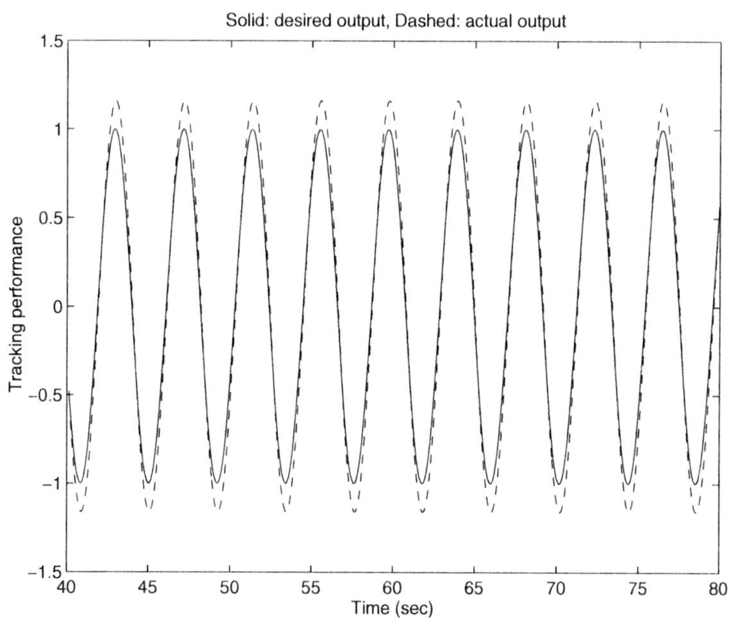

Figure 4.2. *The profile of the tracking performance of the closed-loop system under the linear controller with $\omega = 1.5$ and $A_m = 1$.*

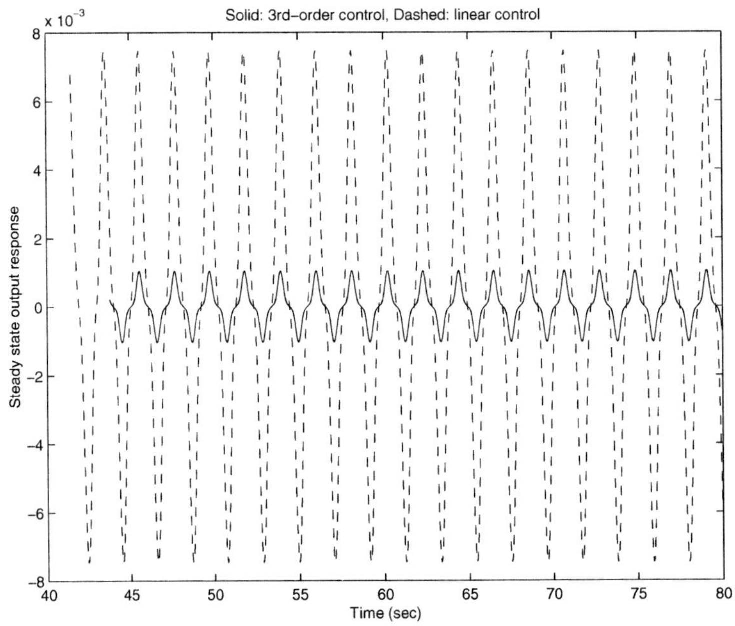

Figure 4.3. *Comparison of the output responses of the closed-loop system under the nonlinear and linear controllers with* $\omega = 1.5$ *and* $A_m = 4$.

where

$$\mathbf{x}^{(1)}(v) = \begin{bmatrix} v_1 \\ \omega v_2 \\ a_{10} v_1 \\ \omega a_{10} v_2 \end{bmatrix},$$

$$\mathbf{u}^{(1)}(v) = (-M\omega^2 + mga_{10})v_1 + b\omega v_2.$$

That is, this linear controller is a linear approximation of the third-order controller. It is seen that, in all cases, the third-order nonlinear controller performs much better than the linear controller. Figures 4.1 and 4.2 show the profiles of the tracking performance of the closed-loop system under the nonlinear controller and the linear controller, respectively, for the case $\omega = 1.5$ and $A_m = 1$. It can be seen that, under the nonlinear controller, no steady-state tracking error is visible, while, under the linear controller, a sizable steady-state tracking error is present. Figure 4.3 further compares the output responses of the closed-loop system under the nonlinear controller and linear controller with $\omega = 1.5$ and $A_m = 4$.

Remark 4.16. We have seen that the coefficients of $v^{[2]}$ of $\mathbf{x}_3(v)$ and $\mathbf{x}_4(v)$ are zero. This is not a coincidence. In fact, it can be seen that if the power series expansion of $\mathbf{x}_3(v)$ and $\mathbf{x}_4(v)$ only contains such terms as $v^{[l]}$ with l an odd integer, so does the power series expansion of the expressions on both sides of (4.73). Thus, we can conclude that the power series solution of equation (4.73) does not contain such terms as $v^{[l]}$ where l is an even integer. ∎

Chapter 5

Nonlinear Robust Output Regulation

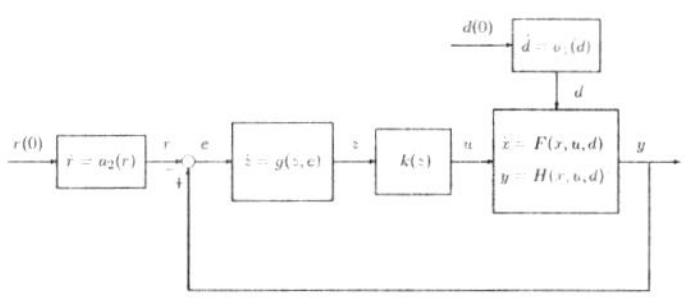

We now turn our attention to the nonlinear robust output regulation problem in which the same objectives as described in Chapters 3 and 4 must be achieved via either dynamic state feedback or output feedback control in the presence of appropriately defined model uncertainties.

Two robust control problems will be defined for a class of general nonlinear systems in this chapter, namely, the robust output regulation problem and the kth-order robust output regulation problem. They are, respectively, the robust enhancement of the output regulation problem studied in Chapter 3 and the kth-order output regulation problem studied in Chapter 4. The chapter is organized as follows. Section 5.1 gives precise descriptions of the two problems and lists some standard assumptions. An equivalent characterization of the robust output regulation property in terms of the solvability of a set of partial differential equations will also be given. Section 5.2 introduces two examples. The first example shows that, when the exogenous signal is constant, the nonlinear robust output regulation problem can still be solved by a linear controller that solves the linear robust output regulation problem of the linearized system of the given nonlinear system. However, this technique does not work when the exogenous signal is time varying, as illustrated by the second example. In Section 5.3, we first reveal why the design method for the linear systems fails to work for the nonlinear system and then proceed to establish the solvability conditions for the kth-order robust output regulation problem. In Section 5.4, we pass to the robust output regulation problem. It is shown that if the solution of the regulator equations is a degree k polynomial in the exogenous signal v, then a controller that solves the kth-order robust output regulation problem also solves the robust output regulation problem. Moreover, by incorporating the feedforward control technique, it is possible to solve the robust output regulation problem for some cases where the solution of the regulator equations is not polynomial. In Section 5.5, we address some computation issues. In Section 5.6, the ball and beam example is used to illustrate the design approach.

5.1 Problem Description

In analogy to the description of the uncertain linear plant given in (1.46) of Chapter 1, we describe an uncertain nonlinear plant as follows:

$$\dot{x}(t) = f(x(t), u(t), v(t), w), \qquad e(t) = h(x(t), u(t), v(t), w), \quad t \geq 0, \tag{5.1}$$

with the same exosystem

$$\dot{v}(t) = a(v(t)), \quad t \geq 0, \tag{5.2}$$

where $x(t)$ is the n-dimensional plant state, $u(t)$ the m-dimensional plant input, $e(t)$ the p-dimensional plant output representing the tracking error, $v(t)$ the q-dimensional exogenous signal representing the disturbance and/or the reference input, and w the n_w-dimensional vector representing the unknown plant parameter. It is assumed that 0 is the nominal value of the uncertain parameter w and that $f(0, 0, 0, w) = 0$ and $h(0, 0, 0, w) = 0$ for all $w \in \mathcal{R}^{n_w}$.

The class of control laws are described by

$$\begin{aligned} u(t) &= k(x(t), v(t), z(t)), \\ \dot{z}(t) &= g(z(t), e(t)), \quad t \geq 0, \end{aligned} \tag{5.3}$$

where $z(t)$ is the compensator state vector of dimension n_z to be specified later. With an abuse of notation, the above controller encompasses three cases:

1. **Dynamic State Feedback Controller:** When $v(t)$ does not appear in (5.3), that is,

$$\begin{aligned} u(t) &= k(x(t), z(t)), \\ \dot{z}(t) &= g(z(t), e(t)). \end{aligned} \tag{5.4}$$

2. **Dynamic Output Feedback Controller:** When $x(t)$ and $v(t)$ do not appear in (5.3), that is,

$$\begin{aligned} u(t) &= k(z(t)), \\ \dot{z}(t) &= g(z(t), e(t)). \end{aligned} \tag{5.5}$$

3. **Dynamic Output Feedback with Feedforward Controller:** When $x(t)$ does not appear in (5.3), that is,

$$\begin{aligned} u(t) &= k(z(t), v(t)), \\ \dot{z}(t) &= g(z(t), e(t)). \end{aligned} \tag{5.6}$$

Letting $x_c = \text{col}(x, z)$, the resulting closed-loop system can be written as

$$\begin{aligned} \dot{x}_c(t) &= f_c(x_c(t), v(t), w), \quad t \geq 0, \\ e(t) &= h_c(x_c(t), v(t), w), \end{aligned} \tag{5.7}$$

where

$$\begin{aligned} f_c(x_c, v, w) &= \begin{bmatrix} f(x, k(x, v, z), v, w) \\ g(z, h(x, k(x, v, z), v, w)) \end{bmatrix}, \\ h_c(x_c, v, w) &= h(x, k(x, v, z), v, w). \end{aligned} \tag{5.8}$$

For simplicity, all the functions involved in this setup are assumed to be sufficiently smooth and defined globally on the appropriate Euclidean spaces, with the value zero at the respective origins. Throughout this chapter, we use V and W to denote some open neighborhoods of the origins of $\mathcal{R}^q$ and $\mathcal{R}^{n_w}$, respectively. For convenience of presentation, we allow V and W to be made arbitrarily small.

For convenience, let us lump the closed-loop system (5.7) and the exosystem (5.2) together as follows:

$$\begin{aligned}\dot{x}_c(t) &= f_c(x_c(t), v(t), w), \quad t \geq 0,\\ \dot{v}(t) &= a(v(t)),\\ e(t) &= h_c(x_c(t), v(t), w),\end{aligned} \tag{5.9}$$

and call (5.9) the closed-loop composite system. It is clear that, for all w, the state $\text{col}(x_c, v) = \text{col}(0, 0)$ is an equilibrium point of the composite system.

Robust Output Regulation Problem (RORP): Find a controller of the form (5.3) such that the closed-loop composite system (5.9) satisfies the following two properties.

Property 5.1. For all sufficiently small $x_c(0)$, $v(0)$, and w, the trajectory $\text{col}(x_c(t), v(t))$ of the closed-loop composite system (5.9) exists and is bounded for all $t \geq 0$, and

Property 5.2. For all sufficiently small $x_c(0)$, $v(0)$, and w, the trajectory $\text{col}(x_c(t), v(t))$ of the closed-loop composite system (5.9) satisfies

$$\lim_{t\to\infty} e(t) = \lim_{t\to\infty} h_c(x_c(t), v(t), w) = 0. \tag{5.10}$$

***k*th-Order Robust Output Regulation Problem (KRORP):** Find a controller of the form (5.3) such that the closed-loop composite system (5.9) satisfies Property 5.1 and the following property.

Property 5.3. For all sufficiently small $x_c(0)$, $v(0)$, and w, the trajectory $\text{col}(x_c(t), v(t))$ of the closed-loop composite system (5.9) satisfies

$$\lim_{t\to\infty} (e(t) - o^k(v(t))) = \lim_{t\to\infty} (h_c(x_c(t), v(t), w) - o^k(v(t))) = 0, \tag{5.11}$$

where k is some given positive integer.

Remark 5.1. It is clear that the robust output regulation problem and the kth-order robust output regulation problem are extensions of the output regulation problem described in Chapter 3 and the kth-order output regulation problem described in Chapter 4, respectively, by further taking into account the uncertain parameter w. On the other hand, the description of the plant (5.1) includes the linear uncertain plant as described in Chapter 1 as a special case. Thus the robust output regulation problem described here is an extension of the linear robust output regulation studied in Chapter 1. Moreover, noting that, for the class of linear systems, Property 5.3 is the same as Property 1.4, the kth-order robust output regulation problem described here is also an extension of the linear robust output regulation studied in Chapter 1. ■

Remark 5.2. The constant parameter w can be viewed as being produced by an autonomous system $\dot{w} = 0, \quad w(0) = w$. Combining this system with the closed-loop composite system gives

$$\begin{aligned} \dot{x}_c &= f_c(x_c, v, w), \\ \dot{v} &= a(v), \\ \dot{w} &= 0, \\ e &= h_c(x_c, v, w). \end{aligned} \tag{5.12}$$

This system takes exactly the same form as (3.13), viewing $\dot{v} = a(v)$ and $\dot{w} = 0$ as the exosystem. Thus, using the same argument as in Remark 3.1, Property 5.1 is guaranteed if the equilibrium point of the system (5.12) at $\text{col}(x_c, v, w) = \text{col}(0, 0, 0)$ is stable in the sense of Lyapunov. Moreover, by Theorem 2.27 and Assumption 3.1, the equilibrium point of the system (5.12) at $\text{col}(x_c, v, 0) = \text{col}(0, 0, 0)$ is stable in the sense of Lyapunov if the closed-loop system has the following property.

Property 5.4. All the eigenvalues of the matrix

$$\frac{\partial f_c}{\partial x_c}(0, 0, 0) \tag{5.13}$$

have negative real parts.

Thus, in this chapter, we will directly impose Property 5.4 instead of Property 5.1 on the closed-loop composite system. ∎

The reason for studying the kth-order robust output regulation problem is at least twofold. First, from a practical point of view, it suffices to require that the steady-state tracking error be sufficiently small, and Property 5.3 is a reasonable measure of smallness of the steady-state tracking error. Second, as will be shown in Section 5.4, under some additional assumption on the solution of the regulator equations, a controller that solves the kth-order robust output regulation problem also solves the robust output regulation problem. In what follows, a controller that solves the robust output regulation problem or the kth-order robust output regulation problem will be called a *robust servoregulator* or *kth-order robust servoregulator*. In particular, (5.4) and (5.5) are called the *(kth-order) state feedback robust servoregulator*, and the *(kth-order) output feedback robust servoregulator*, respectively.

The following result is an extension of Lemmas 3.6 and 4.2 to the case where the model uncertainty is taken into account. The proof of this lemma is exactly the same as those of Lemmas 3.6 and 4.2, viewing w as produced by $\dot{w} = 0$, and is thus omitted.

Lemma 5.3. *Under Assumption 3.1′, suppose the closed-loop system (5.7) has Property 5.4. Then*

(i) *The closed-loop composite system (5.9) has Property 5.2 if and only if it has the following two properties:*

Property 5.5. There exists a sufficiently smooth function $\mathbf{x}_c(v, w)$ with $\mathbf{x}_c(0, 0) = 0$ that satisfies, for $v \in V$ and $w \in W$, the following partial differential equations:

$$\frac{\partial \mathbf{x}_c(v, w)}{\partial v} a(v) = f_c(\mathbf{x}_c(v, w), v, w), \tag{5.14}$$

$$0 = h_c(\mathbf{x}_c(v, w), v, w). \tag{5.15}$$

(ii) *The closed-loop composite system (5.9) has Property 5.3 if and only if*

Property 5.6. There exists a sufficiently smooth function $\mathbf{x}_c^{(k)}(v, w)$ with $\mathbf{x}_c^{(k)}(0, 0) = 0$ that satisfies, for $v \in V$ and $w \in W$, the following partial differential equations:

$$\frac{\partial \mathbf{x}_c^{(k)}(v, w)}{\partial v} a(v) = f_c(\mathbf{x}_c^{(k)}(v, w), v, w), \tag{5.16}$$

$$o^k(v) = h_c(\mathbf{x}_c^{(k)}(v, w), v, w). \tag{5.17}$$

Various assumptions needed for the solvability of the above two problems are listed as follows.

Assumption 5.1. There exist sufficiently smooth functions $\mathbf{x}(v, w)$ and $\mathbf{u}(v, w)$ with $\mathbf{x}(0, 0) = 0$ and $\mathbf{u}(0, 0) = 0$ satisfying, for $v \in V$ and $w \in W$, the following equations:

$$\begin{aligned} \frac{\partial \mathbf{x}(v, w)}{\partial v} a(v) &= f(\mathbf{x}(v, w), \mathbf{u}(v, w), v, w), \\ 0 &= h(\mathbf{x}(v, w), \mathbf{u}(v, w), v, w). \end{aligned} \tag{5.18}$$

Assumption 5.2. The pair $\left(\frac{\partial f}{\partial x}(0, 0, 0, 0), \quad \frac{\partial f}{\partial u}(0, 0, 0, 0)\right)$ is stabilizable.

Assumption 5.3. The pair $\left(\frac{\partial h}{\partial x}(0, 0, 0, 0), \quad \frac{\partial f}{\partial x}(0, 0, 0, 0)\right)$ is detectable.

Assumption 5.4. For $l = 1, 2, \dots,$

$$\operatorname{rank} \begin{bmatrix} \frac{\partial f}{\partial x}(0, 0, 0, 0) - \lambda I & \frac{\partial f}{\partial u}(0, 0, 0, 0) \\ \frac{\partial h}{\partial x}(0, 0, 0, 0) & \frac{\partial h}{\partial u}(0, 0, 0, 0) \end{bmatrix} = n + p \tag{5.19}$$

for all λ given by

$$\{ \lambda \mid \lambda = l_1\lambda_1 + \cdots + l_q\lambda_q, l_1 + \cdots + l_q = l, \ l_1, \dots, l_q = 0, 1, 2, \dots, l \}, \tag{5.20}$$

where $\lambda_1, \dots, \lambda_q$ are the eigenvalues of the matrix $\frac{\partial a}{\partial v}(0)$.

Remark 5.4. Clearly, equations (5.18) are the extension of equations (3.30) and are thus called the regulator equations of the uncertain nonlinear systems (5.1). Using the same argument as that in Theorem 3.8, it can be shown that, under Assumption 3.1′, the solvability of the regulator equations is necessary for the solvability of the robust output regulation problem for the uncertain system (5.1). However, the solvability of the regulator equations does not guarantee the solvability of the robust output regulation problem for the uncertain system (5.1). As will be shown in Section 5.3, an additional condition has to be imposed on the solution of the regulator equations (5.18). Assumption 5.4 is made to guarantee the existence of the formal Taylor series solution of the regulator equations (5.18). It is noted that this assumption does not guarantee the existence of the solution of the regulator equations (5.18). ∎

Remark 5.5. If the functions $\mathbf{x}(v, w)$ and $\mathbf{u}(v, w)$ described in Assumption 5.1 are defined for all $v \in \mathcal{R}^q$ and all $w \in \mathcal{R}^{n_w}$ and satisfy equations (5.18) for all $v \in \mathcal{R}^q$ and all $w \in \mathcal{R}^{n_w}$, then the functions $\mathbf{x}(v, w)$ and $\mathbf{u}(v, w)$ are called the *global solution of the regulator equations.* ∎

5.2 Two Case Studies

By Lemma 5.3, a controller that solves the robust output regulation problem must be able to induce a center manifold defined by the solution $\mathbf{x}_c(v, w)$ of (5.14), and, on the center manifold, the output of the system is identically zero; that is, $\mathbf{x}_c(v, w)$ also satisfies equation (5.15). For the class of linear systems, (5.14) reduces to the Sylvester equation given in the first equation of (1.53). A controller that incorporates a p-copy internal model of the exosystem can solve the robust output regulation problem because the employment of the internal model guarantees that the solution of the first equation of (1.53) also satisfies the second equation of (1.53). For the class of nonlinear systems, due to the Center Manifold Theorem, if a controller can make the closed-loop system satisfy Property 5.4, then equation (5.14) is solvable for a sufficiently smooth function $\mathbf{x}_c(v, w)$ with $\mathbf{x}_c(0, 0) = 0$. The issue is whether or not $\mathbf{x}_c(v, w)$ also satisfies (5.15) if the controller is such that it solves the robust output regulation problem for the linearization of the nonlinear plant (5.1).

Case 1: Let us first take a look at a special case where the exogenous signal v is constant, that is, where v is generated by the exosystem $\dot{v} = 0$. Assume that a linear state feedback controller of the form

$$u = K_1 x + K_2 z, \quad \dot{z} = e, \tag{5.21}$$

makes the closed-loop system (5.7) satisfy Property 5.4. Under this controller, equations (5.14) and (5.15) become

$$0 = f_c(\mathbf{x}_c(v, w), v, w), \tag{5.22}$$

$$0 = h_c(\mathbf{x}_c(v, w), v, w). \tag{5.23}$$

Since f_c has Property 5.4, the existence of a sufficiently smooth function $\mathbf{x}_c(v, w)$ with $\mathbf{x}_c(0, 0) = 0$ satisfying (5.22) is guaranteed by the Implicit Function Theorem. Since $g(z, e) = e$, satisfaction of equation (5.22) by $\mathbf{x}_c(v, w)$ implies the satisfaction of equation (5.23) by $\mathbf{x}_c(v, w)$. That is, the controller also solves the robust output regulation problem for the nonlinear system.

Clearly, the controller given by (5.21) is simply a linear robust controller based on the Jacobian linearization of the nonlinear system (5.1). The robustness is achieved by having the controller incorporate the p-copy internal model of the exosystem. Unfortunately, such a technique only works for the spatial case where the exogenous signal is constant. The following example shows that the above technique is no longer effective for nonlinear systems subject to time-varying exogenous signals.

Case 2: Consider the following one-dimensional plant:

$$\begin{aligned}\dot{x} &= -(1+w_1)x + (1+w_2)x^2 + u,\\ y &= x,\\ e &= y - v_1 = x - v_1,\end{aligned} \tag{5.24}$$

where the exosystem is given by

$$\begin{bmatrix} \dot{v}_1 \\ \dot{v}_2 \end{bmatrix} = \begin{bmatrix} 0 & 1 \\ -1 & 0 \end{bmatrix}\begin{bmatrix} v_1 \\ v_2 \end{bmatrix} \stackrel{def}{=} A_1 v \tag{5.25}$$

and w_1 and w_2 are two unknown parameters with their nominal values being zero.

The Jacobian linearization of (5.24) is given by

$$\begin{aligned}\dot{x} &= -(1+w_1)x + u,\\ y &= x,\\ e &= y - v_1 = x - v_1.\end{aligned} \tag{5.26}$$

It can be verified that the robust output regulation problem for the linear system (5.26) is solvable by either state or output feedback control. A simple output feedback controller is given by

$$\begin{aligned}u &= -z_1,\\ \dot{z}_1 &= z_2,\\ \dot{z}_2 &= -z_1 + e = -z_1 + (y - v_1).\end{aligned} \tag{5.27}$$

However, this controller does not solve the robust output regulation problem for (5.24). To see this point, it suffices to show that there exists no sufficiently smooth function $\mathbf{x}_c(v, w)$ that satisfies both (5.14) and (5.15). In fact, assume $\mathbf{x}(v, w)$, $\mathbf{z}_1(v, w)$, and $\mathbf{z}_2(v, w)$ satisfy (5.14) and (5.15), that is,

$$\begin{bmatrix} \frac{\partial \mathbf{x}(v,w)}{\partial v} \\ \frac{\partial \mathbf{z}_1(v,w)}{\partial v} \\ \frac{\partial \mathbf{z}_2(v,w)}{\partial v} \end{bmatrix} A_1 v = \begin{bmatrix} -(1+w_1)\mathbf{x}(v,w) + (1+w_2)(\mathbf{x}(v,w))^2 - \mathbf{z}_1(v,w) \\ \mathbf{z}_2(v,w) \\ -\mathbf{z}_1(v,w) + \mathbf{x}(v,w) - v_1 \end{bmatrix} \tag{5.28}$$

and

$$\mathbf{x}(v, w) = v_1. \tag{5.29}$$

Then, necessarily, we have, from equation (5.29) and the first two equations of (5.28),

$$\begin{aligned}\mathbf{x}(v, w) &= v_1,\\ \mathbf{z}_1(v, w) &= -v_2 - (1 + w_1)v_1 + (1 + w_2)v_1^2,\\ \mathbf{z}_2(v, w) &= v_1 - (1 + w_1)v_2 + 2(1 + w_2)v_1 v_2.\end{aligned} \tag{5.30}$$

However, the left-hand and right-hand sides of the third equation of (5.28) are given by

$$\frac{\partial \mathbf{z}_2(v, w)}{\partial v}\begin{bmatrix} v_2 \\ -v_1 \end{bmatrix} = v_2 + (1 + w_1)v_1 + 2(1 + w_2)(v_2^2 - v_1^2)$$

and

$$-\mathbf{z}_1(v, w) + \mathbf{x}(v, w) - v_1 = -\mathbf{z}_1(v, w) = v_2 + (1 + w_1)v_1 - (1 + w_2)v_1^2,$$

respectively, so that the third equation in (5.28) does not hold. This gives a contradiction.

Nevertheless, since the controller (5.27) renders the closed-loop system composed of (5.24) and (5.27) into Property 5.4, the center manifold equation (5.14) associated with the closed-loop system has a solution $\mathbf{x}_c(v, w)$. Moreover, it is possible to show that this solution will annihilate the linear term of the right-hand side of (5.15) for all sufficiently small w. However, the solution of (5.14) may not satisfy equation (5.15), as the right-hand side of (5.15) is in general a nonlinear function of v.

5.3 Solvability of the *k*th-Order Robust Output Regulation Problem

To pursue the problem a little further, let us first introduce the following notations:

$$\begin{aligned} f(x, u, v, w) &= A(w)x + B(w)u + E(w)v + f_2(x, u, v, w),\\ h(x, u, v, w) &= C(w)x + D(w)u + F(w)v + h_2(x, u, v, w),\\ a(v) &= A_1 v + a_2(v),\\ f(x, k(x, v, z), v, w) &= A_c(w)x + B_c(w)z + E_c(w)v + f_{c2}(x_c, v, w),\\ h(x, k(x, v, z), v, w) &= C_c(w)x + D_c(w)z + F_c(w)v + h_{c2}(x_c, v, w),\end{aligned}$$

where $A_1 = \frac{\partial a}{\partial v}(0)$ and $A(w)$, $B(w)$, and so forth are

$$A(w) = \frac{\partial f}{\partial x}(0, 0, 0, w), \quad B(w) = \frac{\partial f}{\partial u}(0, 0, 0, w), \ldots .$$

For convenience, in what follows, we will use the shorthand notation A, B, and so forth to denote $A(0)$, $B(0)$, and so forth.

Now assume that a control law of the form (5.3) with $g(z, e) = \mathcal{G}_1 z + \mathcal{G}_2 e$ renders the closed-loop system (5.7) into Property 5.4. Then Theorem 2.25 ensures the existence of a locally defined sufficiently smooth function $\bar{\mathbf{x}}_c(v, w)$ with $\bar{\mathbf{x}}_c(0, 0) = 0$ such that, for $v \in V$, $w \in W$,

$$\frac{\partial \bar{\mathbf{x}}_c(v, w)}{\partial v} a(v) = f_c(\bar{\mathbf{x}}_c(v, w), v, w). \tag{5.31}$$

By partitioning $\bar{\mathbf{x}}_c(v, w) = \text{col}(\bar{\mathbf{x}}(v, w), \bar{\mathbf{z}}(v, w))$ with $\bar{\mathbf{x}}(v, w) \in \mathcal{R}^n$, (5.31) becomes

$$\begin{aligned}
\frac{\partial \bar{\mathbf{x}}(v, w)}{\partial v} a(v) &= f(\bar{\mathbf{x}}(v, w), k(\bar{\mathbf{x}}(v, w), v, \bar{\mathbf{z}}(v, w)), v, w), \\
\frac{\partial \bar{\mathbf{z}}(v, w)}{\partial v} a(v) &= \mathcal{G}_1 \bar{\mathbf{z}}(v, w) + \mathcal{G}_2 \mathbf{e}(v, w),
\end{aligned} \tag{5.32}$$

where

$$\mathbf{e}(v, w) = h(\bar{\mathbf{x}}(v, w), k(\bar{\mathbf{x}}(v, w), v, \bar{\mathbf{z}}(v, w)), v, w). \tag{5.33}$$

For any $k \geq 1$, $\bar{\mathbf{x}}(v, w)$, $\bar{\mathbf{z}}(v, w)$, and $\mathbf{e}(v, w)$ can be uniquely expressed as

$$\begin{aligned}
\bar{\mathbf{x}}(v, w) &= \sum_{l=1}^{k} X_{lw} v^{[l]} + o^k(v), \\
\bar{\mathbf{z}}(v, w) &= \sum_{l=1}^{k} Z_{lw} v^{[l]} + o^k(v), \\
\mathbf{e}(v, w) &= \sum_{l=1}^{k} Y_{lw} v^{[l]} + o^k(v),
\end{aligned} \tag{5.34}$$

where (X_{lw}, Z_{lw}, Y_{lw}) are constant matrices of appropriate dimensions depending, perhaps, on w. In analogy to the derivation of equation (4.29), substituting (5.34) into (5.32) and (5.33), expanding (5.32) and (5.33) into power series in $v^{[l]}$, and identifying the coefficients of $v^{[l]}$ yield, for $l = 1, 2, \ldots, k$,

$$\begin{aligned}
X_{lw} A^{[l]} &= A_c(w) X_{lw} + B_c(w) Z_{lw} + E_{lw}, \\
Z_{lw} A^{[l]} &= \mathcal{G}_1 Z_{lw} + \mathcal{G}_2 (C_c(w) X_{lw} + D_c(w) Z_{lw} + F_{lw}),
\end{aligned} \tag{5.35}$$

and

$$Y_{lw} = C_c(w) X_{lw} + D_c(w) Z_{lw} + F_{lw}, \tag{5.36}$$

where $A^{[l]}$ is as defined in (4.47) and is repeated below:

$$A^{[l]} = M_l \left[\sum_{i=1}^{l} I_q^{(i-1)} \otimes A_1 \otimes I_q^{(l-i)} \right] N_l,$$

$(E_{1w}, F_{1w}) = (E(w), F(w))$, and, for $l = 2, 3, \ldots$, (E_{lw}, F_{lw}) depend only on $X_{1w}, \ldots, X_{(l-1)w}$ and $Z_{1w}, \ldots, Z_{(l-1)w}$.

Now we can invoke Lemma 1.27 to yield the following result.

Lemma 5.6. *Under Assumption 3.1, assume that a control law of the form (5.3) with $g(z, e) = \mathcal{G}_1 z + \mathcal{G}_2 e$ makes the closed-loop system (5.7) satisfy Property 5.4. Then*

(i) *For some integer $l > 0$, let $Y_{lw} v^{[l]}$ be the lth-order term of the Taylor series expansion of $h_c(\mathbf{x}_c(v, w), v, w)$ as a function of v. Then $Y_{lw} = 0$ for all $w \in W$ if the pair $(\mathcal{G}_1, \mathcal{G}_2)$ incorporates a p-copy internal model of the matrix $A^{[l]}$.*

(ii) *The kth-order robust output regulation problem is solvable if the pair* $(\mathcal{G}_1, \mathcal{G}_2)$ *incorporates a p-copy internal model of the matrix* $\mathcal{A}_{kf}$, *where*

$$\mathcal{A}_{kf} = \begin{bmatrix} A^{[1]} & 0 & \cdots & 0 \\ 0 & A^{[2]} & \cdots & 0 \\ \vdots & \vdots & \vdots & \vdots \\ 0 & 0 & \cdots & A^{[k]} \end{bmatrix}. \tag{5.37}$$

Proof. (i) Since, for the given l, equations (5.35) and (5.36) take the same form as (1.70) and (1.71), the fact that the closed-loop system has Property 5.4 means that the matrix

$$\begin{bmatrix} A_c & B_c \\ \mathcal{G}_2 C_c & \mathcal{G}_1 + \mathcal{G}_2 D_c \end{bmatrix}$$

is Hurwitz. Thus, by Lemma 1.27, $Y_{lw} = 0$ for all $w \in W$ if the pair $(\mathcal{G}_1, \mathcal{G}_2)$ incorporates a p-copy internal model of the matrix $A^{[l]}$.

(ii) By the definition of $\mathcal{A}_{kf}$, if the pair $(\mathcal{G}_1, \mathcal{G}_2)$ incorporates a p-copy internal model of the matrix $\mathcal{A}_{kf}$, it also incorporates a p-copy internal model of all the matrices $A^{[l]}$ for $l = 1, \dots, k$. Therefore, the control law makes $Y_{lw} = 0$ for all $l = 1, \dots, k$, thereby solving the kth-order robust output regulation problem. □

As pointed out in Remark 1.23, given any matrix $\mathcal{A}_{kf}$, it is always possible to find a pair of matrices $(\mathcal{G}_1, \mathcal{G}_2)$ such that it is a p-copy internal model of the matrix $\mathcal{A}_{kf}$. Thus we can define an augmented system as follows:

$$\begin{aligned} \dot{x} &= f(x, u, v, w), \\ \dot{z} &= \mathcal{G}_1 z + \mathcal{G}_2 e, \\ e &= h(x, u, v, w), \end{aligned} \tag{5.38}$$

where the pair $(\mathcal{G}_1, \mathcal{G}_2)$ incorporates a p-copy internal model of the matrix $\mathcal{A}_{kf}$. By Lemma 5.6, the kth-order robust output regulation problem is solvable by a control law of the form (5.3) with $g(z, e) = \mathcal{G}_1 z + \mathcal{G}_2 e$ if the static feedback control law of the form $u = k(x, z, v)$ can exponentially stabilize the equilibrium point of the augmented system (5.38). Indeed, such control laws can be found in the linear form under the assumptions listed in Section 5.1.

Theorem 5.7.

(i) *Under Assumptions 3.1, 5.2, and 5.4, for any positive integer k, the kth-order robust output regulation problem is solvable by a linear state feedback control of the form*

$$u = K_1 x + K_2 z, \qquad \dot{z} = G_1 z + G_2 e, \tag{5.39}$$

where (G_1, G_2) *incorporates a p-copy internal model of the matrix* $\mathcal{A}_{kf}$ *with* G_1 *satisfying Property 1.5, i.e.,*

$$\operatorname{rank} \begin{bmatrix} A - \lambda I & B \\ C & D \end{bmatrix} = n + p$$

for all $\lambda \in \sigma(G_1)$.

(ii) *Under Assumptions 3.1 and 5.2 to 5.4, for any positive integer k, the kth-order robust output regulation problem is solvable by a linear output feedback control of the form*

$$u = Kz, \qquad \dot{z} = \mathcal{G}_1 z + \mathcal{G}_2 e, \tag{5.40}$$

where $(\mathcal{G}_1, \mathcal{G}_2)$ incorporates a p-copy internal model of the matrix $\mathcal{A}_{kf}$, where $(\mathcal{G}_1, \mathcal{G}_2)$ takes the form (1.57) with G_1 satisfying Property 1.5.

Proof. (i) Recall from Chapter 4 that the eigenvalues of the matrix $A^{[l]}$ are given by

$$\{ \lambda \mid \lambda = l_1\lambda_1 + \cdots + l_q\lambda_q, l_1 + \cdots + l_q = l,\ l_1, \ldots, l_q = 0, 1, \ldots, l \},$$

where $\lambda_1, \ldots, \lambda_q$ are eigenvalues of A_1. Therefore, Assumption 5.4 guarantees that, for any fixed integer $k > 0$, there exists a pair (G_1, G_2) that incorporates a p-copy internal model of $\mathcal{A}_{kf}$ with G_1 satisfying Property 1.5. By Lemma 1.26, under Assumptions 3.1 and 5.2, the pair

$$\begin{bmatrix} A & 0 \\ G_2C & G_1 \end{bmatrix}, \begin{bmatrix} B \\ G_2D \end{bmatrix} \tag{5.41}$$

is stabilizable. Thus there exist feedback gains K_1 and K_2 such that the eigenvalues of the matrix

$$\begin{bmatrix} A + BK_1 & BK_2 \\ G_2(C + DK_1) & G_1 + G_2DK_2 \end{bmatrix} \tag{5.42}$$

have negative real parts. Thus, under the control law (5.39), the closed-loop system satisfies Property 5.4. It follows from part (ii) of Lemma 5.6 that the control law (5.39) solves the kth-order robust output regulation problem.

(ii) Let (K_1, K_2, G_1, G_2) be what was obtained from part (i). Under Assumption 5.3, there exists L such that $A - LC$ is stable. Let $K = [K_1, K_2]$,

$$\mathcal{G}_1 = \begin{bmatrix} A + BK_1 - L(C + DK_1) & (B - LD)K_2 \\ 0 & G_1 \end{bmatrix}, \quad \mathcal{G}_2 = \begin{bmatrix} L \\ G_2 \end{bmatrix}.$$

Clearly, the pair $(\mathcal{G}_1, \mathcal{G}_2)$ incorporates a p-copy internal model of the matrix $\mathcal{A}_{kf}$. Moreover, under the control law (5.40), the Jacobian matrix of the closed-loop system is given by

$$\begin{bmatrix} A & BK_1 & BK_2 \\ LC & A + BK_1 - LC & BK_2 \\ G_2C & G_2DK_1 & G_1 + G_2DK_2 \end{bmatrix}. \tag{5.43}$$

Subtracting the first row from the second row and then adding the second column to the first column shows that the spectrum of (5.43) is given by those of (5.42) and $A - LC$. Thus, the closed-loop system satisfies Property 5.4. Again, it follows from Lemma 5.6 that the control law (5.40) solves the kth-order robust output regulation problem. □

Remark 5.8.

(i) It is interesting to know that if v satisfies $\dot{v} = A_1 v$, then $v^{[l]}$ satisfies $\frac{dv^{[l]}}{dt} = A^{[l]} v^{[l]}$. Let

$$v_{kf} = \begin{bmatrix} v^{[1]} \\ v^{[2]} \\ \vdots \\ v^{[k]} \end{bmatrix}. \tag{5.44}$$

Then the matrix $\mathcal{A}_{kf}$ is such that

$$\frac{dv_{kf}}{dt} = \mathcal{A}_{kf} v_{kf}. \tag{5.45}$$

The system (5.45) can be considered as a generalized exosystem which generates not only the exogenous signal v (when $a(v) = A_1 v$), but also the higher order terms of the exogenous signal v up to order k. For convenience, we will call the system (5.45) a k-fold exosystem.

(ii) For linear systems, the right-hand side of equation (5.15) is a linear function of v. Thus, in order to solve the robust output regulation problem, it suffices to require a linear control law to incorporate a p-copy internal model of the matrix A_1. For nonlinear systems, the right-hand side of equation (5.15) is a nonlinear function of v. A linear control law that incorporates the p-copy internal model of the matrix $\mathcal{A}_{kf}$ is able to render the right-hand side zero up to order k in v. But the control law cannot solve the robust output regulation problem in general.

(iii) Effectively, Lemma 5.6 asserts that designing a kth-order robust controller for a nonlinear system (5.1) is equivalent to designing a linear robust servoregulator for the linear system consisting of the linear approximation of (5.1) and the k-fold exosystem (5.45). ∎

Remark 5.9. The solvability conditions of the kth-order output regulation problem studied in Chapter 4 and the kth-order robust output regulation problem are basically the same, but the design philosophy of the control laws are completely different. The controller that solves the former problem relies on the approximate solution of the regulator equations, thus demanding the complete knowledge of the plant. On the other hand, the kth-order robust servoregulator is designed completely based on the linearization of the given nonlinear plant at the origin. Regardless of the variations of the uncertain parameter w, the controller can guarantee the zero steady-state tracking error up to order k of the exogenous signal v. ∎

Remark 5.10. Assumption 5.4 is an extension of the transmission zero assumption, that is, Assumption 1.4. For linear systems, the solvability of the linear robust output regulation problem will necessitate the condition Assumption 1.4. However, Assumption 5.4 may not be necessary for the solvability of the kth-order robust output regulation problem. This is because our description of the plant uncertainty does not make the matrices E_{lw} and F_{lw} change arbitrarily in an open neighborhood of E_{l0} and F_{l0} as w varies arbitrarily in W. Thus, even though Assumption 5.4 fails to hold, the linear equation (5.35) may still have a solution. ∎

5.4 Solvability of the Robust Output Regulation Problem

In this section, we will further show that, under some additional assumptions on the solution of the regulator equations, a controller that solves the kth-order robust output regulation problem for the composite system (5.1) and (5.2) also solves the robust output regulation problem for the same system. Let us begin by characterizing the control law of the form (5.3) that solves the robust output regulation problem.

Lemma 5.11. *Under Assumption 3.1′, assume a control law of the form (5.3) is such that the closed-loop system has Property 5.4; then the control law also solves the robust output regulation problem if and only if there exist sufficiently smooth functions* $(\mathbf{x}(v, w), \mathbf{u}(v, w), \mathbf{z}(v, w))$ *locally defined in* $v \in V$, $w \in W$ *with* $(\mathbf{x}(0,0), \mathbf{u}(0,0), \mathbf{z}(0,0)) = (0,0,0)$ *such that* $\mathbf{x}(v, w)$ *and* $\mathbf{u}(v, w)$ *are the solution of the nonlinear regulator equations (5.18), and* $\mathbf{z}(v, w)$ *satisfies*

$$\mathbf{u}(v, w) = k(\mathbf{x}(v, w), v, \mathbf{z}(v, w)), \tag{5.46}$$

$$\frac{\partial \mathbf{z}(v, w)}{\partial v} a(v) = g(\mathbf{z}(v, w), 0). \tag{5.47}$$

Proof. *Necessity.* By Lemma 5.3, there exists a sufficiently smooth function $\mathbf{x}_c(v, w)$ with $\mathbf{x}_c(0,0) = 0$ that satisfies (5.14) and (5.15). Partition $\mathbf{x}_c(v, w)$ as

$$\mathbf{x}_c(v, w) = \begin{bmatrix} \mathbf{x}(v, w) \\ \mathbf{z}(v, w) \end{bmatrix}, \tag{5.48}$$

where $\mathbf{x}(v, w) \in \mathcal{R}^n$. Since $(f_c(x_c, v, w), h_c(x_c, v, w))$ is given by (5.8), expanding (5.14) and (5.15) gives

$$\begin{aligned} \frac{\partial \mathbf{x}(v, w)}{\partial v} a(v) &= f(\mathbf{x}(v, w), k(\mathbf{x}(v, w), v, \mathbf{z}(v, w)), v, w), \\ \frac{\partial \mathbf{z}(v, w)}{\partial v} a(v) &= g(\mathbf{z}(v, w), h(\mathbf{x}(v, w), k(\mathbf{x}(v, w), v, \mathbf{z}(v, w)), v, w)), \\ 0 &= h(\mathbf{x}(v, w), k(\mathbf{x}(v, w), v, \mathbf{z}(v, w)), v, w). \end{aligned} \tag{5.49}$$

Letting $\mathbf{u}(v, w) = k(\mathbf{x}(v, w), v, \mathbf{z}(v, w))$ gives (5.46), and using (5.46) in the first and third equations of (5.49) shows that $\mathbf{x}(v, w)$ and $\mathbf{u}(v, w)$ satisfy the regulator equations (5.18). Finally, using the third equation of (5.49) in the second equation of (5.49) gives (5.47).

Sufficiency. By Lemma 5.3, we only need to show that there exists a sufficiently smooth function $\mathbf{x}_c(v, w)$ with $\mathbf{x}_c(0,0) = 0$ that satisfies (5.14) and (5.15). To this end, define $\mathbf{x}_c(v, w) = \text{col}(\mathbf{x}(v, w), \mathbf{z}(v, w))$. Using (5.8) yields

$$h_c(\mathbf{x}_c(v, w), v, w) = h(\mathbf{x}(v, w), k(\mathbf{x}(v, w), v, \mathbf{z}(v, w)), v, w), \tag{5.50}$$

$$f_c(\mathbf{x}_c(v, w), v, w) = \begin{bmatrix} f(\mathbf{x}(v, w), k(\mathbf{x}(v, w), v, \mathbf{z}(v, w)), v, w) \\ g(\mathbf{z}(v, w), h_c(\mathbf{x}_c(v, w), v, w)) \end{bmatrix}. \tag{5.51}$$

Using (5.46) in (5.50) and (5.51) gives

$$h_c(\mathbf{x}_c(v, w), v, w) = h(\mathbf{x}(v, w), \mathbf{u}(v, w), v, w), \tag{5.52}$$

$$f_c(\mathbf{x}_c(v, w), v, w) = \begin{bmatrix} f(\mathbf{x}(v, w), \mathbf{u}(v, w), v, w) \\ g(\mathbf{z}(v, w), h_c(\mathbf{x}_c(v, w), v, w)) \end{bmatrix}. \tag{5.53}$$

Using the second equation of (5.18) in (5.52) gives

$$h_c(\mathbf{x}_c(v, w), v, w) = 0,$$

that is, equations (5.15) hold. Using (5.15) in (5.53) gives

$$f_c(\mathbf{x}_c(v, w), v, w) = \begin{bmatrix} f(\mathbf{x}(v, w), \mathbf{u}(v, w), v, w) \\ g(\mathbf{z}(v, w), 0) \end{bmatrix}. \tag{5.54}$$

Finally, using the first equation of (5.18) and (5.47) in (5.54) gives

$$f_c(\mathbf{x}_c(v, w), v, w) = \begin{bmatrix} \frac{\partial \mathbf{x}(v,w)}{\partial v} a(v) \\ \frac{\partial \mathbf{z}(v,w)}{\partial v} a(v) \end{bmatrix} = \frac{\partial \mathbf{x}_c(v, w)}{\partial v} a(v);$$

that is, equations (5.14) hold. □

To solve the robust output regulation problem, we need to impose an additional restriction on the exosystem (5.2).

Assumption 5.5. The exosystem (5.2) is linear, that is, $\dot{v} = A_1 v$, for some matrix A_1. Further, all the eigenvalues of A_1 are simple with zero real parts.

Theorem 5.12.

(i) *Under Assumptions 5.1, 5.2, 5.4, and 5.5, assume the solutions* $\mathbf{x}(v, w)$ *and* $\mathbf{u}(v, w)$ *of the regulator equations are degree k polynomials in v. Then if the state feedback controller (5.39) solves the kth-order robust output regulation problem, it also solves the robust output regulation problem.*

(ii) *Under Assumptions 5.1 to 5.5, assume the solution* $\mathbf{u}(v, w)$ *of the regulator equations is degree k polynomial in v. Then if the output feedback controller (5.40) solves the kth-order robust output regulation problem, it also solves the robust output regulation problem.*

Proof. (i) Assume that the controller (5.39) solves the kth-order robust output regulation problem. By Lemma 5.11, we need to show the existence of a sufficiently smooth function $\mathbf{z}(v, w)$ that satisfies

$$\mathbf{u}(v, w) = K_1 \mathbf{x}(v, w) + K_2 \mathbf{z}(v, w), \tag{5.55}$$

$$\frac{\partial \mathbf{z}(v, w)}{\partial v} A_1 v = G_1 \mathbf{z}(v, w). \tag{5.56}$$

To this end, note that since the closed-loop system has Property 5.4, there exist sufficiently smooth functions $\bar{\mathbf{x}}(v, w)$ and $\bar{\mathbf{z}}(v, w)$ satisfying (5.32). Let $\mathbf{e}(v, w)$ be as defined in (5.33). Again, express $\bar{\mathbf{x}}(v, w)$, $\bar{\mathbf{z}}(v, w)$, and $\mathbf{e}(v, w)$ as in (5.34). Then, since the controller (5.39) solves the kth-order robust output regulation problem for $l = 1, \ldots, k$, X_{lw} and Z_{lw} must satisfy (5.35) and (5.36), where $A_c(w) = A(w) + B(w)K_1(w)$, $B_c(w) = B(w)K_2(w)$,

$C_c(w) = C(w) + D(w)K_1(w)$, $D_c(w) = D(w)K_2(w)$. Let $U_{lw} = K_1 X_{lw} + K_2 Z_{lw}$. Then (5.35) and (5.36) imply, for $l = 1, \dots, k$,

$$\begin{aligned} X_{lw} A^{[l]} &= A(w) X_{lw} + B(w) U_{lw} + E_{lw}, \\ 0 &= C(w) X_{lw} + D(w) U_{lw} + F_{lw}. \end{aligned} \tag{5.57}$$

By Lemma 4.7, there exist sufficiently smooth functions $\mathbf{x}_k(v, w) = o^k(v)$ and $\mathbf{u}_k(v, w) = o^k(v)$ such that

$$\mathbf{x}(v, w) = \sum_{l=1}^{k} X_{lw} v^{[l]} + \mathbf{x}_k(v, w), \tag{5.58}$$

$$\mathbf{u}(v, w) = \sum_{l=1}^{k} (K_1 X_{lw} + K_2 Z_{lw}) v^{[l]} + \mathbf{u}_k(v, w). \tag{5.59}$$

But, by assumption of this theorem, $\mathbf{x}(v, w)$ and $\mathbf{u}(v, w)$ are degree k polynomials in v, thus

$$\mathbf{x}(v, w) = \sum_{l=1}^{k} X_{lw} v^{[l]}, \quad \mathbf{u}(v, w) = \sum_{l=1}^{k} U_{lw} v^{[l]}.$$

Let

$$\mathbf{z}(v, w) = \sum_{l=1}^{k} Z_{lw} v^{[l]}. \tag{5.60}$$

Clearly (5.55) is satisfied. Now using (5.35) and (5.36) and noting that $Y_{lw} = 0$ for $l = 1, \dots, k$ gives

$$Z_{lw} A^{[l]} = G_1 Z_{lw}, \quad l = 1, 2, \dots, k. \tag{5.61}$$

Multiplying both sides of (5.61) from the right by $v^{[l]}$ and then summarizing from $l = 1$ to k gives

$$\sum_{1}^{k} Z_{lw} A^{[l]} v^{[l]} = \sum_{1}^{k} G_1 Z_{lw} v^{[l]}. \tag{5.62}$$

Using

$$\frac{\partial v^{[l]}}{\partial v} A_1 v = \dot{v}^{[l]} = A^{[l]} v^{[l]}$$

in (5.62) gives

$$\sum_{l=1}^{k} Z_{lw} \frac{\partial v^{[l]}}{\partial v} A_1 v = G_1 \left(\sum_{l=1}^{k} Z_{lw} v^{[l]} \right),$$

which is the same as (5.56) upon using (5.60).

(ii) The proof of part (ii) is almost the same as that of part (i). Assume that a controller of the form (5.40) solves the kth-order robust output regulation problem. By Lemma 5.11, we need to show the existence of a sufficiently smooth function $\mathbf{z}(v, w)$ with $\mathbf{z}(0, 0) = 0$ that satisfies

$$\mathbf{u}(v, w) = K\mathbf{z}(v, w), \tag{5.63}$$

$$\frac{\partial \mathbf{z}(v, w)}{\partial v} A_1 v = \mathcal{G}_1 \mathbf{z}(v, w). \tag{5.64}$$

Let $\bar{\mathbf{x}}(v, w)$ and $\bar{\mathbf{z}}(v, w)$ be sufficiently smooth functions satisfying (5.32), and $\mathbf{e}(v, w)$ be as defined in (5.33). Again, express $\bar{\mathbf{x}}(v, w)$, $\bar{\mathbf{z}}(v, w)$, and $\mathbf{e}(v, w)$ as in (5.34). Then, for $l = 1, \dots, k$, X_{lw} and Z_{lw} satisfy (5.35) and (5.36), where $A_c(w) = A(w)$, $B_c(w) = B(w)K(w)$, $C_c(w) = C(w)$, $D_c(w) = D(w)K(w)$. Let $U_{lw} = KZ_{lw}$. Then (5.35) and (5.36) imply, for $l = 1, \dots, k$,

$$\begin{aligned} X_{lw} A^{[l]} &= A(w) X_{lw} + B(w) U_{lw} + E_{lw}, \\ 0 &= C(w) X_{lw} + D(w) U_{lw} + F_{lw}. \end{aligned} \tag{5.65}$$

By Lemma 4.7, there exist sufficiently smooth functions $\mathbf{x}_k(v, w) = o^k(v)$ and $\mathbf{u}_k(v, w) = o^k(v)$ such that

$$\mathbf{x}(v, w) = \sum_{l=1}^{k} X_{lw} v^{[l]} + \mathbf{x}_k(v, w), \tag{5.66}$$

$$\mathbf{u}(v, w) = \sum_{l=1}^{k} K Z_{lw} v^{[l]} + \mathbf{u}_k(v, w). \tag{5.67}$$

But, by assumption of this theorem, $\mathbf{u}(v, w)$ is a degree k polynomial in v; thus

$$\mathbf{u}(v, w) = \sum_{l=1}^{k} K Z_{lw} v^{[l]}.$$

Let

$$\mathbf{z}(v, w) = \sum_{l=1}^{k} Z_{lw} v^{[l]}. \tag{5.68}$$

Clearly (5.63) is satisfied. The proof of the satisfaction of (5.64) is exactly the same as that of (5.56) and is omitted. □

Example 5.13. Consider

$$\begin{aligned} \dot{x}_1 &= x_2 + (1 + w_1) v_1^2, \\ \dot{x}_2 &= -x_2 + (1 + w_2) x_1^2 + u, \\ e &= x_1 - v_1, \\ \dot{v}_1 &= v_2, \\ \dot{v}_2 &= -v_1. \end{aligned}$$

With $w = (w_1, w_2)$, we have

$$\mathbf{x}_1(v, w) = v_1, \quad \mathbf{x}_2(v, w) = v_2 - (1 + w_1)v_1^2,$$
$$\mathbf{u}(v, w) = v_2 - (1 + w_1)v_1^2 - (1 + w_2)v_1^2 - v_1 - 2(1 + w_1)v_1 v_2.$$

It is clear that the system satisfies Assumptions 5.1 to 5.5. Moreover, both $\mathbf{x}(v, w)$ and $\mathbf{u}(v, w)$ are polynomials in v with $k = 2$. By Theorem 5.12, the robust output regulation problem for this system is solvable by either state feedback or output feedback. As a matter of fact, a simple calculation gives

$$A^{[1]} = A_1 = \begin{bmatrix} 0 & 1 \\ -1 & 0 \end{bmatrix}, \quad A^{[2]} = M_2(A_1 \otimes I_2 + I_2 \otimes A_1)N_2 = \begin{bmatrix} 0 & 2 & 0 \\ -1 & 0 & 1 \\ 0 & -2 & 0 \end{bmatrix}.$$

Also, the minimal polynomials of $A^{[1]}$ and $A^{[2]}$ are $(\lambda^2 + 1)$ and $\lambda(\lambda^2 + 4)$, respectively. Thus the minimal polynomial of the matrix $\mathcal{A}_{2f}$ is

$$\lambda(\lambda^2 + 1)(\lambda^2 + 4) = \lambda^5 + 5\lambda^3 + 4\lambda.$$

The minimal p-copy internal model for the matrix $\mathcal{A}_{2f}$ is thus given by

$$G_1 = \begin{bmatrix} 0 & 1 & 0 & 0 & 0 \\ 0 & 0 & 1 & 0 & 0 \\ 0 & 0 & 0 & 1 & 0 \\ 0 & 0 & 0 & 0 & 1 \\ 0 & -4 & 0 & -5 & 0 \end{bmatrix}, \quad G_2 = \begin{bmatrix} 0 \\ 0 \\ 0 \\ 0 \\ 1 \end{bmatrix}.$$

This pair of matrices together with a pair of feedback gains $K_1 \in \mathcal{R}^2$, $K_2 \in \mathcal{R}^5$ that makes the matrix (5.42) Hurwitz constitutes a state feedback robust servoregulator. ∎

The polynomial requirement on the solution of the regulator equations is obviously too restrictive. It is possible to somehow relax this requirement if the exogenous signal v is available for control.

Theorem 5.14.

(i) *Under Assumptions 5.1, 5.2, 5.4 and 5.5, suppose there exists some integer $k > 0$ such that the solution of the regulator equations takes the following form:*

$$\mathbf{x}(v, w) = \mathbf{x}^{[k]}(v, w) + \mathbf{x}_{hk}(v),$$
$$\mathbf{u}(v, w) = \mathbf{u}^{[k]}(v, w) + \mathbf{u}_{hk}(v), \tag{5.69}$$

where $\mathbf{x}^{[k]}(v, w)$ and $\mathbf{u}^{[k]}(v, w)$ are degree k polynomials in v with coefficients depending on w, and $\mathbf{x}_{hk}(v)$ and $\mathbf{u}_{hk}(v)$ are some sufficiently smooth functions of v, independent of w, vanishing at the origin together with their derivatives up to order *k. Then if a state feedback controller of the form (5.39) solves the kth-order robust output regulation problem, then the following controller*

$$u = K_1(x - \mathbf{x}_{hk}(v)) + K_2 z + \mathbf{u}_{hk}(v),$$
$$\dot{z} = G_1 z + G_2 e, \tag{5.70}$$

solves the robust output regulation problem.

(ii) *Under Assumptions 5.1 to 5.5, suppose there exists some integer $k > 0$ such that the solution $\mathbf{u}(v, w)$ of the regulator equations takes the following form:*

$$\mathbf{u}(v, w) = \mathbf{u}^{[k]}(v, w) + \mathbf{u}_{hk}(v), \tag{5.71}$$

where $\mathbf{u}^{[k]}(v, w)$ is a degree k polynomial in v with coefficients depending on w, and $\mathbf{u}_{hk}(v)$ is some sufficiently smooth function of v, independent of w, vanishing at the origin together with its derivatives up to order k. Then if an output feedback controller of the form (5.40) solves the kth-order robust output regulation problem, the following controller:

$$u = Kz + \mathbf{u}_{hk}(v), \qquad \dot{z} = \mathcal{G}_1 z + \mathcal{G}_2 e \tag{5.72}$$

solves the robust output regulation problem.

Proof. We will only prove part (i) since the proof of part (ii) is almost the same as the proof of part (i). Let $\mathbf{x}(v, w)$ and $\mathbf{u}(v, w)$ be the solution of the regulator equations associated with (5.1). Let $\mathbf{x}_{hk}(v)$ and $\mathbf{u}_{hk}(v)$ be as defined in (5.69). Applying a state and input transformation $x = \hat{x} + \mathbf{x}_{hk}(v)$, $u = \hat{u} + \mathbf{u}_{hk}(v)$ to (5.1) gives

$$\dot{\hat{x}} = f(\hat{x} + \mathbf{x}_{hk}(v), \hat{u} + \mathbf{u}_{hk}(v), v, w) - \frac{\partial \mathbf{x}_{hk}(v)}{\partial v} A_1 v,$$
$$e = h(\hat{x} + \mathbf{x}_{hk}(v), \hat{u} + \mathbf{u}_{hk}(v), v, w). \tag{5.73}$$

It can be verified that $\mathbf{x}(v, w) - \mathbf{x}_{hk}(v)$ and $\mathbf{u}(v, w) - \mathbf{u}_{hk}(v)$ are the solution of the regulator equations associated with the system (5.73). System (5.73) is still in the form of (5.1) and satisfies Assumptions 5.2, 5.4, and 5.5, and $\hat{\mathbf{x}}(v, w) = \mathbf{x}(v, w) - \mathbf{x}_{hk}(v)$ and $\hat{\mathbf{u}}(v, w) = \mathbf{u}(v, w) - \mathbf{u}_{hk}(v)$ are degree k polynomials in v. By Theorem 5.12, there exists a state feedback controller of the form (5.39) that solves the robust output regulation problem for system (5.73). Thus, a controller of the form (5.70) solves the robust output regulation problem for system (5.1). □

Example 5.15. Consider the system

$$\dot{x}_1 = f_1(x, w) + \frac{g_1(x)}{g_2(x)} u,$$
$$\dot{x}_2 = f_2(x, v),$$
$$e = x_1 - v_1,$$
$$\dot{v}_1 = v_2,$$
$$\dot{v}_2 = -v_1, \tag{5.74}$$

where

$$f_1(x, w) = -x_1 + x_2 + \sum_{l=1}^{k_1} a_l x_1^l + c(\cos x_2 - 1), \quad f_2(x, v) = x_1 + d(\cos x_1)v_2,$$

$$g_1(x) = 1, \quad g_2(x) = \sum_{l=1}^{k_2} b_l x_1^l + 1, \tag{5.75}$$

and $w = (a_1, \ldots, a_{k_1})$, that is, a_i's are the only uncertain parameters. Simple computation gives

$$\mathbf{x}_1(v, w) = v_1, \quad \mathbf{x}_2(v, w) = -v_2 + d \sin v_1,$$

$$\mathbf{u}(v, w) = g_2(\mathbf{x}_1(v, w), \mathbf{x}_2(v, w))(v_2 - f_1(\mathbf{x}_1(v, w), \mathbf{x}_2(v, w), w)). \tag{5.76}$$

Clearly, the solution of the regulator equations satisfies the condition (5.69). ∎

5.5 Computational Issues

To synthesize a kth-order robust servoregulator, we need to compute the minimal polynomial of the matrix $\mathcal{A}_{kf}$. Thanks to Assumption 5.5, this seemingly tedious work can be easily handled due to the following result.

Theorem 5.16. *Under Assumption 5.5, the matrix $\mathcal{A}_{kf}$ is similar to a diagonal matrix. Therefore, the roots of the minimal polynomial of $\mathcal{A}_{kf}$ are precisely given by all the distinct members of the following set:*

$$\Lambda_k = \{ \lambda \mid \lambda = l_1\lambda_1 + \cdots + l_q\lambda_q, \ l_1 + \cdots + l_q = l,$$
$$l = 1, 2, \ldots, k, \ l_1, \ldots, l_q = 0, 1, \ldots l\}, \tag{5.77}$$

where $\lambda_1, \ldots, \lambda_q$ are eigenvalues of the matrix A_1.

Proof. As pointed out in the proof of Lemma 4.8, $(A^{[l]})^T$ is the matrix of the linear mapping $L_{A_1 v} : P^l \to P^l$ as defined in (4.49) under the ordered basis given by (4.50). Therefore, we only need to show that this linear mapping has C^l_{q+l-1} linearly independent eigenvectors since the dimension of $A^{[l]}$ is C^l_{q+l-1}. To this end, let the row eigenvectors of A_1 corresponding to eigenvalues λ_i be ζ_i, $i = 1, \ldots, q$. By Assumption 5.5, ζ_i, $i = 1, \ldots, q$, are linearly independent. Therefore, the following set:

$$\left\{ (\zeta_1 v)^{l_1} (\zeta_2 v)^{l_2} \cdots (\zeta_q v)^{l_q} \;\middle|\; \sum_{j=1}^{q} l_j = l, \ l_1, \ldots, l_q = 0, 1, \ldots, l \right\} \tag{5.78}$$

consists of C^l_{q+l-1} linearly independent vectors. Moreover, noting $L_{A_1 v}((\zeta_i v)^s) = s\lambda_i(\zeta_i v)^s$ gives

$$L_{A_1 v}((\zeta_1 v)^{l_1} \cdots (\zeta_q v)^{l_q}) = \left(\sum_{j=1}^{q} l_j \lambda_j\right) (\zeta_1 v)^{l_1} \cdots (\zeta_q v)^{l_q}.$$

Thus, $(\zeta_1 v)^{l_1}(\zeta_2 v)^{l_2} \cdots (\zeta_q v)^{l_q}$ is the eigenvector of $L_{A_1 v}$ associated with the eigenvalue $\lambda = l_1\lambda_1 + \cdots + l_q\lambda_q$. □

Theorem 5.16 leads to a straightforward way to calculate the minimal polynomial of $\mathcal{A}_{kf}$ as follows. Consider the following two cases:

(i) The total number of the distinct members in Λ_k is an even number. Then there exist a positive integer n_k and positive distinct real numbers $\hat{\omega}_1, \ldots, \hat{\omega}_{n_k}$ such that

$$\Lambda_k = \{\pm j\hat{\omega}_1, \ \pm j\hat{\omega}_2, \ldots, \pm j\hat{\omega}_{n_k}\},$$

where $j = \sqrt{-1}$. Thus the minimal polynomial of block diag$\{A^{[1]}, \ldots, A^{[k]}\}$ is given by

$$\alpha^k(\lambda) = \prod_{i=1}^{n_k} (\lambda^2 + \hat{\omega}_i^2).$$

Let

$$G_1 = \text{block diag } \underbrace{[\beta_1, \ldots, \beta_p]}_{p\text{-tuple}}, \ G_2 = \text{block diag } \underbrace{[\sigma_1, \ldots, \sigma_p]}_{p\text{-tuple}},$$

where $(\beta_i, \ \sigma_i)$, $i = 1, \ldots, p$, is any controllable pair with σ_i a column vector and $\alpha^k(\lambda) = |\lambda I - \beta_i|$. For example,

$$\beta_i = \text{block diag} \left(\begin{bmatrix} 0 & \hat{\omega}_1 \\ -\hat{\omega}_1 & 0 \end{bmatrix}, \ldots, \begin{bmatrix} 0 & \hat{\omega}_{n_k} \\ -\hat{\omega}_{n_k} & 0 \end{bmatrix} \right), \quad \sigma_i = \begin{bmatrix} 0 \\ 1 \\ \vdots \\ 0 \\ 1 \end{bmatrix}.$$

Clearly, (β_i, σ_i) is controllable and the minimal polynomial of β_i is equal to $\alpha^k(\lambda)$. Thus the pair (G_1, G_2) is the minimal p-copy internal model of $\mathcal{A}_{kf}$.

(ii) The total number of the distinct values of (5.77) is an odd number; then there exist a positive integer n_k and positive distinct real numbers $\hat{\omega}_1, \ldots, \hat{\omega}_{n_k}$ such that

$$\Lambda_k = \{0, \ \pm j\hat{\omega}_1, \ \pm j\hat{\omega}_2, \ldots, \pm j\hat{\omega}_{n_k}\}.$$

Then the minimal polynomial of block diag$\{A^{[1]}, \ldots, A^{[k]}\}$ is given by

$$\alpha^k(\lambda) = \lambda \prod_{i=1}^{n_k} (\lambda^2 + \hat{\omega}_i^2).$$

Thus, letting

$$\beta_i = \text{block diag}\left(0, \begin{bmatrix} 0 & \hat{\omega}_1 \\ -\hat{\omega}_1 & 0 \end{bmatrix}, \dots, \begin{bmatrix} 0 & \hat{\omega}_{n_k} \\ -\hat{\omega}_{n_k} & 0 \end{bmatrix}\right), \quad \sigma_i = \begin{bmatrix} 1 \\ 0 \\ 1 \\ \vdots \\ 0 \\ 1 \end{bmatrix}$$

leads to a minimal p-copy internal model of $\mathcal{A}_{kf}$.

Example 5.17. Let

$$A_1 = \begin{bmatrix} 0 & 0 & 0 \\ 0 & 0 & \omega \\ 0 & -\omega & 0 \end{bmatrix}.$$

Then the minimal polynomial of block diag$\{A^{[1]}, A^{[2]}, A^{[3]}\}$ is

$$\lambda(\lambda^2 + \omega^2)(\lambda^2 + (2\omega)^2)(\lambda^2 + (3\omega)^2).$$

Note that the degree of the minimal polynomial of block diag$\{A^{[1]}, A^{[2]}, A^{[3]}\}$ is 7, while the degree of the characteristic polynomial of block diag$\{A^{[1]}, A^{[2]}, A^{[3]}\}$ is 19. ∎

5.6 The Ball and Beam System Example

We will consider the approximate asymptotic tracking problem for the ball and beam system described in Section 2.8. For convenience, let us duplicate equation (2.113) as follows:

$$\begin{aligned}
\dot{x}_1(t) &= x_2(t), \\
\dot{x}_2(t) &= Hx_1(t)x_4^2(t) - GH \sin x_3(t), \\
\dot{x}_3(t) &= x_4(t), \\
\dot{x}_4(t) &= -\frac{2Mx_1(t)x_2(t)x_4(t) + MGx_1(t)\cos x_3(t)}{Mx_1^2(t) + J + J_b} + \frac{\tau}{Mx_1^2(t) + J + J_b}, \\
y(t) &= x_1(t),
\end{aligned} \tag{5.79}$$

where $x = \text{col}(x_1, x_2, x_3, x_4) = \text{col}(r, \dot{r}, \theta, \dot{\theta})$, $y = r$, $H = M/(J_b/R^2 + M)$.

The objective is to design a state-feedback controller such that the position r of the ball asymptotically tracks a sinusoidal reference input $A_m \sin\omega t$, where ω is fixed.

As before, we first define the exosystem as follows:

$$\begin{bmatrix} \dot{v}_1(t) \\ \dot{v}_2(t) \end{bmatrix} = \begin{bmatrix} \omega v_2(t) \\ -\omega v_1(t) \end{bmatrix}, \quad \begin{bmatrix} v_1(0) \\ v_2(0) \end{bmatrix} = \begin{bmatrix} 0 \\ A_m \end{bmatrix}, \tag{5.80}$$

which yields $v_1(t) = A_m \sin\omega t$. Thus the error equation is given by

$$e(t) = x_1(t) - v_1(t). \tag{5.81}$$

Assume the ball mass M and the moment inertia of the beam J in (5.79) are uncertain parameters. Let us write

$$J = J_0 + \Delta J, \quad M = M_0 + \Delta M,$$

where J_0 and M_0 denote the nominal values of J and M, and ΔJ and ΔM the perturbed values of J and M. Perturbation of M will also cause the variation of H, which can be written as $H = H_0 + \Delta H$, with H_0 being the nominal value and ΔH the perturbed value. Let $w = (\Delta M, \Delta J)$. Then clearly, (5.79) is in the form of (5.1). Our design will be based on the nominal plant, that is, the plant (5.79) with $\Delta J = 0$ and $\Delta M = 0$. For this nominal plant, we can simplify the system by performing an input transformation

$$\tau = 2M_0 x_1 x_2 x_4 + M_0 G x_1 \cos x_3 + (M_0 x_1^2 + J_0 + J_b)u, \tag{5.82}$$

which leads to the following:

$$\begin{aligned}
\dot{x}_1(t) &= x_2(t), \\
\dot{x}_2(t) &= H_0 x_1(t) x_4^2(t) - G H_0 \sin x_3(t), \\
\dot{x}_3(t) &= x_4(t), \\
\dot{x}_4(t) &= u, \\
y(t) &= x_1(t).
\end{aligned} \tag{5.83}$$

Recall from Section 2.8 that the system (5.79) does not have a well-defined relative degree at the origin; therefore we cannot assure the existence of the solution of the regulator equations. Nevertheless, it is easy to verify that this system satisfies Assumptions 5.2 and 5.4. Therefore, for any integer $k > 0$, the kth-order robust output regulation problem for this system is solvable. Since the kth-order output regulation problem is the special case of the kth-order robust output regulation problem, the kth-order output regulation problem for this system is also solvable for any integer k assuming all the plant parameters are precisely known. In what follows, we will design both a third-order state feedback servoregulator and a third-order state feedback robust servoregulator for this system.

A third-order controller for this plant can be designed as follows. First, let us use the approach described in Chapter 4 to obtain a third-order solution of the regulator equations associated with the ball and beam system. The scalar form of the regulator equations associated with the above tracking problem takes the following form:

$$\begin{aligned}
\frac{\partial \mathbf{x}_1(v)}{\partial v} A_1 v &= \mathbf{x}_2(v), \\
\frac{\partial \mathbf{x}_2(v)}{\partial v} A_1 v &= H_0 \mathbf{x}_1(v) \mathbf{x}_4^2(v) - H_0 G \sin \mathbf{x}_3(v), \\
\frac{\partial \mathbf{x}_3(v)}{\partial v} A_1 v &= \mathbf{x}_4(v), \\
\frac{\partial \mathbf{x}_4(v)}{\partial v} A_1 v &= \mathbf{u}(v), \\
\mathbf{x}_1(v) &= v_1.
\end{aligned}$$

By inspection, we can obtain the partial solution as follows:

$$\begin{aligned}
\mathbf{x}_1(v) &= v_1, \\
\mathbf{x}_2(v) &= \omega v_2, \\
\mathbf{u}(v) &= \frac{\partial \mathbf{x}_4(v)}{\partial v} A_1 v,
\end{aligned} \tag{5.84}$$

with two undetermined functions $\mathbf{x}_3(v)$ and $\mathbf{x}_4(v)$ satisfying

$$-\omega^2 v_1 = H_0 v_1 (\mathbf{x}_4(v))^2 - H_0 G \sin \mathbf{x}_3(v), \tag{5.85}$$

$$\frac{\partial \mathbf{x}_3(v)}{\partial v} A_1 v = \mathbf{x}_4(v). \tag{5.86}$$

Again, by the reason given in Remark 4.16, we can assume that the Taylor series solution of (5.85) and (5.86) can be expressed as follows:

$$\mathbf{x}_3(v) = a_{10} v_1 + a_{01} v_2 + a_{30} v_1^3 + a_{21} v_1^2 v_2 + a_{12} v_1 v_2^2 + a_{03} v_2^3 + \cdots, \tag{5.87}$$

$$\mathbf{x}_4(v) = b_{10} v_1 + b_{01} v_2 + b_{30} v_1^3 + b_{21} v_1^2 v_2 + b_{12} v_1 v_2^2 + b_{03} v_2^3 + \cdots. \tag{5.88}$$

Substituting (5.87) and (5.88) into (5.85) and (5.86) and identifying the coefficients gives a third-order approximation of $\mathbf{x}_3(v)$ and $\mathbf{x}_4(v)$ as follows:

$$\begin{aligned}
\mathbf{x}_3^{(3)}(v) &= a_{10} v_1 + a_{30} v_1^3 + a_{12} v_1 v_2^2, \\
\mathbf{x}_4^{(3)}(v) &= b_{01} v_2 + b_{21} v_1^2 v_2 + b_{03} v_2^3,
\end{aligned} \tag{5.89}$$

where

$$\begin{aligned}
a_{10} &= \frac{\omega^2}{G H_0}, \\
a_{30} &= \frac{\omega^6}{6 G^3 H_0^3}, \\
a_{12} &= \frac{\omega^6}{G^3 H_0^2}, \\
b_{01} &= \frac{\omega^3}{G H_0}, \\
b_{21} &= \frac{\omega^7 - 4 H_0 \omega^7}{2 G^3 H_0^3}, \\
b_{03} &= \frac{\omega^7}{G^3 H_0^2}.
\end{aligned}$$

Using the last equation of (5.84) gives the third-order approximation of $\mathbf{u}(v)$ as follows:

$$\begin{aligned}
\mathbf{u}^{(3)}(v_1, v_2) &= \frac{\partial \mathbf{x}_4^{(3)}(v)}{\partial v} A_1 v \\
&= -\frac{\omega^4}{G H_0} v_1 - \frac{\omega^8 - 4 H_0 \omega^8}{2 G^3 H_0^3} v_1^3 + \frac{(1 - 7 H_0)\omega^8}{G^3 H_0^3} v_1 v_2^2.
\end{aligned} \tag{5.90}$$

Thus the third-order approximation of the solution of the regulator equations of the ball and beam system is given by (5.90) and

$$\mathbf{x}^{(3)}(v_1, v_2) = \begin{bmatrix} v_1 \\ \omega v_2 \\ \frac{\omega^2}{GH_0} v_1 + \frac{\omega^6}{6G^3H_0^3} v_1^3 + \frac{\omega^6}{G^3H_0^2} v_1 v_2^2 \\ \frac{\omega^3}{GH_0} v_2 + \frac{\omega^7 - 4H_0\omega^7}{2G^3H_0^3} v_1^2 v_2 + \frac{\omega^7}{G^3H_0^2} v_2^3 \end{bmatrix}.$$

The Jacobian linearization of the nominal plant is given by

$$A = \begin{bmatrix} 0 & 1 & 0 & 0 \\ 0 & 0 & -H_0G & 0 \\ 0 & 0 & 0 & 1 \\ 0 & 0 & 0 & 0 \end{bmatrix}, \quad B = \begin{bmatrix} 0 \\ 0 \\ 0 \\ 1 \end{bmatrix},$$

$$C = [1 \ 0 \ 0 \ 0], \quad D = 0, \quad A_1 = \begin{bmatrix} 0 & \omega \\ -\omega & 0 \end{bmatrix}.$$

It can be verified that the pair (A, B) is controllable. Thus a feedback gain K_x that renders the matrix $A + BK_x$ Hurwitz can be found. To be more specific, letting $K_x = [-0.2826, -1.1604, 6.8783, 3.1500]$ will place the eigenvalues of $A + BK_x$ at

$$-0.6360 \pm 1.8945j, -0.9390 \pm 0.6212j,$$

which is based on the ITAE criterion with the cutoff frequency equal to 1.5.

Next we consider the design of a third-order robust servoregulator. For this purpose, we need to find a pair of matrices (G_1, G_2) that incorporates a minimal p-copy ($p = 1$) internal model of $\mathcal{A}_{3f}$. But as pointed out above, since the solution of the regulator equations does not contain the second-order term, the output equation $h_c(\mathbf{x}_c(v, w), v) = \mathbf{x}(v, w) - v_1$ of the closed-loop system for any state feedback control law of the form (5.39) will not contain the second-order term either. Thus, it suffices to find a pair of matrices (G_1, G_2) that incorporates a minimal 1-copy internal model of $A^{[1]}$ and $A^{[3]}$. The minimal polynomials of $A^{[1]}$ and $A^{[3]}$ are computed as follows:

$$\alpha_1(\lambda) = \lambda^2 + \omega^2, \quad \alpha_3(\lambda) = (\lambda^2 + \omega^2)(\lambda^2 + 9\omega^2).$$

The minimal polynomial of block diag$\{A^{[1]}, A^{[3]}\}$ is thus

$$(\lambda^2 + \omega^2)(\lambda^2 + 9\omega^2).$$

Therefore,

$$G_1 = \begin{bmatrix} 0 & \omega & 0 & 0 \\ -\omega & 0 & 0 & 0 \\ 0 & 0 & 0 & 3\omega \\ 0 & 0 & -3\omega & 0 \end{bmatrix}, \quad G_2 = \begin{bmatrix} 0 \\ 1 \\ 0 \\ 1 \end{bmatrix}.$$

The compensator together with the plant forms an eight-dimensional system. The feedback gain (K_1, K_2) is chosen such that the eigenvalues of the linearized closed-loop system are

$$-1.0013, -3.1173, -0.3046 \pm 1.7661j, -0.5917 \pm 1.1218j, -0.9445 \pm 0.8351j,$$

Amp	First order	Third order	Third-order robust
3.0000	0.0180	0.0001	0.0000
5.0000	0.0877	0.0021	0.0003
6.0000	0.1585	0.0058	0.0008

Table 5.1. *Maximal steady-state tracking error of nominal system with* $\omega = \frac{\pi}{5}$.

Case	ΔM	ΔJ	First order	Third order	Third-order robust
1	0	0	0.0877	0.0021	0.0003
2	0.0100	0.0100	2.6586	2.6882	0.0333
3	0.0150	0.0100	6.4298	6.6523	0.0527
4	-0.0200	0.0100	2.9178	2.8305	0.0417
5	-0.0250	0.0100	Unstable	Unstable	0.0484

Table 5.2. *Maximal steady-state tracking error of the perturbed system with* $A_m =$ 5 *and* $\omega = \frac{\pi}{5}$.

which again are obtained based on the ITAE prototype design with the cutoff frequency equal to 1.5 rad/sec. The resulting feedback gains are

$$K_1 = [-4.4018, -6.0091, 24.8522, 7.8000],$$
$$K_2 = [1.1226, -1.4605, 0.0144, 2.6865].$$

Computer simulation is conducted to compare the performance of the two controllers. The nominal values of the various system parameters are given as follows: $M_0 = 0.05$ kg, $R = 0.01$ m, $J_0 = 0.02$ kg m^2, $J_b = 2 \times 10^{-6}$ kg m^2, and $G = 9.81$ m/s^2. As a result, $H_0 = 0.7134$. It is assumed that the initial states of the plant and compensator are zero. The frequency of the reference input is fixed at $\omega = \frac{\pi}{5}$, while the amplitude A_m of the reference input takes the values of 3, 5, and 6. Five cases are presented:

- Nominal case: $\Delta M = 0.0$ kg, and $\Delta J = 0.0$ kg m^2.
- $\Delta M = 0.010$ kg, $\Delta J = 0.01$ kg m^2.
- $\Delta M = 0.015$ kg, $\Delta J = 0.01$ kg m^2.
- $\Delta M = -0.02$ kg, $\Delta J = 0.01$ kg m^2.
- $\Delta M = -0.025$ kg, $\Delta J = 0.01$ kg m^2.

Comparison is first made for the nominal case. Table 5.1 shows the maximal steady-state tracking errors of the closed-loop systems under the linear controller, third-order controller, and third-order robust controller for $\omega = \pi/5$ and $A_m = 3,\ 5,\ 6$. It is seen that, in every case, the performance of the various controllers is quite good, though the third-order robust controller is superior to the third-order controller, while the third-order controller is superior to the linear controller. Next, we compare the performance of the various controllers in the presence of the parameter uncertainty with $A_m = 5$ and $\omega = \frac{\pi}{5}$. As shown in Table 5.2, the third-order robust controller is quite capable of tolerating the parametric uncertainties. In various cases of the parametric uncertainty, the maximal steady-state tracking errors are kept within the order of 10^{-2}. In contrast, the tracking performance of both the linear and the third-order controller severely deteriorates when the parametric uncertainties

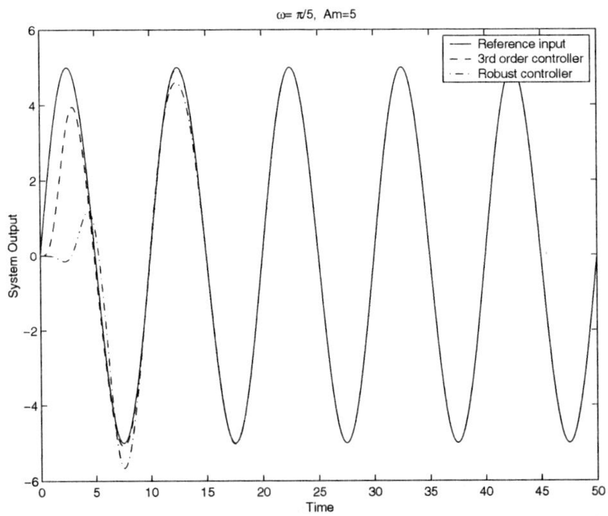

Figure 5.1. *Tracking performance: Nominal case $A_m = 5$ and $\omega = \frac{\pi}{5}$.*

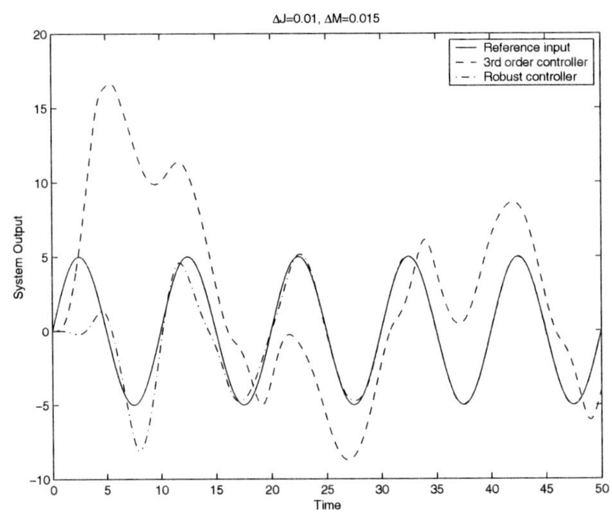

Figure 5.2. *Tracking performance: Perturbed system with $A_m = 5$ and $\omega = \frac{\pi}{5}$.*

are present. For example, in case 3, the maximal steady-state tracking errors of the linear and third-order controllers are over 100 times that of the third-order robust controller. Moreover, in case 5, neither the linear controller nor the third-order controller can stabilize the system. Also note that while the third-order controller performs much better than the linear controller in the nominal case, it has no advantage over the linear controller when the parameter uncertainties are present. Figures 5.1 and 5.2 show the tracking performance of the closed-loop system resulting from the third-order controller and third-order robust controllers with $\omega = \pi/5$ and $A_m = 5$.

Chapter 6

From Output Regulation to Stabilization

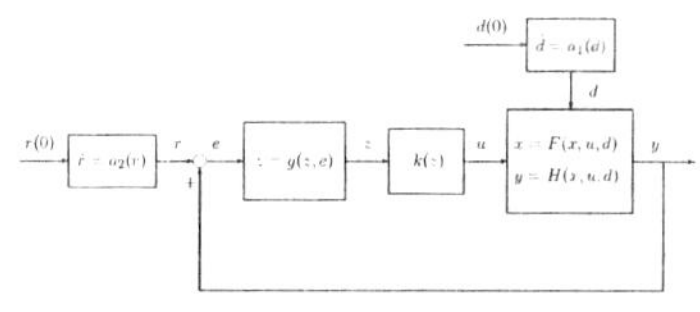

The approach described in Chapter 5 employs an extended version of the internal model principle introduced in Chapter 1 to handle the robust output regulation problem for nonlinear systems. The design approach consists of two steps. First, augment the given plant by a linear dynamic system that incorporates a p-copy internal model of the k-fold exosystem of the given system. Second, stabilize the linear approximation of the augmented system. This design method has two fundamental limitations. First, the linearity of the internal model is incapable of handling nonlinear systems whose regulator equations have nonpolynomial solution. Second, the linear stabilization method employed is incapable of achieving global stability of the closed-loop system. In this chapter, we introduce a new design framework to deal with the robust output regulation problem. This design framework aims to systematically convert the robust output regulation problem for a given system into a robust stabilization problem for an appropriately augmented system.

This new framework, on one hand, removes the polynomial assumption on the solution of the regulator equations, and on the other hand, offers greater flexibility in incorporating global stabilization techniques, thus setting the stage for studying a robust output regulation problem with global stability in Chapter 7.

This chapter is organized as follows. In Section 6.1, the notion of the steady-state generator is introduced which is a dynamic system that can reproduce the solution or partial solution of the regulator equations of the given plant. The notion of the steady-state generator leads to a new definition of the internal model. The composition of the given plant and the internal model is called the augmented system. It is shown that the stabilizing solution of the augmented system will lead to the solution of the robust output regulation problem of the original system. In Section 6.2, the existence conditions of the steady-state generator are established. These conditions in turn lead to the construction of a nonlinear internal model. Section 6.3 shows that, due to the employment of the nonlinear internal model, it is possible to design a dynamic output feedback controller to solve the robust output regulation problem for a nonlinear system whose regulator equations admit a nonpolynomial solution. In Section 6.4, the new framework is applied to solve the robust disturbance rejection problem of the RTAC system.

The notation defined in Chapter 5 will be used freely in this chapter. In particular, we define

$$A = \frac{\partial f}{\partial x}(0,0,0,0), \quad B = \frac{\partial f}{\partial u}(0,0,0,0), \quad C = \frac{\partial h}{\partial x}(0,0,0,0), \quad D = \frac{\partial h}{\partial u}(0,0,0,0).$$

Also, for convenience, we will lump the plant (5.1) and the exosystem (5.2) together as follows:

$$\begin{aligned} \dot{x}(t) &= f(x(t), u(t), v(t), w), \\ \dot{v}(t) &= a(v(t)), \\ e(t) &= h(x(t), u(t), v(t), w). \end{aligned} \tag{6.1}$$

We will refer to (6.1) as a composite system.

6.1 A New Design Framework

As pointed out in Remark 3.12, the output regulation problem can be viewed as a stabilization problem about an invariant manifold defined by the solution of the regulator equations. When the solution of the regulator equations is available for feedback control, one can convert the output regulation problem into a stabilization problem about the equilibrium point at the origin of a translated system, as was done in Chapter 3. However, when the plant contains unknown parameters, the solution of the regulator equations cannot be used for feedback. One wonders if the solution of the regulator equations can be obtained by some other means so that the robust output regulation problem can also be converted into the stabilization problem of some related system. This idea motivates a new design framework to tackle the robust output regulation problem. This framework includes the following three steps. First, introduce the concept of the steady-state generator for the system (6.1), which is some dynamic system that can produce a partial or whole solution of the regulator equations. Second, define a generalized internal model based on the steady-state generator which, together with the plant, is called the augmented system. Third, show that, after a suitable coordinate and input transformation, the stabilizability of the equilibrium at the origin of the augmented system implies the solvability of the robust output regulation problem of the original system.

Definition 6.1. *Let $F : V \times W \to \mathcal{R}^l$, where V and W are some open neighborhoods of the origins of $\mathcal{R}^q$ and $\mathcal{R}^{n_w}$, respectively, and l is some integer, be a smooth function vanishing at the origin. The function F is said to have a* generator *if, for some integer s, there exists a triple $\{\theta, \alpha, \beta\}$, where $\theta : V \times W \to \mathcal{R}^s$, $\alpha : \mathcal{R}^s \to \mathcal{R}^s$, and $\beta : \mathcal{R}^s \to \mathcal{R}^l$ are sufficiently smooth functions vanishing at the origin, such that, for all trajectories $v(t) \in V$ of the exosystem (5.2) and all $w \in W$,*

$$\begin{aligned} \frac{d\theta(v,w)}{dt} &= \alpha(\theta(v,w)), \\ F(v,w) &= \beta(\theta(v,w)). \end{aligned} \tag{6.2}$$

If $V = \mathcal{R}^q$, $W = \mathcal{R}^{n_w}$, then the triple $\{\theta, \alpha, \beta\}$ is called a global generator of $F(v,w)$.

Let the triple $\{\theta, \alpha, \beta\}$ be a (global) generator of $F(v, w)$. If, in addition, the linearization of the pair $\{\beta(\theta), \alpha(\theta)\}$ at the origin is observable, then the triple $\{\theta, \alpha, \beta\}$ is called a linearly observable *(global) generator of $F(v, w)$.*

Definition 6.2. *Let $g_o : \mathcal{R}^{n+m} \rightarrow \mathcal{R}^l$ be a mapping for some integer $1 \leq l \leq n+m$. Under Assumptions 3.1 and 5.1, the composite system (6.1) is said to have a (global) steady-state generator with output $g_o(x, u)$ if the function $g_o(\mathbf{x}(v, w), \mathbf{u}(v, w))$ has a (global) generator. The system (6.1) is said to have a (global) steady-state generator with output $g_o(x, u)$ with linear observability if the function $g_o(\mathbf{x}(v, w), \mathbf{u}(v, w))$ has a (global) generator with linear observability.*

Remark 6.3. Existence of a steady-state generator with output $g_o(x, w)$ means that some function of the solution of the regulator equations can be reproduced by an autonomous system of the form

$$\dot{\tau} = \alpha(\tau), \qquad y = \beta(\tau), \tag{6.3}$$

which is independent of the model uncertainty w and exogenous signal v. As will be seen later, it is possible to use the information provided by $g_o(\mathbf{x}(v(t), w), \mathbf{u}(v(t), w))$ to design a controller. In particular, when $g_o(x, u) = \text{col}(x, u)$, the steady-state generator reproduces the whole solution $\mathbf{x}(u, w)$ and $\mathbf{u}(v, w)$ of the regulator equations, and when $g_o(x, u) = u$, the steady-state generator reproduces the partial solution $\mathbf{u}(v, w)$ of the regulator equations. In what follows, we will assume that $g_o(x, u) = \text{col}(x_{i_1}, x_{i_2}, \ldots, x_{i_d}, u)$, where $1 \leq i_1 < i_2 < \cdots < i_d \leq n$ for some integer d satisfying $0 \leq d \leq n$. Without loss of generality, we can always assume $i_j = j$ for $j = 1, \ldots, d$, since the index of the state variable can be relabelled to have this assumption satisfied. ∎

Remark 6.4. The motivation of introducing the notion of the steady-state generator will be briefly elucidated in Remark 6.11. Here let us first connect this notion to the previous results obtained in Chapter 5. By Lemma 5.11, under Assumptions 3.1′ and 5.2, if there is an output feedback control law of the form (5.5) that solves the robust output regulation problem for system (6.1), then there exists a sufficiently smooth function $\mathbf{z}(v, w)$ defined for $v \in V$, $w \in W$ with $\mathbf{z}(0, 0) = 0$ such that $\mathbf{z}(v, w)$ satisfies

$$\begin{aligned} \mathbf{u}(v, w) &= k(\mathbf{z}(v, w)), \\ \frac{\partial \mathbf{z}(v, w)}{\partial v} a(v) &= g(\mathbf{z}(v, w), 0). \end{aligned} \tag{6.4}$$

Let $\theta(v, w) = \mathbf{z}(v, w), \alpha(\theta) = g(\theta, 0), \beta(\theta) = k(\theta)$. Then clearly, the triple $\{\theta, \alpha(\theta), \beta(\theta)\}$ is a steady-state generator of system (6.1) with output $g_o(x, u) = u$. Moreover, denote the linearization at the origin of the control law (5.5) by the triple $(K, \mathcal{G}_1, \mathcal{G}_2)$. Then

$$\frac{\partial f_c}{\partial x_c}(0, 0, 0) = \begin{bmatrix} A & BK \\ \mathcal{G}_2 C & \mathcal{G}_1 + \mathcal{G}_2 DK \end{bmatrix} = \begin{bmatrix} A & BK \\ 0 & \mathcal{G}_1 \end{bmatrix} + \begin{bmatrix} 0 \\ \mathcal{G}_2 \end{bmatrix} \begin{bmatrix} C & DK \end{bmatrix}.$$

The fact that the matrix $\frac{\partial f_c}{\partial x_c}(0, 0, 0)$ is Hurwitz implies that the pair

$$\left(\begin{bmatrix} C & DK \end{bmatrix}, \begin{bmatrix} A & BK \\ 0 & \mathcal{G}_1 \end{bmatrix} \right)$$

is detectable. Hence, the following decomposition:

$$\begin{bmatrix} A-\lambda I & BK \\ 0 & \mathcal{G}_1-\lambda I \\ C & DK \end{bmatrix} = \begin{bmatrix} A-\lambda I & 0 & B \\ 0 & I & 0 \\ C & 0 & D \end{bmatrix} \begin{bmatrix} I & 0 \\ 0 & \mathcal{G}_1-\lambda I \\ 0 & K \end{bmatrix}$$

further shows that the pair $(K, \mathcal{G}_1)$ is detectable, too. In particular, when all the eigenvalues of $\mathcal{G}_1$ have zero real parts, then the pair $(K, \mathcal{G}_1)$ is observable. Thus, if the robust output regulation for the system (6.1) is solvable by an output feedback controller of the form (5.5), then the system (6.1) must have a steady-state generator whose linearization at the origin is detectable. ∎

Remark 6.5. It is known that, under Assumptions 5.1 to 5.5, if the solution of the regulator equations of the system (6.1) is a degree k polynomial in v, then the robust output regulation for the system (6.1) is solvable by a linear output feedback control law of the form $u = Kz$, $\dot{z} = \mathcal{G}_1 z + \mathcal{G}_2 e$, where the pair $(\mathcal{G}_1, \mathcal{G}_2)$ incorporates a p-copy internal model of the matrix $\mathcal{A}_{kf}$. Moreover, by Lemma 5.11, there exists a sufficiently smooth function $\mathbf{z}(v, w)$ locally defined in $v \in V$, $w \in W$ with $\mathbf{z}(0, 0) = 0$ such that $\mathbf{z}(v, w)$ satisfies

$$\begin{aligned} \mathbf{u}(v, w) &= K\mathbf{z}(v, w), \\ \frac{\partial \mathbf{z}(v, w)}{\partial v} A_1 v &= \mathcal{G}_1 \mathbf{z}(v, w). \end{aligned} \tag{6.5}$$

Let $\theta(v, w) = \mathbf{z}(v, w)$, $\alpha(\theta) = \mathcal{G}_1\theta$, $\beta(\theta) = K\theta$. Then, clearly, the triple is a steady-state generator of (6.1) with output $g_o(x, u) = u$. Nevertheless, the polynomial assumption on the solution of the regulator equations is restrictive. We will show in the next section that the steady-state generator may exist even when the solution of the regulator equations is not polynomial. Before doing this, let us first give a more general characterization of the concept of the internal model as follows. ∎

Definition 6.6. *Under Assumptions 3.1 and 5.1, suppose the composite system (6.1) has a (global) steady-state generator with output $g_o(x, u)$. Let $\gamma : \mathcal{R}^{s+d+m+p} \to \mathcal{R}^s$ be a sufficiently smooth function vanishing at the origin. Then we call the following system:*

$$\dot{\eta} = \gamma(\eta, g_o(x, u), e) \tag{6.6}$$

an internal model *of (6.1) with output $g_o(x, u)$ if*

$$\gamma(\theta(v, w), g_o(\mathbf{x}(v, w), \mathbf{u}(v, w)), 0) = \alpha(\theta(v, w)). \tag{6.7}$$

For convenience at the price of the abuse of the notation, in what follows, we will always use the notation $\gamma(\eta, x, u, e)$ to stand for $\gamma(\eta, g_o(x, u), e)$.

Remark 6.7. The reason for defining the internal model this way will be given in Remark 6.11. At this stage, let us first note that the characterization of the internal model here, on one hand, contains the one described in Chapter 5 as a special case. As pointed out in

Remark 6.5, if a linear output feedback control law of the form $u = Kz$, $\dot{z} = \mathcal{G}_1 z + \mathcal{G}_2 e$ solves the robust output regulation problem for system (6.1), then there exists a function $\mathbf{z}(v, w)$ such that $\mathbf{u}(v, w) = K\mathbf{z}(v, w)$ and $\frac{\partial \mathbf{z}(v,w)}{\partial v} A_1 v = \mathcal{G}_1 \mathbf{z}(v, w)$. Moreover, let $\theta(v, w) = \mathbf{z}(v, w)$, $\alpha(\theta) = \mathcal{G}_1\theta$, $\beta(\theta) = K\theta$. Then the triple $\{\theta, \alpha(\theta), \beta(\theta)\}$ is a steady-state generator of (6.1) with output $g_o(x, u) = u$. Now, let $\gamma(\eta, x, u, e) = \mathcal{G}_1\eta + \mathcal{G}_2 e$. Then $\gamma(\theta(v, w), \mathbf{x}(v, w), \mathbf{u}(v, w), 0) = \mathcal{G}_1\theta(v, w) = \alpha(\theta(v, w))$. Thus, the internal model described here is an extension of what is described in Chapter 5. On the other hand, this characterization is much more general than the existing one in two aspects. First, it can be a *nonlinear* system, and second, the system dynamics can be coupled to the given system not only through the tracking error e, but also through the state x and input u. We will see in Section 6.3 that this generality can be used to construct a particularly useful nonlinear internal model. For the time being, we will first show that an internal model defined this way leads to an augmented system with the property that the *stabilizability* of this augmented system implies the solvability of the robust output regulation problem of the original system (6.1). ∎

Attaching the internal model (6.6) to the given plant yields the following augmented system:

$$\begin{aligned} \dot{x} &= f(x, u, v, w), \\ \dot{\eta} &= \gamma\,(\eta, x, u, e)\,, \\ e &= h(x, u, v, w). \end{aligned} \tag{6.8}$$

Performing on (6.8) the following coordinate and input transformation:

$$\begin{aligned} \bar{x}_i &= x_i - \beta_i\,(\eta)\,, \quad i = 1, \dots, d, \\ \bar{x}_i &= x_i - \mathbf{x}_i\,(v, w)\,, \quad i = d+1, \dots, n, \\ \bar{\eta} &= \eta - \theta\,(v, w)\,, \\ \bar{u} &= u - \left[\beta_{d+1}\,(\eta)\,, \dots, \beta_{d+m}\,(\eta)\right]^T = u - \beta_u\,(\eta) \end{aligned} \tag{6.9}$$

gives a new system denoted by

$$\begin{aligned} \dot{\bar{x}} &= \bar{f}(\bar{x}, \bar{\eta}, \bar{u}, v, w), \\ \dot{\bar{\eta}} &= \bar{\gamma}(\bar{x}, \bar{\eta}, \bar{u}, v, w), \\ e &= \bar{h}(\bar{x}, \bar{\eta}, \bar{u}, v, w), \end{aligned} \tag{6.10}$$

where

$$\begin{aligned} \bar{f}_i(\bar{x}, \bar{\eta}, \bar{u}, v, w) &= f_i(x, u, v, w) - \frac{\partial \beta_i(\eta)}{\partial \eta}\gamma(\eta, x, u, e), \quad i = 1, \dots, d, \\ \bar{f}_i(\bar{x}, \bar{\eta}, \bar{u}, v, w) &= f_i(x, u, v, w) - f_i(\mathbf{x}(v, w), \mathbf{u}(v, w), v, w), \quad i = d+1, \dots, n, \\ \bar{\gamma}(\bar{x}, \bar{\eta}, \bar{u}, v, w) &= \gamma(\eta, x, u, e) - \alpha(\theta(v, w)), \\ \bar{h}(\bar{x}, \bar{\eta}, \bar{u}, v, w) &= h(x, u, v, w). \end{aligned}$$

The system (6.10) has the following property.

Proposition 6.8. *Suppose the composite plant (6.1) satisfies Assumptions 3.1 and 5.1 and has a steady-state generator with output* $g_o(x, u) = \mathrm{col}(x_1, \ldots, x_d, u)$ *and an internal model described by (6.6). Then the augmented system in the new coordinates and input described by (6.10) has the property that, for all trajectories* $v(t) \in V$ *of the exosystem and all* $w \in W$*, where* V *and* W *are some open neighborhoods of the origins of* $\mathcal{R}^q$ *and* $\mathcal{R}^{n_w}$*, respectively,*

$$\begin{aligned} \bar{f}(0,0,0,v,w) &= 0, \\ \bar{\gamma}(0,0,0,v,w) &= 0, \\ \bar{h}(0,0,0,v,w) &= 0. \end{aligned} \tag{6.11}$$

Proof. Consider the augmented system (6.8). Since $\mathbf{x}(v, w)$ and $\mathbf{u}(v, w)$ are the solution of the regulator equations, and $\theta(v, w)$ satisfies (6.2), the hypersurface $\{(x, \eta, v) \mid x = \mathbf{x}(v, w),\ \ \eta = \theta(v, w)\}$ is an output zeroing manifold of the composite system consisting of (6.8) and the exosystem (5.2) rendered by the feedback control $u = \mathbf{u}(v, w)$. Therefore, the hypersurface $\{(\bar{x}, \bar{\eta}, v) \mid \bar{x} = 0,\ \bar{\eta} = 0\}$ is the output zeroing manifold of the composite system (6.10) and (5.2) rendered by the feedback control $\bar{u} = 0$. This is, the origin $(\bar{x}, \bar{\eta}) = (0, 0)$ is the equilibrium point of the unforced augmented system for all trajectories $v(t) \in V$ of the exosystem, and any $w \in W$, and the error output equation is identically zero at $(\bar{x}, \bar{\eta}, \bar{u}) = (0, 0, 0)$ for all trajectories $v(t) \in V$ of the exosystem and for any $w \in W$. Thus the proof is completed. □

Consider a controller of the form

$$\begin{aligned} \bar{u} &= k\left(\bar{x}_1, \ldots, \bar{x}_d, \xi, e\right), \\ \dot{\xi} &= g_\xi\left(\bar{x}_1, \ldots, \bar{x}_d, \xi, e\right), \end{aligned} \tag{6.12}$$

where $\xi \in \mathcal{R}^{n_\xi}$ for some integer n_ξ, and k and g_ξ are sufficiently smooth functions vanishing at their respective origins. Let $\bar{x}_c = \mathrm{col}(\bar{x}, \bar{\eta}, \xi)$ be the state of the closed-loop system composed of the augmented system (6.10) and the controller (6.12). Then this closed-loop system takes the following form:

$$\dot{\bar{x}}_c = \bar{f}_c(\bar{x}_c, v, w), \qquad e = \bar{h}_c(\bar{x}_c, v, w). \tag{6.13}$$

It is possible to show that if (6.13) satisfies Property 5.4, then the following controller:

$$\begin{aligned} u &= \beta_u(\eta) + k\left(x_1 - \beta_1(\eta), \ldots, x_d - \beta_d(\eta), \xi, e\right), \\ \dot{\eta} &= \gamma\left(\eta, x, u, e\right), \\ \dot{\xi} &= g_\xi\left(x_1 - \beta_1(\eta), \ldots, x_d - \beta_d(\eta), \xi, e\right) \end{aligned} \tag{6.14}$$

solves the robust output regulation problem for the original plant (6.1).

Corollary 6.9. *If the controller (6.12) is such that the closed-loop system (6.13) satisfies Property 5.4, then the controller (6.14) solves the robust output regulation problem for the original system (6.1).*

Proof. Consider the closed-loop system composed of the plant (6.1) and the controller (6.14) and denote its state by $x_c = \text{col}(x, \eta, \xi)$. Then

$$x_c = \bar{x}_c + \text{col}(\beta_1(\bar{\eta} + \theta(v, w)), \ldots, \beta_d(\bar{\eta} + \theta(v, w)),$$
$$\mathbf{x}_{d+1}(v, w), \ldots, \mathbf{x}_n(v, w), \theta(v, w), 0).$$

Thus, when $v = 0$ and $w = 0$, the state $\bar{x}_c$ of the closed-loop system (6.13) and the state x_c of the closed-loop system composed of (6.1) and (6.14) are related by a diffeomorphism $x_c = \bar{x}_c + \text{col}(\beta_1(\bar{\eta}), \ldots, \beta_d(\bar{\eta}), 0, \ldots, 0, 0)$. Thus the closed-loop system composed of (6.1) and (6.14) also satisfies Property 5.4. Next let $\mathbf{x}_c(v, w) = \text{col}(\mathbf{x}(v, w), \theta(v, w), 0)$. Then it can be verified that $\mathbf{x}_c(v, w)$ is a zero error center manifold of the closed-loop system (6.1) and (6.14). The proof is thus completed by using part (i) of Lemma 5.3. □

Remark 6.10. Corollary 6.9 concludes that if a controller solves the stabilization problem for system (6.10), then this controller together with the internal model solves the output regulation problem for the original system (6.1). Thus the robust output regulation problem for (6.1) is converted into a robust stabilization problem for (6.10). In particular, the controller (6.14) can take the output feedback form when $d = 0$ and the function γ is independent of x, or the full state feedback form when $d = n$, or the partial state feedback form when $0 < d < n$. The number d depends on how many components of the state x can be reproduced by the steady-state generator and /or how many components of the state x are needed for feedback. ■

Remark 6.11. Having established Proposition 6.8 and Corollary 6.9, it is possible to further elaborate the concepts of the steady-state generator and the internal model as well as their relationship. For convenience, first assume $g_o(x, u) = u$. Then, from Proposition 6.8 and Corollary 6.9, it can be seen that, in order to convert a robust output regulation problem for the composite system (6.1) into a robust stabilization problem for the augmented system (6.8), the dynamic compensator as defined by (6.6) should have two properties:

(i) The augmented system (6.8) together with the exosystem (5.2) has an output zeroing manifold $\{(x, \eta, v) \mid x = \mathbf{x}(v, w), \ \ \eta = \theta(v, w)\}$ rendered by the feedback control $u = \mathbf{u}(v, w)$.

(ii) The output zeroing manifold of the augmented system (6.8) and the exosystem (5.2) can be made attractive by a feedback control independent of x, v, and w.

Once the coordinate and input transformation (6.9) are introduced, then the second property can be translated into saying that the equilibrium point of the augmented system (6.10) with $d = 0$ can be stabilized by an output feedback controller.

In other words, the first property of the dynamic compensator (6.6) guarantees that the robust output regulation problem for the original system (6.1) can be converted into a robust stabilization problem of the equilibrium point of the augmented system (6.10), and the second property of the dynamic compensator (6.6) guarantees that the equilibrium point of the augmented system (6.10) is stabilizable by an output feedback controller.

Clearly, the internal model as defined in Definition 6.6 renders the dynamic compensator (6.6) the first property explicitly. However, Definition 6.6 does not say anything about the second property of the dynamic compensator (6.6). The reason is that there is no uniform concept of the stabilizability for nonlinear systems due to the varieties of the stability concepts and the complexity of nonlinear systems. The construction of the internal model (6.6) depends not only on the systems under consideration, but also on the specific stability requirements on the augmented system (6.10). Nevertheless, the generality of Definition 6.6 has offered the functional flexibility for constructing an internal model with the second property for a given class of nonlinear systems with a specific stability requirement. When it comes to local asymptotic stabilizability of the equilibrium point of the augmented system, it is possible to synthesize a generic internal model having this property. This model will be shown in Section 6.3. In Chapter 7, we will further address the global robust output regulation problem. In this case, the stabilizability of nonlinear systems is intractable in general. We have to address this issue on a case by case basis when the specific form of the nonlinear systems is available.

It is quite clear that the steady-state generator itself can be viewed as a dynamic compensator of the form (6.6) with property (i); that is, it can be viewed as an internal model. However, the steady-state generator can never have property (ii) since the dynamics $\dot{\tau} = \alpha(\tau)$ of the steady-state generator is not coupled with the plant (6.1), and the equilibrium point of $\dot{\eta} = \alpha(\eta)$ at the origin is not asymptotically stable, as will be shown in the next section. Thus a more general characterization of the internal model has to be introduced in Definition 6.6.

The above description also applies to the case where $g_o(x, u)$ also depends on the state x or part of the components of the state x. In this case, the state x or part of the components of the state x is assumed to be available for feedback control. ∎

6.2 Existence of the Steady-State Generator and the Internal Model

Let us first show that the steady-state generator exists when the solution of the regulator equations satisfies certain differential equations.

Proposition 6.12. *Assume the exosystem satisfies Assumption 5.5, and let $\pi : V \times W \to \mathcal{R}$ be a sufficiently smooth function vanishing at the origin. Then $\pi(v, w)$ has a generator with linear observability if there exists some set of r real numbers $a_1, a_2, \ldots, a_r$ such that*

$$\frac{d^r \pi(v(t), w)}{dt^r} - a_1 \pi(v(t), w) - a_2 \frac{d\pi(v(t), w)}{dt} - \cdots - a_r \frac{d^{(r-1)} \pi(v(t), w)}{dt^{(r-1)}} = 0 \tag{6.15}$$

for all trajectories $v(t) \in V$ of the exosystem and all $w \in W$.

Proof. Let T be any nonsingular matrix of dimension r,

$$\theta(v,w) = T \begin{bmatrix} \pi(v,w) \\ \dot{\pi}(v,w) \\ \vdots \\ \dfrac{d^{(r-1)}\pi(v,w)}{dt^{(r-1)}} \end{bmatrix},$$
$$\alpha(\theta) = T\Phi T^{-1}\theta,$$
$$\beta(\theta) = \Psi T^{-1}\theta, \tag{6.16}$$

where

$$\Phi = \begin{bmatrix} 0 & 1 & 0 & \cdots & 0 \\ 0 & 0 & 1 & \cdots & 0 \\ \vdots & \vdots & \vdots & \cdots & \vdots \\ 0 & 0 & 0 & \cdots & 1 \\ a_1 & a_2 & a_2 & \cdots & a_r \end{bmatrix}, \quad \Psi = \begin{bmatrix} 1 & 0 & \cdots & 0 \end{bmatrix}. \tag{6.17}$$

It can be readily verified that the triple defined by (6.16) is a generator of $\pi(v, w)$ with linear observability. □

Corollary 6.13. *Let $g_o : \mathcal{R}^{n+m} \to \mathcal{R}^l$ for some positive integer $1 \le l \le n+m$ be a sufficiently smooth function vanishing at the origin. Under Assumptions 5.1 and 5.5, for $i = 1, \ldots, l$, let $\pi_i(v,w) = g_{oi}(\mathbf{x}(v,w), \mathbf{u}(v,w))$. Then the system (6.1) has a steady-state generator with output $g_o(x,u)$ with linear observability if, for each $i = 1, \ldots, l$, there exist positive integers r_i and real numbers $a_{i,1}, \ldots, a_{i,r_i}$ such that*

$$\frac{d^{r_i}\pi_i(v(t),w)}{dt^{r_i}} - a_{i,1}\pi_i(v(t),w) - a_{i,2}\frac{d\pi_i(v(t),w)}{dt} - \cdots - a_{i,r_i}\frac{d^{(r_i-1)}\pi_i(v(t),w)}{dt^{(r_i-1)}} = 0. \tag{6.18}$$

Proof. For each i, let $\{\theta_i, \alpha_i, \beta_i\}$ be a generator of $\pi_i(v, w)$ with linear observability. Let

$$\theta(v,w) = \begin{bmatrix} \theta_1(v,w) \\ \theta_2(v,w) \\ \vdots \\ \theta_l(v,w) \end{bmatrix}, \quad \alpha(\theta) = \begin{bmatrix} \alpha_1(\theta_1) \\ \alpha_2(\theta_2) \\ \vdots \\ \alpha_l(\theta_l) \end{bmatrix}, \quad \beta(\theta) = \begin{bmatrix} \beta_1(\theta_1) \\ \beta_2(\theta_2) \\ \vdots \\ \beta_l(\theta_l) \end{bmatrix}. \tag{6.19}$$

Then it is possible to verify that the triple $\{\theta, \alpha, \beta\}$ is a steady-state generator with linear observability of the system (6.1) with output $g_o(x,u) = \text{col}(g_{o1}(x,u), \ldots, g_{ol}(x,u))$. □

Equation (6.15) is a linear differential equation. It is interesting to find the class of functions that satisfy this equation. For this purpose, without loss of generality, we assume that the dimension q of the matrix A_1 is an odd integer and A_1 takes the following form:

$$A_1 = \text{block diag}\{S_0, S_1, \ldots, S_k\}, \tag{6.20}$$

where $S_0 = 0$, and

$$S_i = \begin{bmatrix} 0 & \omega_i \\ -\omega_i & 0 \end{bmatrix}, \quad \omega_i > 0, \quad i = 1, \ldots, k. \tag{6.21}$$

Proposition 6.14. *Assume that the exosystem satisfies Assumption 5.5, and let $\pi : V \times W \to \mathcal{R}$ be an analytic function vanishing at the origin. Then the following are equivalent:*

(i) *There exists some set of r real numbers $a_1, a_2, \ldots, a_r$ such that*

$$\frac{d^r\pi(v(t), w)}{dt^r} - a_1\pi(v(t), w) - a_2\frac{d\pi(v(t), w)}{dt} - \cdots - a_r\frac{d^{(r-1)}\pi(v(t), w)}{dt^{(r-1)}} = 0 \tag{6.22}$$

for all trajectories $v(t) \in V$ of the exosystem and all $w \in W$.

(ii) *Let $\Omega = \{l_1\omega_1 + \cdots + l_k\omega_k \geq 0, l_1, \ldots, l_k = 0, \pm 1, \pm 2, \ldots, \}$. Then, there exist $0 = \hat{\omega}_0$ and $\hat{\omega}_1, \ldots, \hat{\omega}_{n_k} \in \Omega$ for some finite integer n_k such that*

$$\pi(v(t), w) = \sum_{l=-n_k}^{n_k} C_l(w, v_0)e^{j\hat{\omega}_l t}, \tag{6.23}$$

where $j = \sqrt{-1}$, and, for $l \geq 0$, $\hat{\omega}_l = -\hat{\omega}_{-l}$, and $C_l^* = C_{-l}$, where C_l^* is the conjugate complex of C_l.

(iii) *There exist some integer n_i and real numbers $\psi_l(w, v_0)$, $l = 1, \ldots, n_i$, such that*

$$\pi(v(t), w) = \sum_{l=1}^{n_i} \psi_l(w, v_0)v^{[l]}(t). \tag{6.24}$$

Proof. For $i = 1, \ldots, k$, let

$$V_i = \sqrt{v_{2i}^2(0) + v_{2i+1}^2(0)}e^{j\tan^{-1}\left(\frac{v_{2i}(0)}{v_{2i+1}(0)}\right)}.$$

Then the solution of the exosystem $\dot{v} = A_1 v$ is $v_1(t) = v_1(0)$, and for $i = 1, \ldots, k$,

$$v_{2i}(t) = \frac{V_i e^{j\omega_i t} - V_i^* e^{-j\omega_i t}}{2j}, \quad v_{2i+1}(t) = \frac{V_i e^{j\omega_i t} + V_i^* e^{-j\omega_i t}}{2}. \tag{6.25}$$

(i) $\longrightarrow$ (ii). Since $\pi(v, w)$ is an analytic function of v, we can expand $\pi(v, w)$ into a power series in v as follows:

$$\pi(v, w) = \sum_{l=1}^{\infty} \Pi_l(w) v^{[l]},$$

where $\Pi_l(w)$ is some real number depending on w and the notation $v^{[l]}$ is as defined in Chapter 4.

Substituting $v_1(t) = v_1(0)$ and (6.25) into the power series expansion of $\pi(v(t), w)$ gives

$$\pi(v(t), w) = \sum_{l=-\infty}^{\infty} C_l(w, v_0) e^{j\hat{\omega}_l t}, \tag{6.26}$$

where $\hat{\omega}_l \in \Omega$ and $C_l^* = C_{-l}$. Using (6.26) gives

$$\frac{d^i \pi(v(t), w)}{dt^i} = \sum_{l=-\infty}^{\infty} (j\hat{\omega}_l)^i C_l(w, v_0) e^{j\hat{\omega}_l t}, \quad i = 1, 2, \ldots, r. \tag{6.27}$$

Thus equation (6.22) implies

$$\begin{aligned} &\frac{d^r \pi(v(t), w)}{dt^r} - a_1 \pi(v(t), w) - a_2 \frac{d\pi(v(t), w)}{dt} - \cdots - a_r \frac{d^{(r-1)} \pi(v(t), w)}{dt^{(r-1)}} \\ &= \sum_{l=-\infty}^{\infty} p(j\hat{\omega}_l) C_l(w, v_0) e^{j\hat{\omega}_l t} \\ &= 0, \end{aligned}$$

where

$$p(\lambda) = \lambda^r - a_1 - a_2\lambda - \cdots - a_r \lambda^{r-1}. \tag{6.28}$$

Since $p(\lambda)$ can only have r roots, there must exist an integer n_k such that $C_l(w, v_0) = 0$ for all $|l| > n_k$. Thus $\pi(v(t), w)$ must take the form (6.23).

(ii) $\longrightarrow$ (i). Let $r = 2n_k + 1$ and $a_1, a_2, \ldots, a_r$ be such that

$$\lambda \Pi_{l=1}^{n_k} (\lambda^2 + \hat{\omega}_l^2) = \lambda^r - a_1 - a_2\lambda - \cdots - a_r \lambda^{r-1}. \tag{6.29}$$

Then $\pi(v(t), w)$ as given by (6.23) satisfies

$$\begin{aligned} &\frac{d^r \pi(v(t), w)}{dt^r} - a_1 \pi(v(t), w) - a_2 \frac{d\pi(v(t), w)}{dt} - \cdots - a_r \frac{d^{(r-1)} \pi(v(t), w)}{dt^{(r-1)}} \\ &= \sum_{l=-n_k}^{n_k} ((j\hat{\omega}_l)^r - a_1 - a_2(j\hat{\omega}_l) - \cdots - a_r (j\hat{\omega}_l)^{r-1}) C_l(w, v_0) e^{j\hat{\omega}_l t}, \end{aligned} \tag{6.30}$$

which shows that $\pi(v(t), w)$ satisfies (6.22) upon using (6.29).

(ii) $\longrightarrow$ (iii). Note that

$$v_{2i+1}(t) + jv_{2i}(t) = V_i e^{j\omega_i t},$$
$$v_{2i+1}(t) - jv_{2i}(t) = V_i^* e^{-j\omega_i t}.$$

Also note that there exist integers $l_1, \ldots, l_k$ such that $\hat{\omega}_l = l_1\omega_1 + \cdots + l_k\omega_k$. For convenience, assume the integers $l_1, \ldots, l_k$ are nonnegative. Then,

$$\begin{aligned}
&C_l(w, v_0)e^{j\hat{\omega}_l t} + C_l^*(w, v_0)e^{-j\hat{\omega}_l t} \\
&\quad = C_l(w, v_0)e^{jl_1\omega_1 t} \cdots e^{jl_k\omega_k t} + C_l^*(w, v_0)e^{-jl_1\omega_1 t} \cdots e^{-jl_k\omega_k t} \\
&\quad = C_l(w, v_0)(e^{j\omega_1 t})^{l_1} \cdots (e^{j\omega_k t})^{l_k} + C_l^*(w, v_0)(e^{-j\omega_1 t})^{l_1} \cdots (e^{-j\omega_k t})^{l_k} \\
&\quad = C_l(w, v_0)(V_1)^{-l_1}(v_3(t) + jv_2(t))^{l_1} \cdots (V_k)^{-l_k}(v_{2k+1}(t) + jv_{2k}(t))^{l_k} \\
&\qquad + C_l^*(w, v_0)(V_1^*)^{-l_1}(v_3(t) - jv_2(t))^{l_1} \cdots (V_k^*)^{-l_k}(v_{2k+1}(t) - jv_{2k}(t))^{l_k}.
\end{aligned}$$

Thus, $\pi(v(t), w)$ must be a polynomial in $v(t)$ with real coefficients depending on both the uncertain parameter w and the initial state v_0. Clearly, the above derivation can be slightly modified to suit the case where some of the integers l_i, $1 \le i \le k$, are negative.

(iii) $\longrightarrow$ (ii). It follows straightforwardly from (6.24) and (6.25). □

Remark 6.15.

(i) Let $\pi(v, w)$ be any sufficiently smooth function in v and w. We call a monic polynomial $P(\lambda) = \lambda^r - a_1 - a_2\lambda - \cdots - a_r\lambda^{r-1}$ a *zeroing polynomial* of $\pi(v, w)$ if, along all trajectories $v(t)$ of the exosystem $\dot{v} = A_1 v$, $\pi(v(t), w)$ satisfies a differential equation of the form (6.22). By Proposition 6.14, if $\pi(v, w)$ has a zeroing polynomial $P(\lambda)$, then $\pi(v(t), w)$ must be a polynomial in the trajectory $v(t)$ of the exosystem or a trigonometric polynomial of the form (6.23). But $\pi(v, w)$ itself does not have to be a polynomial in v. In fact, consider a function of the form $\delta(v_2^2 + v_3^2)\pi(v, w)$, where $\pi(v, w)$ satisfies (6.22) and $\delta(\cdot)$ is any sufficiently smooth scalar function. Since $v_2^2(t) + v_3^2(t)$ is actually a constant equal to $v_2^2(0) + v_3^2(0)$, $\delta(v_2^2(t) + v_3^2(t))\pi(v(t), w) = \delta(v_2^2(0) + v_3^2(0))\pi(v(t), w)$. Clearly, $\delta(v_2^2 + v_3^2)\pi(v, w)$ also satisfies (6.22).

(ii) If $\pi(v, w)$ is a degree k polynomial in v, we can write

$$\pi(v, w) = \sum_{l=1}^{k} \Pi_l(w)v^{[l]}.$$

Let $P(\lambda) = \lambda^r - a_1 - a_2\lambda - \cdots - a_r\lambda^{r-1}$ be the minimal polynomial of the matrix $\mathcal{A}_{kf}$. Then it follows from the Cayley–Hamilton theorem and equation (5.45) that $P(\lambda)$ is a zeroing polynomial of $\pi(v, w)$.

(iii) A monic polynomial $P(\lambda)$ is called a *minimal zeroing polynomial* of $\pi(v, w)$ if $P(\lambda)$ is a zeroing polynomial of $\pi(v, w)$ of least degree. Now assume that $\pi(v(t), w)$ takes the form (6.23) and $C_l \neq 0$, $l = 0, 1, 2, \ldots, n_k$; then, clearly, $P(\lambda) = \lambda\Pi_{l=1}^{n_k}(\lambda^2 + \hat{\omega}_l^2)$ is the minimal zeroing polynomial of $\pi(v, w)$. It is noted that, if $P(\lambda)$ is the minimal zeroing polynomial of $\pi(v, w)$, then all the zeros of $P(\lambda)$ are simple and pure imaginary. This property will be useful later when the stabilizability of the augmented system (6.10) is considered. ■

By Propositions 6.12 and 6.14, the existence of the steady-state generator of the form (6.16) requires that the solution of the regulator equations be a polynomial function of $v(t)$ which is still quite restrictive as it essentially requires that the nonlinear systems contain only polynomial nonlinearity. We now propose a more general steady-state generator as follows.

Definition 6.16. *Let $\pi_i(v(t), w)$, $i = 1, \ldots, I$, for some positive integer I, be I trigonometric polynomials of t or polynomials in $v(t)$. They are called* pairwise coprime *if their minimal zeroing polynomials $P_1(\lambda), \ldots, P_I(\lambda)$ are pairwise coprime.*

Lemma 6.17. *Let $g_o : \mathcal{R}^{n+m} \to \mathcal{R}^{d+m}$ for some integer $0 \leq d \leq n$ be a sufficiently smooth function vanishing at the origin. Under Assumptions 5.1 and 5.5, assume, for $i = 1, \ldots, d+m$, that there exist pairwise coprime polynomials $\pi_i^1(v, w), \ldots, \pi_i^{I_i}(v, w)$, with $r_i^1, \ldots, r_i^{I_i}$ being the degrees of their minimal zeroing polynomials $P_i^1(\lambda), \ldots, P_i^{I_i}(\lambda)$ and a sufficiently smooth function $\Gamma_i : \mathcal{R}^{r_i^1+\cdots+r_i^{I_i}} \to \mathcal{R}$ vanishing at the origin such that, for all trajectories $v(t) \in V$ of the exosystem, and $w \in W$,*

$$g_{oi}(\mathbf{x}(v, w), \mathbf{u}(v, w)) = \Gamma_i\Big(\pi_i^1(v, w), \dot{\pi}_i^1(v, w), \ldots, \frac{d^{(r_i^1-1)}\pi_i^1(v, w)}{dt^{(r_i^1-1)}}, \ldots, \pi_i^{I_i}(v, w), \dot{\pi}_i^{I_i}(v, w), \ldots, \frac{d^{(r_i^{I_i}-1)}\pi_i^{I_i}(v, w)}{dt^{(r_i^{I_i}-1)}}\Big). \quad (6.31)$$

Then

(i) *For $i = 1, \ldots, d+m$, $j = 1, \ldots, I_i$, let*

$$\theta_i^j(v, w) = \begin{bmatrix} \pi_i^j(v, w) \\ \dot{\pi}_i^j(v, w) \\ \vdots \\ \dfrac{d^{(r_i^j-1)}\pi_i^j(v, w)}{dt^{(r_i^j-1)}} \end{bmatrix}, \quad \theta_i(v, w) = T_i \begin{bmatrix} \theta_i^1(v, w) \\ \vdots \\ \theta_i^{I_i}(v, w) \end{bmatrix},$$

with T_i being any nonsingular matrix of dimension $r_i^1 + \cdots + r_i^{I_i}$, Φ_i^j the companion matrix of $P_i^j(\lambda)$ satisfying $\dot{\theta}_i^j(v, w) = \Phi_i^j \theta_i^j(v, w)$, $\alpha_i(\theta_i) = T_i \Phi_i T_i^{-1} \theta_i$ with $\Phi_i =$ block diag$(\Phi_i^1, \ldots, \Phi_i^{I_i})$, and $\beta_i(\theta_i) = \Gamma_i(T_i^{-1}\theta_i)$. Then the system (6.1) has a steady-state generator $\{\theta, \alpha, \beta\}$ with output $g_o(x, u) = \text{col}(g_{o1}(x, u), \ldots, g_{o(d+m)}(x, u))$ as follows:

$$\theta(v, w) = \begin{bmatrix} \theta_1(v, w) \\ \vdots \\ \theta_{d+m}(v, w) \end{bmatrix}, \quad \alpha(\theta) = \begin{bmatrix} \alpha_1(\theta_1) \\ \vdots \\ \alpha_{d+m}(\theta_{d+m}) \end{bmatrix} = T\Phi T^{-1}\theta,$$

$$\beta(\theta) = \begin{bmatrix} \beta_1(\theta_1) \\ \vdots \\ \beta_{d+m}(\theta_{d+m}) \end{bmatrix}, \quad (6.32)$$

where $\Phi =$ block diag$(\Phi_1, \ldots, \Phi_{d+m})$ and $T =$ block diag$(T_1, \ldots, T_{d+m})$.

(ii) *For* $i = 1, \dots, d+m$, *let* $\Psi_i = [\Psi_i^1, \dots, \Psi_i^{I_i}]$ *be the Jacobian of* Γ_i *at the origin, where* $\Psi_i^j \in \mathcal{R}^{1\times r_i^j}$. *Then the pair* (Ψ_i, Φ_i) *is observable, hence the generator (6.32) is linearly observable if*

$$\text{the pair } (\Psi_i^j, \Phi_i^j) \text{ is observable, } i = 1, \dots, d+m, \quad j = 1, \dots, I_i. \tag{6.33}$$

Proof. (i) The triple $(\theta_i, \alpha_i, \beta_i)$ is clearly a steady-state generator of (6.1) with output $g_{oi}(x, u)$. Thus, the triple defined in (6.32) satisfies $\dot{\theta} = T\Phi T^{-1}\theta = \alpha(\theta)$ and $\beta(\theta(v, w)) = g_o(\mathbf{x}(v, w), \mathbf{u}(v, w))$; that is, it is a steady-state generator of (6.1) with output $g_o(x, u)$.

(ii) To verify the observability of (Ψ_i, Φ_i), it suffices to show, by the PBH test, that, for any λ,

$$\begin{aligned}
&\operatorname{rank}\begin{bmatrix} \lambda I - \Phi_i \\ \Psi_i \end{bmatrix} \\
&= \operatorname{rank}\begin{bmatrix} \lambda I - \Phi_i^1 & 0 & \cdots & 0 \\ 0 & \lambda I - \Phi_i^2 & \cdots & 0 \\ \vdots & \vdots & \vdots & \vdots \\ 0 & 0 & 0 & \lambda I - \Phi_i^{I_i} \\ \Psi_i^1 & \Psi_i^2 & \cdots & \Psi_i^{I_i} \end{bmatrix} \\
&= r_i^1 + \cdots + r_i^{I_i}.
\end{aligned} \tag{6.34}$$

It is clear that (6.34) holds for any $\lambda \notin \sigma(\Phi_i)$. For any $\lambda \in \sigma(\Phi_i)$, there exists $1 \leq k \leq I_i$ such that $\lambda \in \sigma(\Phi_i^k)$ and $\lambda \notin \sigma(\Phi_i^j)$, $j \neq k$ and $1 \leq j \leq I_i$, since, for any $j \neq k$, $P_i^j(\lambda)$ and $P_i^k(\lambda)$ are coprime. Thus

$$\begin{aligned}
\operatorname{rank}\begin{bmatrix} \lambda I - \Phi_i \\ \Psi_i \end{bmatrix} &= \operatorname{rank}\begin{bmatrix} \lambda I - \Phi_i^k \\ \Psi_i^k \end{bmatrix} \\
&+ \sum_{\substack{j=1 \\ j\neq k}}^{I_i} \operatorname{rank}\left[\lambda I - \Phi_j^i\right] = r_i^1 + \cdots + r_i^{I_i},
\end{aligned}$$

since (Ψ_i^k, Φ_i^k) is observable. Moreover, the linearization of $\alpha(\theta)$ and $\beta(\theta)$ at the origin is $T\Phi T^{-1}$ and ΨT^{-1} with $\Psi = \operatorname{diag}(\Psi_1, \dots, \Psi_{d+m})$. As a result, the generator (6.32) is linearly observable. The proof is thus completed. □

Remark 6.18. Denote the eigenvalues of A_1 by $\lambda_1, \dots, \lambda_q$. Then, by Proposition 6.14, the collection of the zeros of all $P_i^j(\lambda)$ or, what is the same, the eigenvalues of Φ take the form

$$l_1\lambda_1 + \cdots + l_q\lambda_q, \quad l_1 + \cdots + l_q = 1, 2, \dots, \quad l_1, \dots, l_q = 0, 1, \dots. \tag{6.35}$$

Thus, all the eigenvalues of the matrix Φ are semisimple with zero real parts. ■

Remark 6.19. From the proof of part (ii) of Lemma 6.17, it can be seen that it is necessary to require that, for each i, the zeroing polynomials $P_i^j(\lambda)$, $j = 1, \dots, I_i$, be pairwise coprime to guarantee the observability of the pair (Ψ_i, Φ_i). But, for $i \neq k$, the polynomials $P_i^j(\lambda)$,

$P_k^l(\lambda)$, $j = 1, \ldots, I_i$, $l = 1, \ldots, I_k$, do not have to be different. As we will see in Section 6.4, for some class of systems, the functions Γ_i and Γ_k, $i \neq k$, may rely on the same set of polynomials. In this case, one can synthesize a reduced-order steady-state generator. ∎

Remark 6.20. Suppose a function $\pi(v, w)$ satisfies equation (6.15), with r as the degree of its minimal zeroing polynomial. Then defining $\Gamma : \mathcal{R}^r \to \mathcal{R}$ as a linear function such that, for all $x \in \mathcal{R}^r$, $\Gamma(x) = [1, 0, \ldots, 0]x$ shows that

$$\Gamma\left(\pi(v, w), \dot{\pi}(v, w), \ldots, \frac{d^{(r-1)}\pi(v, w)}{dt^{r-1}}\right) = \pi(v, w).$$

Thus, the class of functions satisfying (6.31) includes the class of polynomial or trigonometric polynomial functions. Moreover, this class of functions is much larger than the class of the polynomial or trigonometric polynomial functions; for example, the regulator equations of the system in Example 6.25 to be introduced later admit the following solution:

$$\mathbf{u}(v, w) = \sin\left(\frac{-(1 + w)}{1 + \omega_1^2}(v_1 + \omega_1 v_2)\right) + (1 - \omega_2^2)v_3.$$

Assume $\omega_2 \neq 1$. Let $\pi_1(v, w) = \frac{-(1+w)}{1+\omega_1^2}(v_1 + \omega_1 v_2)$, and $\pi_2(v, w) = (1 - \omega_2^2)v_3$. Then $P_1(\lambda) = (\lambda^2 + \omega_1^2)$ and $P_2(\lambda) = (\lambda^2 + \omega_2^2)$ are the minimal zeroing polynomials of $\pi_1(v, w)$ and $\pi_2(v, w)$, respectively. Thus, defining

$$\Gamma\left(\pi_1(v, w), \dot{\pi}_1(v, w), \pi_2(v, w), \dot{\pi}_2(v, w)\right) = \sin(\pi_1(v, w)) + \pi_2(v, w)$$

gives

$$\mathbf{u}(v, w) = \Gamma\left(\pi_1(v, w), \dot{\pi}_1(v, w), \pi_2(v, w), \dot{\pi}_2(v, w)\right).$$

Thus, the system has a steady-state generator with output $g_o(x_1, x_2, u) = u$. Moreover, it is easy to see that

$$\Phi = \begin{bmatrix} 0 & 1 & 0 & 0 \\ -\omega_1^2 & 0 & 0 & 0 \\ 0 & 0 & 0 & 1 \\ 0 & 0 & -\omega_2^2 & 0 \end{bmatrix}$$

and $\Psi = [1, 0, 1, 0]$. Thus this steady-state generator is linearly observable. Note that the selection of $\pi(v, w)$ and Γ is not unique. Taking the same example as above with $\omega_2 \neq 1$ and letting $\pi_1(v, w) = \frac{-(1+w)}{1+\omega_1^2}(v_1 + \omega_1 v_2) + (1 - \omega_2^2)v_3$ shows that $P_1(\lambda) = (\lambda^2 + \omega_1^2)(\lambda^2 + \omega_2^2)$ and

$$\begin{aligned} \mathbf{u}(v, w) &= \sin\left(\frac{\omega_2^2\pi_1(v, w) + \ddot{\pi}_1(v, w)}{\omega_2^2 - \omega_1^2}\right) + \frac{\omega_1^2\pi_1(v, w) + \ddot{\pi}_1(v, w)}{\omega_1^2 - \omega_2^2} \\ &\stackrel{def}{=} \Gamma\left(\pi_1(v, w), \dot{\pi}_1(v, w), \ddot{\pi}_1(v, w), \frac{d^3\pi_1(v, w)}{dt^3}\right), \end{aligned}$$

which in turn gives

$$\Phi = \begin{bmatrix} 0 & 1 & 0 & 0 \\ 0 & 0 & 1 & 0 \\ 0 & 0 & 0 & 1 \\ -\omega_1^2\omega_2^2 & 0 & -\omega_1^2 - \omega_2^2 & 0 \end{bmatrix}$$

and $\Psi = [1, 0, 0, 0]$. Thus, the function $\mathbf{u}(v, w)$ has another generator with linear observability. ∎

Corresponding to a steady-state generator of the form (6.32), we can construct a nonlinear internal model as follows. Let $M_i \in \mathcal{R}^{(r_i^1+\cdots+r_i^{l_i})\times(r_i^1+\cdots+r_i^{l_i})}$ and $N_i \in \mathcal{R}^{(r_i^1+\cdots+r_i^{l_i})\times 1}$, $i = 1, \ldots, d+m$, be such that M_i is Hurwitz, and (M_i, N_i) is controllable. Since, for each $i = 1, \ldots, d+m$, the spectra of the matrices Φ_i and M_i are disjoint and the pair (Ψ_i, Φ_i) is observable, there exists a unique, nonsingular matrix $T_i \in \mathcal{R}^{(r_i^1+\cdots+r_i^{l_i})\times(r_i^1+\cdots+r_i^{l_i})}$ that satisfies the Sylvester equation (Appendix A)

$$T_i\Phi_i - M_iT_i = N_i\Psi_i.$$

Let

$$\begin{aligned} M &= \text{block diag}(M_1, \ldots, M_{d+m}), \\ N &= \text{block diag}(N_1, \ldots, N_{d+m}), \\ T &= \text{block diag}(T_1, \ldots, T_{d+m}). \end{aligned}$$

Then we have the following result.

Proposition 6.21. *Under the same assumptions as those of Lemma 6.17, for $i = 1, \ldots, d+m$, the following dynamic system:*

$$\dot{\eta} = \gamma(\eta, x, u, e) = M\eta + N(g_o(x, u) - \beta(\eta) + \Psi T^{-1}\eta) \tag{6.36}$$

is an internal model of the system (6.1) with output $g_o(x, u)$.

Proof. Let $\gamma_i(\eta_i, x, u, e) = M_i\eta_i + N_i(g_{oi}(x, u) - \beta_i(\eta_i) + \Psi_iT_i^{-1}\eta_i)$. Then

$$\begin{aligned} &\gamma_i\,(\theta_i(v, w), \mathbf{x}(v, w), \mathbf{u}(v, w), 0) \\ &\quad = M_i\theta_i(v, w) + N_i(g_{oi}(\mathbf{x}(v, w), \mathbf{u}(v, w)) - \beta_i(\theta_i(v, w)) + \Psi_iT_i^{-1}\theta_i(v, w)) \\ &\quad = M_i\theta_i(v, w) + N_i\Psi_iT_i^{-1}\theta_i(v, w) \\ &\quad = T_i\Phi_iT_i^{-1}\theta_i(v, w) = \alpha_i(\theta_i(v, w)), \quad i = 1, \ldots, d+m. \end{aligned} \tag{6.37}$$

Putting these equations together with $\eta = \text{col}(\eta_1, \ldots, \eta_{d+m})$ gives

$$\gamma\,(\theta(v, w), \mathbf{x}(v, w), \mathbf{u}(v, w), 0) = \alpha(\theta(v, w)). \tag{6.38}$$

Thus (6.36) is an internal model of (6.1) with output $g_o(x, u)$, where $g_o = \text{col}(g_{o1}, \ldots, g_{o(d+m)})$. □

Remark 6.22. In the next section, we will give conditions for the composite system (6.1) under which the internal model given by (6.36) will render the augmented system (6.10) the local asymptotic stabilizability property by the output feedback. We also note that, in the special case where the solution of the regulator equations is polynomial in the solution $v(t)$ of the exosystem $\dot{v} = A_1 v$, the function $\beta(\theta)$ is linear with $\beta(\theta) = \Psi T^{-1}\theta$. Thus the internal model (6.36) reduces to a linear internal model of the form $\dot{\eta} = M\eta + Ng_o(x, u)$. ∎

6.3 Robust Output Regulation with the Nonlinear Internal Model

In this section, we will apply the framework described in Section 6.1 to establish the solvability of the robust output regulation problem without assuming that the solution of the regulator equations is a polynomial.

Theorem 6.23. *Consider composite system (6.1). Let Assumptions 5.1 to 5.3 and 5.5 hold and the conditions (6.31) and (6.33) be satisfied with $g_o(x, u) = u$. Further, assume*

$$\operatorname{rank} \begin{bmatrix} A - \lambda I & B \\ C & D \end{bmatrix} = n + m \tag{6.39}$$

for all λ such that $P_i^j(\lambda) = 0$ for some $i = 1, \ldots, m$, and some $j = 1, \ldots, l_i$.

Then, the robust output regulation problem is solvable by an output feedback control law.

Proof. Under the assumptions of this theorem, the system (6.1) has a linearly observable steady-state generator of the form (6.32) with output $g_o(x, u) = u$. Corresponding to this steady-state generator, define the internal model as given by (6.36) with output $g_o(x, u) = u$, and a transformation of the form (6.9) with $d = 0$. This transformation converts the augmented system (6.8) into the form (6.10), where

$$\begin{aligned}
\bar{f}(\bar{x}, \bar{\eta}, \bar{u}, v, w) &= f(x, u, v, w) - f(\mathbf{x}, \mathbf{u}, v, w) \\
&= f(\bar{x} + \mathbf{x}, \bar{u} + \beta(\eta), v, w) - f(\mathbf{x}, \mathbf{u}, v, w) \\
&= f(\bar{x} + \mathbf{x}, \bar{u} + \beta(\bar{\eta} + \theta), v, w) - f(\mathbf{x}, \mathbf{u}, v, w), \\
\bar{\gamma}(\bar{x}, \bar{\eta}, \bar{u}, v, w) &= M\eta + N(u - \beta(\eta) + \Psi T^{-1}\eta) - T\Phi T^{-1}\theta \\
&= M\eta + N\Psi T^{-1}\eta - T\Phi T^{-1}\theta + N\bar{u} \\
&= \left(M + N\Psi T^{-1}\right)(\bar{\eta} + \theta) - T\Phi T^{-1}\theta + N\bar{u} \\
&= \left(M + N\Psi T^{-1}\right)\bar{\eta} + N\bar{u}, \\
\bar{h}(\bar{x}, \bar{\eta}, \bar{u}, v, w) &= h(x, u, v, w) \\
&= h(\bar{x} + \mathbf{x}, \bar{u} + \beta(\bar{\eta} + \theta), v, w).
\end{aligned}$$

By Corollary 6.9, it suffices to (locally) stabilize the equilibrium point at the origin of (6.10) with $v = 0$ and $w = 0$. To this end, linearizing (6.10) at $\bar{x} = 0$, $\bar{\eta} = 0$, $\bar{u} = 0$ with v and w

being set to zero gives

$$
\begin{aligned}
\dot{\bar{x}} &= A\bar{x} + B\bar{u} + B\Psi T^{-1}\bar{\eta}, \\
\dot{\bar{\eta}} &= (M + N\Psi T^{-1})\bar{\eta} + N\bar{u}, \\
e_l &= C\bar{x} + D\bar{u} + D\Psi T^{-1}\bar{\eta}.
\end{aligned}
\tag{6.40}
$$

Consider the decomposition

$$
\begin{aligned}
&\begin{bmatrix} A - \lambda I & B\Psi T^{-1} & B \\ 0 & M + N\Psi T^{-1} - \lambda I & N \end{bmatrix} \\
&\quad = \begin{bmatrix} A - \lambda I & 0 & B \\ 0 & M - \lambda I & N \end{bmatrix} \begin{bmatrix} I & 0 & 0 \\ 0 & I & 0 \\ 0 & \Psi T^{-1} & I \end{bmatrix}.
\end{aligned}
$$

From Assumption 5.2 as well as the fact that M is Hurwitz, we conclude that (6.40) is stabilizable using the PBH test. To show that (6.40) is detectable, first note that $M + N\Psi T^{-1} = T\Phi T^{-1}$ and all the eigenvalues of Φ have zero real parts. Thus, under Assumption 5.3, the following matrix:

$$
\begin{bmatrix} A - \lambda I & B\Psi T^{-1} \\ 0 & M + N\Psi T^{-1} - \lambda I \\ C & D\Psi T^{-1} \end{bmatrix}
$$

has full rank for all $\lambda \notin \sigma(\Phi)$ and $\mathrm{Re}\{\lambda\} \geq 0$. Next, using the decomposition

$$
\begin{bmatrix} A - \lambda I & B\Psi T^{-1} \\ 0 & M + N\Psi T^{-1} - \lambda I \\ C & D\Psi T^{-1} \end{bmatrix} = \begin{bmatrix} A - \lambda I & 0 & B \\ 0 & M - \lambda I & N \\ C & 0 & D \end{bmatrix} \begin{bmatrix} I & 0 \\ 0 & I \\ 0 & \Psi T^{-1} \end{bmatrix}
$$

and condition (6.39), we conclude that the matrix also has full rank for all $\lambda \in \sigma(\Phi)$. The detectability of (6.40) then follows from the PBH test. As a result, let K and L be such that

$$
\begin{bmatrix} A & B\Psi T^{-1} \\ 0 & M + N\Psi T^{-1} \end{bmatrix} + \begin{bmatrix} B \\ N \end{bmatrix} K
$$

and

$$
\begin{bmatrix} A & B\Psi T^{-1} \\ 0 & M + N\Psi T^{-1} \end{bmatrix} - L \begin{bmatrix} C & D\Psi T^{-1} \end{bmatrix}
$$

are Hurwitz. Then, system (6.40) can be stabilized by a linear feedback control law as follows:

$$
\begin{aligned}
\bar{u} &= K\xi, \\
\dot{\xi} &= \begin{bmatrix} A & B\Psi T^{-1} \\ 0 & M + N\Psi T^{-1} \end{bmatrix} \xi + \begin{bmatrix} B \\ N \end{bmatrix} \bar{u} \\
&\quad + L(e_l - C\xi_1 - D\bar{u} - D\Psi T^{-1}\xi_2),
\end{aligned}
\tag{6.41}
$$

where $\xi = \mathrm{col}(\xi_1, \xi_2)$ with $\xi_1 \in \mathcal{R}^n$ and $\xi_2 \in \mathcal{R}^s$. Note that the variable e_l in the control law (6.41) is not the true error output of the original plant and may not be measurable. Nevertheless, replacing e_l in (6.41) by the true error output of the plant $e = h(\bar{x} + \mathbf{x}, \bar{u} + \beta(\bar{\eta} + \theta), v, w)$ gives an output feedback control law as follows:

$$\begin{aligned} \bar{u} &= K\xi, \\ \dot{\xi} &= \begin{bmatrix} A & B\Psi T^{-1} \\ 0 & M + N\Psi T^{-1} \end{bmatrix} \xi + \begin{bmatrix} B \\ N \end{bmatrix} \bar{u} \\ &\quad + L(h(\bar{x} + \mathbf{x}, \bar{u} + \beta(\bar{\eta} + \theta), v, w) - C\xi_1 - D\bar{u} - D\Psi T^{-1}\xi_2). \end{aligned} \tag{6.42}$$

Clearly, the linear approximation of the closed-loop system composed of the composite system (6.1) and this control law at the origin is the same as that of the closed-loop system composed of the composite system (6.1) and the control law (6.41). Thus (6.42) also solves the robust output regulation problem of the composite system (6.1).

The control law (6.42) can be written as follows:

$$\begin{aligned} \bar{u} &= K\xi, \\ \dot{\xi} &= Q\xi + Le, \end{aligned} \tag{6.43}$$

where

$$Q = \begin{bmatrix} A & B\Psi T^{-1} \\ 0 & M + N\Psi T^{-1} \end{bmatrix} + \left(\begin{bmatrix} B \\ N \end{bmatrix} - LD \right) K - L \begin{bmatrix} C & D\Psi T^{-1} \end{bmatrix}.$$

Finally, using (6.14) with $d = 0$ shows that the following output feedback control law:

$$\begin{aligned} u &= \beta(\eta) + K\xi, \\ \dot{\eta} &= M\eta + N(K\xi + \Psi T^{-1}\eta), \\ \dot{\xi} &= Q\xi + Le, \end{aligned} \tag{6.44}$$

solves the robust output regulation problem of the original system (6.1). □

Remark 6.24. In the special case where the system (6.1) is linear, the solvability conditions of Theorem 6.23 are basically the same as those given in Theorem 1.31, and the controller also takes a linear form. However, the design method illustrated here is quite different from that described in Chapter 1. In particular, the dimension of the output feedback controller given in Chapter 1 is $n_q \times m + n$ (assuming $m = p$), where n_q is the degree of the minimal polynomial of A_1, but the dimension of the output feedback controller (6.44) is $2n_q \times m + n$. This difference is caused by the need to estimate the state $\mathrm{col}(\bar{x}, \bar{\eta})$ of the system (6.40). Next consider the nonlinear system (6.1) and assume the solution of the regulator equations of (6.1) is a degree k polynomial in v. In this case, the dimension of the output feedback controller given in Chapter 5 is $n_k \times m + n$ (assuming $m = p$), where n_k is the degree of the minimal polynomial of the matrix $\mathcal{A}_{kf}$, while the dimension of the output feedback controller (6.44) is $2 \times K + n$ (assuming $m = p$), where K is the dimension of the matrix Φ. K can be much smaller than n_k. For example, given some hypothetical nonlinear system with $m = 1$, $n_w = 1$, and $q = 2$, suppose, for some $k \geq 1$, that $\mathbf{u}(v, w) = \sum_{l_1+l_2=1}^{k} a_{l_1 l_2}(wv_1)^{l_1}(wv_2)^{l_2}$, where $\dot{v}_1 = v_2$, $\dot{v}_2 = -v_1$, and $a_{l_1 l_2}$ are known real scalars. Let $\pi(v, w) = wv_1$. Then $\dot{\pi}(v, w) = wv_2$. Defining $\Gamma(\pi(v, w), \dot{\pi}(v, w)) = \sum_{l_1+l_2=1}^{k} a_{l_1 l_2}(\pi(v, w))^{l_1}(\dot{\pi}(v, w))^{l_2}$ gives $\mathbf{u}(v, w) = \Gamma(\pi(v, w), \dot{\pi}(v, w))$. Thus, for this system $K = 2$ regardless of k, but $n_k = 2k + 1$ when $k > 1$. ∎

Example 6.25. Consider the following nonlinear system:

$$\begin{aligned}\dot{x}_1 &= x_1 + e\cos(x_3) + d,\\ \dot{x}_2 &= x_3,\\ \dot{x}_3 &= -x_2 - \sin((1 + w)x_1) + u,\\ e &= x_2 - y_d,\end{aligned} \tag{6.45}$$

where d and y_d are produced by

$$\begin{aligned}&\dot{v}_1 = \omega_1 v_2, \quad \dot{v}_2 = -\omega_1 v_1, \quad d = v_1,\\ &\dot{v}_3 = \omega_2 v_4, \quad \dot{v}_4 = -\omega_2 v_3, \quad y_d = v_3\end{aligned}$$

with $\omega_1 \neq \omega_2$. The robust output regulation problem of system (6.45) reflects the objective of asymptotic tracking of a sinusoidal reference input y_d and rejection of a sinusoidal disturbance d. The system is clearly nonminimum phase. Therefore, none of the inversion-based control approaches can handle this problem.

By inspection, the solution of the regulator equations is

$$\begin{aligned}\mathbf{x}_1(v, w) &= \frac{-1}{1 + \omega_1^2}(v_1 + \omega_1 v_2),\\ \mathbf{x}_2(v, w) &= v_3,\\ \mathbf{x}_3(v, w) &= \omega_2 v_4,\\ \mathbf{u}(v, w) &= \sin\left(\frac{-(1 + w)}{1 + \omega_1^2}(v_1 + \omega_1 v_2)\right) + (1 - \omega_2^2)v_3.\end{aligned}$$

Two different steady-state generators with linear observability with output $g_o(x, u) = u$ have been constructed in Remark 6.20. Here we will further construct an output feedback controller to solve the robust output regulation problem. To give a specific solution, we suppose $\omega_1 = 1$, $\omega_2 = 2$. As described in Remark 6.20, letting

$$\pi_1(v, w) = \frac{-(1 + w)}{2}(v_1 + v_2), \quad \pi_2(v, w) = -3v_3$$

gives

$$\begin{aligned}\mathbf{u}(v, w) &= \Gamma\left(\pi_1(v, w), \dot{\pi}_1(v, w), \pi_2(v, w), \dot{\pi}_2(v, w)\right)\\ &= \sin(\pi_1(v, w)) + \pi_2(v, w)\end{aligned}$$

and $\Psi = [1, 0, 1, 0]$. Thus, we can obtain a generator $\{\theta, \alpha(\theta), \beta(\theta)\}$ for $\mathbf{u}(v, w)$ with $\theta(v, w) = T\,[\pi_1, \dot{\pi}_1, \pi_2, \dot{\pi}_2]^T$, $\alpha(\theta) = T\Phi T^{-1}\theta$, and $\beta(\theta) = \Gamma(T^{-1}\theta)$ for any nonsingular matrix T and

$$\Phi = \text{block diag}\left([\Phi_1, \Phi_2]\right) = \text{block diag}\left(\begin{bmatrix} 0 & 1\\ -1 & 0\end{bmatrix}, \begin{bmatrix} 0 & 1\\ -4 & 0\end{bmatrix}\right).$$

To obtain an internal model, let

$$M = \text{block diag}\left(\begin{bmatrix} 0 & 1 \\ -3 & -2 \end{bmatrix}, \begin{bmatrix} 0 & 1 \\ -3 & -2 \end{bmatrix}\right), N = \begin{bmatrix} 0 \\ 1 \\ 0 \\ 1 \end{bmatrix}.$$

Solving the Sylvester equations $T_i\Phi_i - M_iT_i = N_i\Psi_i$, $i = 1, 2$, gives

$$\begin{aligned} T &= \text{block diag}\,(T_1, T_2) \\ &= \text{block diag}\left(\begin{bmatrix} 0.25 & -0.25 \\ 0.25 & 0.25 \end{bmatrix}, \begin{bmatrix} -0.0588 & -0.1176 \\ 0.4706 & -0.0588 \end{bmatrix}\right). \end{aligned}$$

The matrices that define the Jacobian linearization of the system (6.45) at the origin are

$$A = \begin{bmatrix} 1 & 1 & 0 \\ 0 & 0 & 1 \\ -1 & -1 & 0 \end{bmatrix}, \ B = \begin{bmatrix} 0 \\ 0 \\ 1 \end{bmatrix}, \ C = \begin{bmatrix} 0 & 1 & 0 \end{bmatrix}, \ D = 0.$$

It can be verified that the pair (A, B) is controllable, and (C, A) observable. Also, the system (A, B, C, D) has only one transmission zero, which is equal to 1, and thus does not coincide with the eigenvalues of Φ. Hence, it is possible to achieve the robust output regulation for this system by Theorem 6.23. ∎

6.4 Robust Asymptotic Disturbance Rejection of the RTAC System

In Chapter 3, we have formulated the disturbance rejection problem of the RTAC system as an output regulation problem and solved the problem with both the static state feedback and the dynamic measurement output feedback controllers. It is seen that while the controller can completely eliminate the effect of the sinusoidal disturbance on the output of the system asymptotically for the nominal case, its performance deteriorates when the system's parameter ϵ is perturbed. In this section, we will further apply the approach introduced in this chapter to design a robust output feedback controller for the asymptotic disturbance rejection of the RTAC system.

Let us write $\epsilon = \epsilon_0 + w$, where ϵ_0 is the nominal value of ϵ and w is the perturbation. Thus the regulator equations of the system can be written as follows:

$$\begin{aligned} \mathbf{x}_1(v, w) &= 0, \\ \mathbf{x}_2(v, w) &= 0, \\ \mathbf{x}_3(v, w) &= \arcsin\left(\frac{-v_1}{(\epsilon_0 + w)\omega^2}\right), \\ \mathbf{x}_4(v, w) &= \frac{-v_2}{(\epsilon_0 + w)\omega}\left(\frac{1}{\sqrt{1 - \left(\frac{-v_1}{(\epsilon_0+w)\omega^2}\right)^2}}\right), \\ \mathbf{u}(v, w) &= \mathbf{x}_4{}^2(v, w)\tan \mathbf{x}_3(v, w) + \frac{v_1}{(\epsilon_0 + w)\cos \mathbf{x}_3(v, w)}. \end{aligned}$$

We note that the solution of the regulation equations is not polynomial in v, and therefore the approach given in Chapter 5 cannot solve the robust output regulation problem of the RTAC system. Nevertheless, assuming that the displacement x_1 of the cart and the angular position x_3 of the proof-mass are measurable output variables, it is possible to design a measurement output feedback control law to solve the asymptotic disturbance rejection problem of the RTAC system in the presence of the variations of the parameter ϵ. Indeed, let $g_o(x,u) = \text{col}(x_1, x_3, u)$, $\pi(v,w) = \frac{v_1}{(\epsilon_0+w)}$, $\dot{\pi}(v,w) = \omega\frac{v_2}{(\epsilon_0+w)}$. Then

$$\mathbf{x}_1(v,w) = 0 \stackrel{def}{=} \Gamma_{x_1}(\pi,\dot{\pi}),$$

$$\mathbf{x}_3(v,w) = \arcsin\left(\frac{-v_1}{(\epsilon_0+w)\omega^2}\right) = \arcsin\left(\frac{-\pi}{\omega^2}\right) \stackrel{def}{=} \Gamma_{x_3}(\pi,\dot{\pi}),$$

$$\mathbf{u}(v,w) = \frac{\dot{\pi}^2}{\omega^4\left(1-\left(\frac{\pi}{\omega^2}\right)^2\right)}\frac{\frac{-\pi}{\omega^2}}{\sqrt{1-\left(\frac{\pi}{\omega^2}\right)^2}} + \frac{\pi}{\sqrt{1-\left(\frac{\pi}{\omega^2}\right)^2}} \stackrel{def}{=} \Gamma_u(\pi,\dot{\pi}).$$

Thus letting

$$\Gamma(\pi(v,w),\dot{\pi}(v,w)) = \begin{bmatrix} \Gamma_{x_1}(\pi(v,w),\dot{\pi}(v,w)) \\ \Gamma_{x_3}(\pi(v,w),\dot{\pi}(v,w)) \\ \Gamma_u(\pi(v,w),\dot{\pi}(v,w)) \end{bmatrix}$$

shows that the solution of the regulation equations satisfies condition (6.31).

Since $\ddot{\pi} = -\frac{\omega^2 v_1}{(\epsilon_0+w)}$, we have $\ddot{\pi} + \omega^2\pi = 0$; that is, $P(\lambda) = \lambda^2 + \omega^2$ is the minimal zeroing polynomial of $\pi(v,w)$.

It is ready to verify that the RTAC system admits a steady-state generator $\{\theta, \alpha, \beta\}$, where

$$\theta = T\begin{bmatrix} \pi \\ \dot{\pi} \end{bmatrix}, \quad \Phi = \begin{bmatrix} 0 & 1 \\ -\omega^2 & 0 \end{bmatrix},$$

$$\alpha(\theta) = T\Phi T^{-1}\theta, \quad \beta(\theta) = \Gamma(T^{-1}\theta) = \begin{bmatrix} \Gamma_{x_1}(T^{-1}\theta) \\ \Gamma_{x_3}(T^{-1}\theta) \\ \Gamma_u(T^{-1}\theta) \end{bmatrix} \stackrel{def}{=} \begin{bmatrix} \beta_{x_1}(\theta) \\ \beta_{x_3}(\theta) \\ \beta_u(\theta) \end{bmatrix},$$

where $T \in \mathcal{R}^{2\times 2}$ is any nonsingular matrix. Clearly, the steady-state generator is linearly observable since the pair (Ψ_u, Φ) is observable, where $\Psi_u = [1 \;\; 0]$ is the Jacobian of Γ_u at the origin. Thus, condition (6.33) is also satisfied

Corresponding to the above steady-state generator, we can obtain an internal model as follows:

$$\dot{\eta} = M\eta + N(u - \beta_u(\eta) + \Psi_u T^{-1}), \tag{6.46}$$

where $M = \begin{bmatrix} 0 & 1 \\ a_1 & a_2 \end{bmatrix}$ with $a_1 < 0$ and $a_2 < 0$, $N = \begin{bmatrix} 0 \\ 1 \end{bmatrix}$, and T is the solution of the Sylvester equation $T\Phi - MT = N\Psi_u$. Since M is Hurwitz and (M, N) is controllable, the Sylvester equation has a unique nonsingular solution T as follows:

$$T = \begin{bmatrix} t_{11} & t_{12} \\ t_{21} & t_{22} \end{bmatrix} = \begin{bmatrix} \frac{-1}{a_1+\omega^2+\frac{a_2^2\omega^2}{a_1+\omega^2}} & \frac{-a_2}{a_1+\omega^2}t_{11} \\ \frac{a_2\omega^2}{a_1+\omega^2}t_{11} & t_{11} \end{bmatrix}.$$

Performing the following coordinate and input transformation:

$$
\begin{aligned}
\bar{x}_1 &= x_1 - \beta_{x_1}(\eta),\\
\bar{x}_2 &= x_2 - \mathbf{x}_2(v, w),\\
\bar{x}_3 &= x_3 - \beta_{x_3}(\eta),\\
\bar{x}_4 &= x_4 - \mathbf{x}_4(v, w),\\
\bar{\eta} &= \eta - \theta(v, w),\\
\bar{u} &= u - \beta_u(\eta)
\end{aligned}
$$

on the augmented system consisting of the RTAC system and the internal model (6.46) gives

$$
\begin{aligned}
\dot{\bar{x}}_1 &= \bar{x}_2,\\
\dot{\bar{x}}_2 &= \frac{-\bar{x}_1 + (\epsilon_0 + w)(\bar{x}_4 + \mathbf{x}_4(v, w))^2 \sin(\bar{x}_3 + \beta_{x_3}(\bar{\eta} + \theta))}{1 - (\epsilon_0 + w)^2 \cos^2(\bar{x}_3 + \beta_{x_3}(\bar{\eta} + \theta))}\\
&\quad + \frac{-(\epsilon_0 + w)\cos(\bar{x}_3 + \beta_{x_3}(\bar{\eta} + \theta))}{1 - (\epsilon_0 + w)^2 \cos^2(\bar{x}_3 + \beta_{x_3}(\bar{\eta} + \theta))}(\bar{u} + \beta_u(\bar{\eta} + \theta))\\
&\quad + \frac{v_1}{1 - (\epsilon_0 + w)^2 \cos^2(\bar{x}_3 + \beta_{x_3}(\bar{\eta} + \theta))},\\
\dot{\bar{x}}_3 &= \dot{x}_3 - \dot{\beta}_{x_3}(\bar{\eta} + \theta) = \bar{x}_4 + \mathbf{x}_4(v, w) - \frac{\partial \beta_{x_3}(\bar{\eta} + \theta)}{\partial \bar{\eta}}\dot{\bar{\eta}} - \frac{\partial \beta_{x_3}(\bar{\eta} + \theta)}{\partial \theta}\dot{\theta}\\
&= \bar{x}_4 + \frac{-v_2}{(\epsilon_0 + w)\omega}\frac{1}{\sqrt{1 - \left(\frac{-v_1}{(\epsilon_0 + w)\omega^2}\right)^2}}\\
&\quad + \frac{[1\ 0]T^{-1}(M + N\Psi_u T^{-1})\bar{\eta} + [1\ 0]T^{-1}N\bar{u} + [1\ 0]\Phi T^{-1}\theta}{\omega^2\sqrt{1 - \left(\frac{[1\ 0]T^{-1}(\bar{\eta} + \theta)}{\omega^2}\right)^2}},\\
\dot{\bar{x}}_4 &= \frac{(\epsilon_0 + w)\cos(\bar{x}_3 + \beta_{x_3}(\bar{\eta} + \theta))}{1 - (\epsilon_0 + w)^2 \cos^2(\bar{x}_3 + \beta_{x_3}(\bar{\eta} + \theta))}\\
&\quad \times \left(\bar{x}_1 - (\epsilon_0 + w)(\bar{x}_4 + \mathbf{x}_4(v, w))^2 \sin(\bar{x}_3 + \beta_{x_3}(\bar{\eta} + \theta))\right)\\
&\quad + \frac{1}{1 - (\epsilon_0 + w)^2 \cos^2(\bar{x}_3 + \beta_{x_3}(\bar{\eta} + \theta))}(\bar{u} + \beta_u(\bar{\eta} + \theta))\\
&\quad + \frac{-(\epsilon_0 + w)\cos(\bar{x}_3 + \beta_{x_3}(\bar{\eta} + \theta))}{1 - (\epsilon_0 + w)^2 \cos^2(\bar{x}_3 + \beta_{x_3}(\bar{\eta} + \theta))}v_1 + \frac{-v_1\omega + \frac{v_1 v_2^2 + v_1^3}{(\epsilon_0 + w)^2\omega^3}}{(\epsilon_0 + w)\omega\left[1 - \frac{v_1^2}{(\epsilon_0 + w)^2\omega^4}\right]^{\frac{3}{2}}},\\
\dot{\bar{\eta}} &= (M + N\Psi_u T^{-1})\bar{\eta} + N\bar{u}.
\end{aligned}
\tag{6.47}
$$

By Corollary 6.9, it suffices to (locally) stabilize the equilibrium point at the origin of (6.47) with $v = 0$ and $w = 0$ by a controller depending on $\bar{x}_1$ and $\bar{x}_3$ only. To this end, linearizing the augmented system (6.47) with v and w being set to zero and noting $\Psi_u = [1, 0]$ gives

$$
\begin{aligned}
\dot{\bar{x}}_1 &= \bar{x}_2, \\
\dot{\bar{x}}_2 &= \frac{-1}{1-\epsilon_0^2}\bar{x}_1 + \frac{-\epsilon_0}{1-\epsilon_0^2}\bar{u} + \frac{-\epsilon_0}{1-\epsilon_0^2}\Psi_u T^{-1}\bar{\eta}, \\
\dot{\bar{x}}_3 &= \bar{x}_4 + \frac{1}{\omega^2}\Psi_u T^{-1}(M + N\Psi_u T^{-1})\bar{\eta} + \frac{1}{\omega^2}\Psi_u T^{-1} N\bar{u}, \\
\dot{\bar{x}}_4 &= \frac{\epsilon_0}{1-\epsilon_0^2}\bar{x}_1 + \frac{1}{1-\epsilon_0^2}\bar{u} + \frac{1}{1-\epsilon_0^2}\Psi_u T^{-1}\bar{\eta}, \\
\dot{\bar{\eta}} &= (M + N\Psi_u T^{-1})\bar{\eta} + N\bar{u}.
\end{aligned}
$$

The above system can be put into the following matrix form as follows:

$$
\begin{aligned}
\dot{\bar{x}} &= A\bar{x} + B_\eta\bar{\eta} + \bar{B}\bar{u}, \\
\dot{\bar{\eta}} &= (M + N\Psi_u T^{-1})\bar{\eta} + N\bar{u},
\end{aligned}
\tag{6.48}
$$

where

$$
A = \begin{bmatrix} 0 & 1 & 0 & 0 \\ \frac{-1}{1-\epsilon_0^2} & 0 & 0 & 0 \\ 0 & 0 & 0 & 1 \\ \frac{\epsilon_0}{1-\epsilon_0^2} & 0 & 0 & 0 \end{bmatrix}, \quad B_\eta = \begin{bmatrix} 0_{1\times 2} \\ \frac{-\epsilon_0}{1-\epsilon_0^2}\Psi_u T^{-1} \\ \frac{1}{\omega^2}\Psi_u T^{-1}(M + N\Psi_u T^{-1}) \\ \frac{1}{1-\epsilon_0^2}\Psi_u T^{-1} \end{bmatrix},
$$

$$
\bar{B} = \begin{bmatrix} 0 \\ \frac{-\epsilon_0}{1-\epsilon_0^2} \\ \frac{1}{\omega^2}\Psi_u T^{-1} N \\ \frac{1}{1-\epsilon_0^2} \end{bmatrix}.
$$

Moreover, let

$$
y_m = C_m \begin{bmatrix} \bar{x} \\ \bar{\eta} \end{bmatrix},
$$

where

$$
C_m = \begin{bmatrix} 1 & 0 & 0 & 0 & 0 & 0 \\ 0 & 0 & 1 & 0 & 0 & 0 \end{bmatrix}.
$$

Then it can be verified that the linear system with $\mathrm{col}(\bar{x}, \bar{\eta})$ as the state, $\bar{u}$ as the input, and y_m as the output is both stabilizable and detectable.

Now let K and L be such that the two matrices

$$\begin{bmatrix} A & B_\eta \\ 0 & M + N\Psi_u T^{-1} \end{bmatrix} + K \begin{bmatrix} \bar{B} \\ N \end{bmatrix} \tag{6.49}$$

and

$$\begin{bmatrix} A & B_\eta \\ 0 & M + N\Psi_u T^{-1} \end{bmatrix} - LC_m \tag{6.50}$$

are Hurwitz. Then a linear output feedback controller that stabilizes (6.47) can be given as follows:

$$\begin{aligned} \bar{u} &= K\xi, \\ \dot{\xi} &= \begin{bmatrix} A & B_\eta \\ 0 & M + N\Psi_u T^{-1} \end{bmatrix} \xi + \begin{bmatrix} \bar{B} \\ N \end{bmatrix} \bar{u} + L \begin{bmatrix} \bar{x}_1 - \xi_1 \\ \bar{x}_3 - \xi_3 \end{bmatrix}. \end{aligned} \tag{6.51}$$

By Corollary 6.9, the controller that solves the robust output regulation problem of the original system is given as follows:

$$\begin{aligned} u &= K\xi + \beta_u(\eta), \\ \dot{\xi} &= \begin{bmatrix} A & B_\eta \\ 0 & M + N\Psi_u T^{-1} \end{bmatrix} \xi + \begin{bmatrix} \bar{B} \\ N \end{bmatrix} K\xi + L \begin{bmatrix} x_1 - \xi_1 \\ x_3 - \beta_{x_3}(\eta) - \xi_3 \end{bmatrix}, \\ \dot{\eta} &= M\eta + N(u - \beta_u(\eta) + \Psi_u T^{-1}\eta). \end{aligned} \tag{6.52}$$

A specific controller has been synthesized with the various parameters as follows:

$$\omega = 3,\ \epsilon_0 = 0.2,\ \Phi = \begin{bmatrix} 0 & 1 \\ -\omega^2 & 0 \end{bmatrix} = \begin{bmatrix} 0 & 1 \\ -9 & 0 \end{bmatrix},\ M = \begin{bmatrix} 0 & 1 \\ -3 & -2 \end{bmatrix},$$

$$N = \begin{bmatrix} 0 \\ 1 \end{bmatrix},\ \text{and } T = \begin{bmatrix} -0.0833 & -0.0278 \\ 0.2500 & -0.0833 \end{bmatrix}.$$

Also,

$$K = \begin{bmatrix} 5.9374 & -3.4198 & -0.9555 & -2.5082 & 5.9333 & -1.7874 \end{bmatrix},$$

which is such that the eigenvalues of the matrix (6.49) are

$$1.2 \times \begin{bmatrix} -0.3099 \pm 1.2634j & -0.5805 \pm 0.7828j & -0.7346 \pm 0.2873j \end{bmatrix}$$

and

$$L = \begin{bmatrix} 12.7755 & 0.7210 \\ 46.6889 & 9.4943 \\ -3.1786 & 5.9745 \\ -203.9327 & -13.7275 \\ -51.6165 & 7.6811 \\ -129.3720 & -54.0792 \end{bmatrix},$$

which is such that the eigenvalues of the matrix (6.50) are given by

$$\begin{bmatrix} -1.50 \pm j1.50 & -2.25 & -3.75 & -4.50 & -5.25 \end{bmatrix}.$$

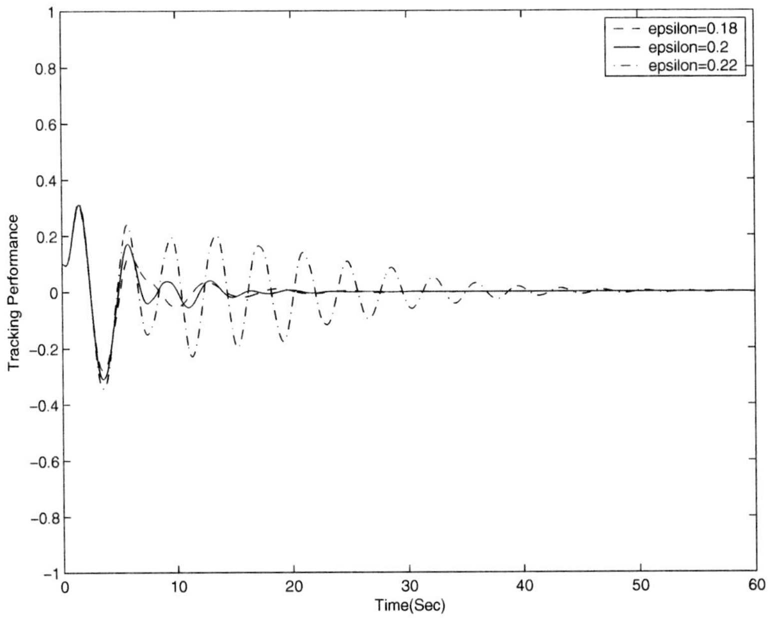

Figure 6.1. *The profiles of the displacement x_1 with $\epsilon = 0.18, 0.2, 0.22$, $\omega = 3$, and $A_m = 0.5$.*

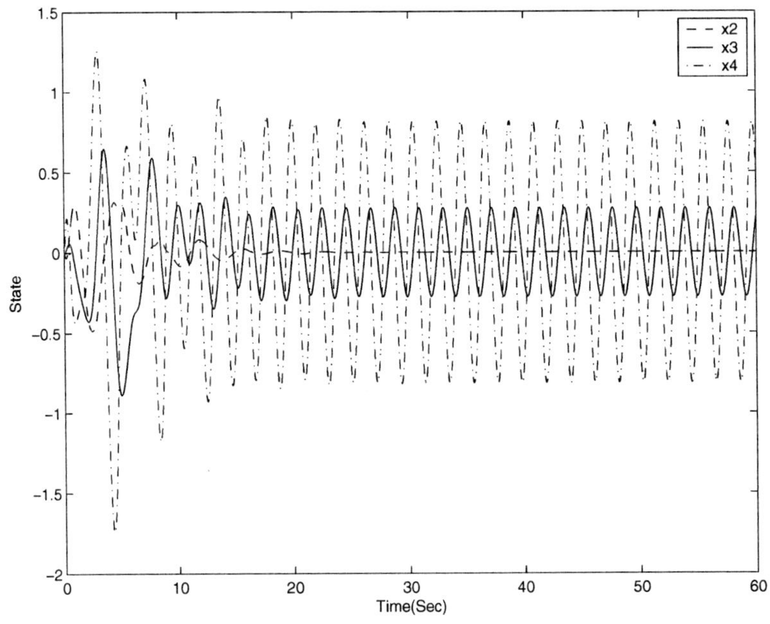

Figure 6.2. *The profiles of the state variables (x_2, x_3, x_4) with $\epsilon = 0.2$, $\omega = 3$, and $A_m = 0.5$.*

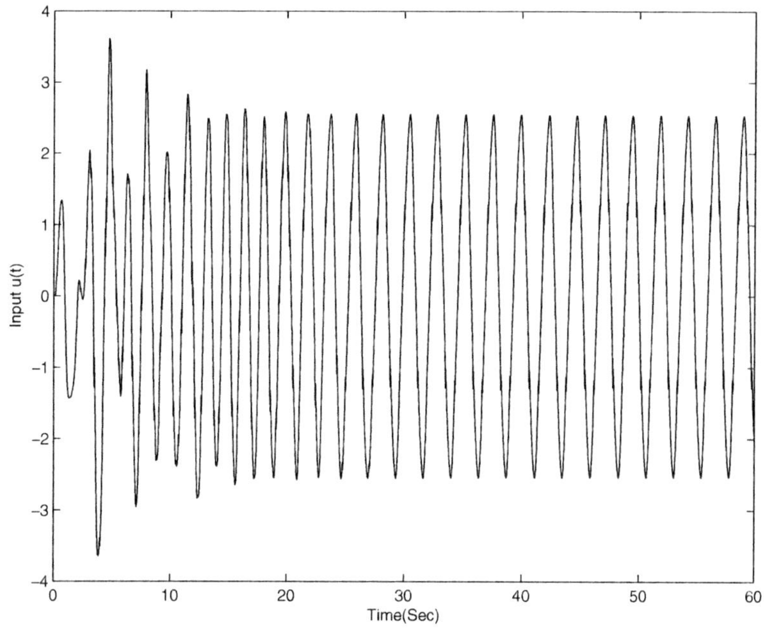

Figure 6.3. *The profile of the control input u with $\epsilon = 0.2$, $\omega = 3$, and $A_m = 0.5$.*

Computer simulation has been used to evaluate the performance of the closed-loop system with the initial state being $x(0) = \text{col}(0.1, 0, 0, 0)$, $\eta(0) = 0$, and $\xi(0) = 0$. Figure 6.1 shows the displacement x_1 of the cart under a sinusoidal disturbance $v_1 = 0.5 \sin \omega t$ for cases where $\epsilon = 0.18, \; 0.2, \; 0.22$. As expected, the parameter variations do not affect the steady-state response of the output. This is in sharp contrast with the nonlinear servoregulator designed in Chapter 3, where the same parameter variations significantly affect the steady-state response of the output. Figure 6.2 shows the profile of the other three state variables x_2, x_3, x_4, and Figure 6.3 shows the profile of the control input $u(t)$.

Chapter 7

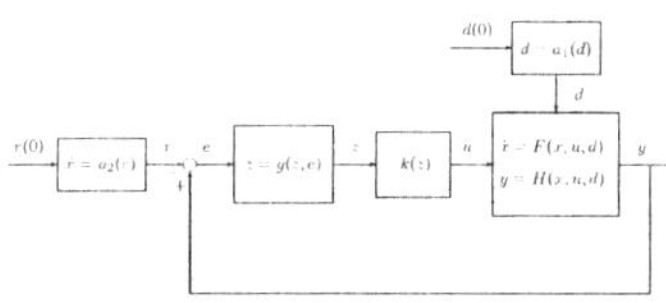

Global Robust Output Regulation

The robust output regulation problem that we studied in previous chapters is local in the sense that Property 5.1 only guarantees the boundedness of the trajectories of the closed-loop system, and Property 5.2 only ensures the asymptotic regulation of the error output of the closed-loop system when the initial state of the plant, the controller, and the exosystem, and the uncertain parameter w are all sufficiently small. In practice, it is desirable to design controllers that render the global boundedness of the trajectories of the closed-loop system, asymptotic regulation of the error output of the closed-loop system for any initial state of the plant, the controller, arbitrarily large exogenous signals, and an arbitrarily large uncertain parameter w. A formal formulation of such a problem is called the global robust output regulation problem and is the topic of this chapter.

We have already known from Chapter 6 that, under some suitable assumptions, the robust output regulation problem for a given plant can be converted into a robust stabilization problem for an augmented system. This design philosophy can also be used to handle the global robust output regulation problem. However, the global robust stabilization problem itself is a challenging topic. Only some limited results are available for handling certain classes of nonlinear systems with special structures. Two such classes of nonlinear systems are called nonlinear systems in output feedback form and nonlinear systems in lower triangular form, respectively. This chapter will give the solvability conditions of the global robust output regulation problem for both of these classes of nonlinear systems.

This chapter is organized as follows. Section 7.1 describes the problem. Section 7.2 presents some stabilization results for nonlinear systems in lower triangular form. Sections 7.3 and 7.4 establish the solvability conditions of the global robust output regulation problem for nonlinear systems in output feedback form and for nonlinear systems in lower triangular form, respectively.

7.1 Problem Description

The plant and exosystem considered in this chapter is described by

$$\dot{x}(t) = f(x(t), u(t), v(t), w), \qquad \dot{v}(t) = A_1 v(t), \qquad e(t) = h(x(t), v(t), w),\ t \geq 0, \tag{7.1}$$

where $x(t)$ is the n-dimensional plant state, $u(t)$ the m-dimensional plant input, $e(t)$ the p-dimensional plant output representing the tracking error, $v(t)$ the q-dimensional exogenous signal representing the disturbance and/or the reference input, and w the n_w-dimensional plant uncertain parameter. The plant is somehow simpler than (5.1) in that the error output equation does not depend on u explicitly. Again, we assume that all the eigenvalues of the matrix A_1 are simple with zero real parts.

The class of control laws considered here is described by

$$u = k_\zeta(x, \zeta, e), \qquad \dot{\zeta} = g_\zeta(x, \zeta, e), \tag{7.2}$$

where ζ is the compensator state vector of dimension n_ζ to be specified later. The above control law is called the dynamic state feedback control law. When x does not explicitly appear in (7.2), that is,

$$u = k_\zeta(\zeta, e), \qquad \dot{\zeta} = g_\zeta(\zeta, e),$$

the control law is called the dynamic output feedback control law. With $x_c = \text{col}(x, \zeta)$, the closed-loop system can be written as

$$\dot{x}_c = f_c(x_c, v, w), \qquad e = h_c(x_c, v, w), \tag{7.3}$$

where

$$f_c(x_c, v, w) = \begin{bmatrix} f(x, k_\zeta(x, \zeta, e), v, w) \\ g_z(x, \zeta, h(x, v, w)) \end{bmatrix},$$

$$h_c(x_c, v, w) = h(x, v, w). \tag{7.4}$$

Again, all the functions involved in this setup are assumed to be sufficiently smooth and defined globally on the appropriate Euclidean spaces, with the value zero at the respective origins. Also it is assumed that 0 is the nominal value of the uncertain parameter w, and $f(0, 0, 0, w) = 0$ and $h(0, 0, w) = 0$ for all $w \in \mathcal{R}^{n_w}$. Note that in (7.2), the feedback control is allowed to rely on the error output e explicitly.

In terms of the closed-loop system, we can describe the problem as follows.

Global Robust Output Regulation Problem (GRORP): For any compact set $V_0 \in \mathcal{R}^q$ with a known bound and any compact set $W \in \mathcal{R}^{n_w}$ with a known bound, find a controller of the form (7.2) such that the closed-loop system (7.3) has the following two properties.

Property 7.1. For all $v(0) \in V_0$ and $w \in W$, the trajectory of the closed-loop system (7.3) starting from any initial states $x_c(0)$ exists and is bounded for all $t > 0$.

Property 7.2.

$$\lim_{t\to\infty} e(t) = 0. \tag{7.5}$$

A few remarks are in order.

Remark 7.1.

(i) By saying the bound of a compact set $X \in \mathcal{R}^n$ is known, we mean that there exists a known number $c > 0$ such that $X \subset \{x \mid \|x\| \leq c,\ x \in \mathcal{R}^n\}$.

(ii) Since $v(t)$ is generated by a stable linear system with $v(0) \in V_0$, where V_0 is some compact set of $\mathcal{R}^q$ with a known bound, there exists a compact set $V \in \mathcal{R}^q$ with a known bound such that $v(t) \in V$ for all $t \geq 0$.

(iii) Unlike the local case, Property 7.1 cannot be guaranteed by requiring the global asymptotic stability of the equilibrium point of the system

$$\dot{x}_c = f_c(x_c, 0, 0)$$

at $x_c = 0$. For example, consider the following system:

$$\begin{aligned} \dot{x}_1 &= -x_1 + v, \\ \dot{x}_2 &= -0.5x_2 + x_1 x_2, \\ \dot{v} &= 0. \end{aligned}$$

The solution of the system is given by

$$x_1(t) = v + (x_1(0) - v)e^{-t}, \quad x_2(t) = x_2(0)e^{(v-0.5)t}e^{[(x_1(0)-v)(1-e^{-t})]}.$$

It can be seen that the equilibrium point of this system is globally asymptotically stable when $v = 0$. Nevertheless, when $v \neq 0$, for example, $v = 1$, and $\text{col}(x_1(0), x_2(0)) = \text{col}(1, 1)$, $x_2(t) = e^{0.5t}$ approaches infinity.

Thus, as will be seen later, in order to guarantee the satisfaction of Property 7.1 by the closed-loop system, we need go farther than rendering $x_c = 0$ a globally asymptotically stable equilibrium point of $\dot{x}_c = f_c(x_c, 0, 0)$. ∎

When dealing with the (local) robust output regulation problem, it suffices to assume that the solution of the regulator equations exists in an arbitrarily small open neighborhood of the origin of $\mathcal{R}^q \times \mathcal{R}^{n_w}$. To handle the global robust output regulation problem, we require that the solution of the regulator equations exist globally. Thus, Assumption 5.1 is modified as follows.

Assumption 7.1. There exist sufficiently smooth functions $\mathbf{x}(v, w)$ and $\mathbf{u}(v, w)$ with $\mathbf{x}(0, 0) = 0$ and $\mathbf{u}(0, 0) = 0$ satisfying, for all $v \in \mathcal{R}^q$ and $w \in \mathcal{R}^{n_w}$, the following equations:

$$\begin{aligned} \frac{\partial \mathbf{x}(v, w)}{\partial v} A_1 v &= f(\mathbf{x}(v, w), \mathbf{u}(v, w), v, w), \\ 0 &= h(\mathbf{x}(v, w), v, w). \end{aligned} \tag{7.6}$$

Remark 7.2. Let $\mathbf{x}(v, w)$ and $\mathbf{u}(v, w)$ be a global solution of the regulator equations (7.6). Assume that system (7.1) has a global steady-state generator and an internal model characterized in (6.6). Then the coordinate transformation (6.9) and the augmented system (6.10) are defined globally. As a result, we can obtain a global version of Proposition 6.8 as follows. ∎

Proposition 7.3. *Suppose Assumption 7.1 and assume that system (7.1) has a global steady-state generator with output* $g_o(x, u) = \text{col}(x_1, \ldots, x_d, u)$ *and an internal model described by (6.6). Then the augmented system in the new coordinates and input described by (6.10) has the property that, for all trajectories* $v(t) \in \mathcal{R}^q$ *of the exosystem, and all* $w \in \mathcal{R}^{n_w}$,

$$\begin{aligned} \bar{f}(0, 0, 0, v, w) &= 0, \\ \bar{\gamma}(0, 0, 0, v, w) &= 0, \\ \bar{h}\,(0, 0, 0, v, w) &= 0. \end{aligned} \tag{7.7}$$

Using this proposition, it is also possible to convert the global robust output regulation problem for the given plant (7.1) into a global robust stabilization problem of the equilibrium point $(\bar{\eta}, \bar{x}) = (0, 0)$ of the augmented system (6.10) for any $v(t) \in V$ and $w \in W$ by the class of controllers of the form (6.12). To this end, recall that the closed-loop system composed of the augmented system (6.10) and the controller (6.12) is denoted by (6.13) and is repeated as follows:

$$\dot{\bar{x}}_c = \bar{f}_c(\bar{x}_c, v, w), \qquad e = \bar{h}_c(\bar{x}_c, v, w), \tag{7.8}$$

where $\bar{x}_c = \text{col}(\bar{x}, \bar{\eta}, \xi)$.

Global Robust Stabilization Problem (GRSP). For any compact set $V_0 \in \mathcal{R}^q$ with a known bound and any compact set $W \in \mathcal{R}^{n_w}$ with a known bound, find a controller of the form (6.12) such that, for any $\bar{x}_c(0)$, any $v(0) \in V_0$, and any $w \in W$, the trajectory of the closed-loop system (7.8) exists for all $t \geq 0$ and satisfies

$$\|\bar{x}_c(t)\| \leq \beta^{kl}(\|\bar{x}_c(0)\|, t), \quad t \geq 0 \tag{7.9}$$

for some class $\mathcal{KL}$ function $\beta^{kl}(\cdot, \cdot)$ independent of v and w.

Corollary 7.4. *Suppose Assumptions 7.1 and 5.5 hold. Given any compact set* $V_0 \in \mathcal{R}^q$ *with a known bound and any compact set* $W \in \mathcal{R}^{n_w}$ *with a known bound, assume that controller (6.12) solves the global robust stabilization problem for the augmented system (6.10). Then a controller of the form (7.2), where* $\zeta = \text{col}(\eta, \xi)$,

$$\begin{aligned} k_\zeta(x, \zeta, e) &= \beta_u(\eta) + k\left(x_1 - \beta_1(\eta), \ldots, x_d - \beta_d(\eta), \xi, e\right), \\ g_\zeta(x, \zeta, e) &= \begin{bmatrix} \gamma(\eta, x, u, e) \\ g_\xi\left(x_1 - \beta_1(\eta), \ldots, x_d - \beta_d(\eta), \xi, e\right) \end{bmatrix} \end{aligned} \tag{7.10}$$

solves the global robust output regulation problem for the original system (7.1).

Proof. Assume that the controller (6.12) solves the global robust stabilization problem of system (6.10) for some given compact sets $V_0 \in \mathcal{R}^q$ and $W \in \mathcal{R}^{n_w}$. Denote the state of the closed-loop system composed of the plant (7.1) and the controller (7.10) by $x_c = \text{col}(x, \eta, \xi)$. Then

$$\begin{aligned} x_c = \bar{x}_c + \\ &\text{col}(\beta_1(\bar{\eta} + \theta(v, w)), \ldots, \beta_d(\bar{\eta} + \theta(v, w)), \mathbf{x}_{d+1}(v, w), \ldots, \mathbf{x}_n(v, w), \theta(v, w), 0). \end{aligned}$$

Let $\mathbf{x}_c(v, w) = \mathrm{col}(\mathbf{x}(v, w), \theta(v, w), 0)$. Then

$$\begin{aligned}
&\|x_c - \mathbf{x}_c(v, w)\| \\
&\quad = \|\mathrm{col}(\bar{x}_1 + (\beta_1(\eta) - \beta_1(\theta)), \ldots, \bar{x}_d + (\beta_d(\eta) - \beta_d(\theta)), \bar{x}_{d+1}, \ldots, \bar{x}_n, \bar{\eta}, \xi)\| \\
&\quad \le \|\bar{x}_c\| + \|\mathrm{col}(\beta_1(\eta) - \beta_1(\theta), \ldots, \beta_d(\eta) - \beta_d(\theta))\|.
\end{aligned}$$

Using inequality (7.9) gives, for all $t \ge 0$,

$$\begin{aligned}
\|x_c(t) - \mathbf{x}_c(v(t), w)\| &\le \beta^{kl}(\|\bar{x}_c(0)\|, t) \\
&\quad + \left\| \begin{bmatrix} \beta_1(\bar{\eta}(t) + \theta(v(t), w)) - \beta_1(\theta(v(t), w)) \\ \vdots \\ \beta_d(\bar{\eta}(t) + \theta(v(t), w)) - \beta_d(\theta(v(t), w)) \end{bmatrix} \right\|.
\end{aligned}$$

Note that, for all $x_c(0)$, all $v(0) \in V_0$, and all $w \in W$, $\bar{\eta}(t)$ and $\theta(v(t), w)$ are bounded for all $t \ge 0$. Therefore, the fact that the functions β and θ are C^1 and vanish at their respective origins guarantees the existence of constants $L_i > 0$, $i = 1, \ldots, d$, such that, for $t \ge 0$,

$$|\beta_i(\bar{\eta}(t) + \theta(v(t), w)) - \beta_i(\theta(v(t), w))| \le L_i \|\bar{\eta}(t)\|.$$

Then we further have, for $t \ge 0$,

$$\|x_c(t) - \mathbf{x}_c(v(t), w)\| \le \beta^{kl}(\|\bar{x}_c(0)\|, t) + L\|\bar{\eta}(t)\|$$

for some positive constant L. As a result, for all $x_c(0)$, all $v(0) \in V_0$, and all $w \in W$, $x_c(t)$ is bounded for all $t \ge 0$ and

$$\lim_{t\to\infty} \|x_c(t) - \mathbf{x}_c(v(t), w)\| = 0. \tag{7.11}$$

Similarly, since the function h is C^1, there exists a constant $L_0 > 0$ such that

$$\|h(x(t), v(t), w) - h(\mathbf{x}(v(t), w), v(t), w)\| \le L_0\|x(t) - \mathbf{x}(v(t))\|, \quad t \ge 0.$$

Thus, using (7.11) gives

$$\begin{aligned}
\lim_{t\to\infty} \|e(t)\| &= \lim_{t\to\infty} \|h(x(t), v(t), w) - h(\mathbf{x}(v(t), w), v(t), w)\| \\
&\le \lim_{t\to\infty} L_0\|x(t) - \mathbf{x}(v(t))\| \\
&\le \lim_{t\to\infty} L_0\|x_c(t) - \mathbf{x}_c(v(t), w)\| = 0. \quad \square
\end{aligned}$$

Due to this corollary, we have also converted the global robust output regulation problem for the given plant (7.1) into a global robust stabilization problem of the augmented system (6.10).

7.2 Stabilization of Systems in Lower Triangular Form

In this section, we will study the class of nonlinear systems in the following form:

$$\begin{aligned}\dot{z}_0 &= f_0\left(z_0, x_1, \mu(t)\right),\\ \dot{z}_i &= Q_i\left(z_0, z_1, \ldots, z_i, x_1, \ldots, x_i, \mu(t)\right),\\ \dot{x}_i &= f_i\left(z_0, z_1, \ldots, z_i, x_1, \ldots, x_i, \mu(t)\right) + b_i(\mu(t))x_{i+1}, \quad t \geq t_0 \geq 0,\\ & i = 1, \ldots, r, \end{aligned} \tag{7.12}$$

where $x = \text{col}(x_1, \ldots, x_r)$ and $z = \text{col}(z_0, z_1, \ldots, z_r)$ are the states with $x_i \in \mathcal{R}$, $i = 1, \ldots, r$, and $z_i \in \mathcal{R}^{n_i}$, $i = 0, \ldots, r$, $u \in \mathcal{R}$ is the input, and $\mu : [t_0, \infty) \to \Sigma \subset \mathcal{R}^{n_\mu}$ is a piecewise continuous function with Σ a prescribed compact set of $\mathcal{R}^{n_\mu}$. The function $f_0 : \mathcal{R}^{n_0+1+n_\mu} \to \mathcal{R}^{n_0}$ and functions $Q_i : \mathcal{R}^{n_0+n_1+\cdots+n_i+i+n_\mu} \to \mathcal{R}^{n_i}$, $f_i : \mathcal{R}^{n_0+n_1+\cdots+n_i+i+n_\mu} \to \mathcal{R}$, and $b_i : \mathcal{R}^{n_\mu} \to \mathcal{R}$ for $i = 1, \ldots, r$ are C^1 satisfying $f_0(0, 0, \mu) = 0$, $Q_i(0, \ldots, 0, \mu) = 0$, and $f_i(0, \ldots, 0, \mu) = 0$ for all $\mu \in \mathcal{R}^{n_\mu}$.

In (7.12), the vector μ represents a set of unknown parameters and/or disturbances and is called the *static* uncertainty. On the other hand, the functions Q_i may not be known precisely and/or the state z_i may not be available for feedback control. Thus the dynamics governing z_i, $i = 1, \ldots, r$, are called the *dynamic* uncertainty of system (7.12) as opposed to the static uncertainty $\mu(t)$. In the special case where system (7.12) involves no dynamic uncertainty, that is, $n_i = 0$, $i = 1, \ldots, r$, the system reduces to the following:

$$\begin{aligned}\dot{z}_0 &= f_0\left(z_0, x_1, \mu(t)\right),\\ \dot{x}_1 &= f_1\left(z_0, x_1, \mu(t)\right) + b_1(\mu(t))x_2,\\ &\vdots\\ \dot{x}_r &= f_r\left(z_0, x_1, \ldots, x_r, \mu(t)\right) + b_r(\mu(t))u. \end{aligned} \tag{7.13}$$

System (7.13) is called a (strictly feedback) lower triangular system. In the more special case where $\mu(t)$ does not appear on the right side of (7.13), the subsystem $\dot{z}_0 = f_0\left(z_0, 0\right)$ is the zero dynamics of system (7.13) viewing x_1 as the output.

In what follows, we will consider the global robust stabilization problem for system (7.12) with respect to both static and dynamic uncertainties using a sufficiently smooth partial state feedback control of the form $u = k(x_1, \ldots, x_r)$ with $k(0, \ldots, 0) = 0$. For this purpose, let us list a few assumptions as follows.

Assumption 7.2. For all $\mu \in \mathcal{R}^{n_\mu}$ and all $i = 1, \ldots, r$, $b_i(\mu) > 0$.

Assumption 7.3. The system $\dot{z}_0 = f\left(z_0, x_1, \mu(t)\right)$, $t \geq t_0$, is RISS with respect to μ with state z_0 and input x_1 and has a known C^1 gain function $\kappa_0(\cdot)$.

Assumption 7.4. For all $i = 1, \ldots, r$, the system $\dot{z}_i = Q_i\left(z_0, z_1, \ldots, z_i, x_1, \ldots, x_i, \mu(t)\right)$, $t \geq t_0 \geq 0$, is RISS with respect to μ with state z_i and input $\text{col}(z_0, z_1, \ldots, z_{i-1}, x_1, \ldots, x_i)$ and has a known C^1 gain function $\kappa_i(\cdot)$.

Remark 7.5. By the definition of RISS for systems of the form (7.12), Assumptions 7.3 and 7.4 mean the existence of some class $\mathcal{KL}$ function $\beta_i^{kl}(\cdot, \cdot)$, some known C^1 class

$\mathcal{K}_\infty$ functions $\kappa_i(\cdot)$, $i = 0, 1, \ldots, r$, which are independent of μ, such that, for all $\mu : [t_0, \infty) \to \Sigma \subset \mathcal{R}^{n_\mu}$, the solution $z_0(t)$ of system $\dot{z}_0 = f(z_0, x_1, \mu(t))$ and the solutions $z_i(t)$ of $\dot{z}_i = Q_i(z_0, z_1, \ldots, z_i, x_1, \ldots, x_i, \mu(t))$, $i = 1, \ldots, r$, exist and satisfy, for all $t \geq t_0 \geq 0$,

$$\|z_0(t)\| \leq \max\left\{\beta_0^{kl}(\|z_0(t_0)\|, t - t_0), \kappa_0\left(\sup_{t_0 \leq \tau \leq t} \|x_1(\tau)\|\right)\right\} \tag{7.14}$$

and

$$\|z_i(t)\| \leq \max\left\{\beta_i^{kl}(\|z_i(t_0)\|, t - t_0), \kappa_i\left(\sup_{t_0 \leq \tau \leq t} \|\mathrm{col}(z_0, \ldots, z_{i-1}, x_1, \ldots, x_i)(\tau)\|\right)\right\} \tag{7.15}$$

for all $z_i(t_0) \in \mathcal{R}^{n_i}$, and all $\mathrm{col}(z_0, \ldots, z_{i-1}, x_1, \ldots, x_i) \in L_\infty^{n_0+\cdots+n_{i-1}+i}$. Also note that, under Assumption 7.2, for any compact set $\Sigma \subset \mathcal{R}^{n_\mu}$, there exist real numbers b_{Mi}, b_{mi}, $i = 1, \ldots, r$, such that $\infty > b_{Mi} \geq b_i(\mu) \geq b_{mi} > 0$ for all $\mu \in \Sigma$. ∎

The main result of this section is given as follows.

Theorem 7.6. *Under Assumptions 7.2 to 7.4, there exists a sufficiently smooth state feedback controller of the form $u = k(x_1, \ldots, x_r)$ satisfying $k(0, \ldots, 0) = 0$ such that the equilibrium point of the closed-loop system at the origin is globally asymptotically stable for all $\mu : [t_0, \infty) \to \Sigma \subset \mathcal{R}^{n_\mu}$.*

We will use a recursive approach to synthesize a state feedback controller to globally stabilize (7.12). The recursive approach will be based on the following proposition, which handles a special case of (7.12) with $r = 1$ and $n_1 = 0$.

Proposition 7.7. *Consider the system*

$$\begin{aligned} \dot{z} &= \varphi(z, x, \mu(t)), \\ \dot{x} &= \phi(z, x, \mu(t)) + \psi(\mu(t))u, \quad t \geq t_0 \geq 0, \end{aligned} \tag{7.16}$$

in which $(z, x) \in \mathcal{R}^m \times \mathcal{R}$, $\mu : [t_0, \infty) \to \Sigma \subset \mathcal{R}^{n_\mu}$ is piecewise continuous with Σ a prescribed compact set of $\mathcal{R}^{n_\mu}$, $\varphi(z, x, \mu)$ and $\phi(z, x, \mu)$ are C^1 functions satisfying $\varphi(0, 0, \mu) = 0$, $\phi(0, 0, \mu) = 0$ for $\mu \in \Sigma \subset \mathcal{R}^{n_\mu}$. Suppose the following:

(i) *The upper subsystem in (7.16) is RISS with respect to μ with state z and input x, and has a known C^1 class $\mathcal{K}_\infty$ gain function $\kappa(\cdot)$.*
(ii) *For all $\mu \in \mathcal{R}^{n_\mu}$, $\psi(\mu) > 0$.*

Then, there exists a smooth function $\rho : \mathcal{R} \to [0, \infty)$ such that, under the controller

$$u = -x\rho(x) + \hat{u}, \tag{7.17}$$

the closed-loop system (7.16) and (7.17) is RISS with respect to μ with state $Z = \mathrm{col}(z, x)$ and input $\hat{u}$ and has a known C^1 class $\mathcal{K}_\infty$ gain function $\tilde{\kappa}(\cdot)$.

Proof. Consider the system composed of the lower subsystem of (7.16) and controller (7.17):

$$\dot{x} = \phi(z, x, \mu) + \psi(\mu)(-x\rho(x) + \hat{u}). \tag{7.18}$$

If $\rho(x)$ can be chosen such that system (7.18) is RISS with respect to μ with state x and input $\text{col}\left(z, \hat{u}\right)$, in particular, the solution of system (7.18) exists and satisfies, for some class $\mathcal{KL}$ function $\beta^{kl}(\cdot, \cdot)$, some known class $\mathcal{K}_\infty$ function $\gamma_z(\cdot)$, and some known C^1 class $\mathcal{K}_\infty$ function $\gamma_u(\cdot)$, independent of $\mu(t)$,

$$|x(t)| \leq \max\left\{\beta^{kl}(|x(t_0)|, t - t_0), \gamma_z\left(\sup_{t_0 \leq \tau \leq t} ||z(\tau)||\right), \gamma_u\left(\sup_{t_0 \leq \tau \leq t} ||\hat{u}(\tau)||\right)\right\}, \quad t \geq t_0 \geq 0, \tag{7.19}$$

for all $x(t_0) \in \mathcal{R}$, $z \in L_\infty^m$, $\hat{u} \in L_\infty$, and $\mu : [t_0, \infty) \to \Sigma \subset \mathcal{R}^{n_\mu}$. Further, if $\rho(x)$ is such that

$$\kappa(\gamma_z(s)) < s, \quad \forall s > 0, \tag{7.20}$$

then the proof is completed upon using Corollary 2.19 (The Small Gain Theorem) with the C^1 gain function $\tilde{\kappa}(s)$ being any C^1 class $\mathcal{K}_\infty$ function satisfying

$$\tilde{\kappa}(s) \geq \max\{2\kappa \circ \gamma_u(s), 2\gamma_u(s)\}. \quad \square \tag{7.21}$$

To complete the proof of Proposition 7.7, we need to establish two more lemmas as follows.

Lemma 7.8. *Let $f : \mathcal{R}^m \times \mathcal{R}^n \times \mathcal{R}^p \to \mathcal{R}$ be a C^1 function satisfying $f(0, 0, \mu) = 0$ for all $\mu \in \Sigma$, with Σ being a compact set of $\mathcal{R}^p$. Then there exist smooth functions $F_1 : \mathcal{R}^m \to \mathcal{R}$ and $F_2 : \mathcal{R}^n \to \mathcal{R}$ satisfying $F_1(0) = 0$ and $F_2(0) = 0$ such that*

$$|f(x, y, \mu)| \leq F_1(x) + F_2(y), \quad \forall x \in \mathcal{R}^m, \ y \in \mathcal{R}^n, \ \mu \in \Sigma. \tag{7.22}$$

Proof. Let

$$f_1(x) = \max_{\{(a,\mu) \mid (a,\mu) \in \mathcal{R}^n \times \Sigma, \ ||a|| \leq ||x||\}} |f(x, a, \mu)|, \quad \forall x \in \mathcal{R}^m$$

and

$$f_2(y) = \max_{\{(b,\mu) \mid (b,\mu) \in \mathcal{R}^m \times \Sigma, \ ||b|| \leq ||y||\}} |f(b, y, \mu)|, \quad \forall y \in \mathcal{R}^n.$$

Then, $|f(x, y, \mu)| \leq f_1(x)$ for all $\mu \in \Sigma$ when $||y|| \leq ||x||$, and $|f(x, y, \mu)| \leq f_2(y)$ for all $\mu \in \Sigma$ when $||x|| \leq ||y||$. Thus, for all $(x, y, \mu) \in \mathcal{R}^m \times \mathcal{R}^n \times \Sigma$,

$$|f(x, y, \mu)| \leq f_1(x) + f_2(y).$$

Clearly $f_1(0) = 0$ and $f_2(0) = 0$. Moreover, since $f(x, y, \mu)$ is C^1 and Σ is compact, there exists a constant $L > 0$, independent of μ, such that $|f(x, y, \mu)| \leq L(||x|| + ||y||)$

for all sufficiently small $x \in \mathcal{R}^m$, $y \in \mathcal{R}^n$, and all $\mu \in \Sigma$. Thus, for all sufficiently small $x \in \mathcal{R}^m$,

$$f_1(x) \leq \max_{\{a \mid a \in \mathcal{R}^n, \|a\| \leq \|x\|\}} L(\|x\| + \|a\|) \leq 2L\|x\|,$$

that is, $f_1(x)$ is linear locally. Similarly, $f_2(y)$ is also linear locally. Therefore, there exist smooth functions $F_1(x)$ and $F_2(y)$ with $F_1(0) = 0$ and $F_2(0) = 0$ such that $f_1(x) \leq F_1(x)$ for all $x \in \mathcal{R}^m$, and $f_2(y) \leq F_2(y)$ for all $y \in \mathcal{R}^n$. □

Lemma 7.9. *There exists a smooth function $\rho : \mathcal{R} \to [0, \infty)$, such that, under the controller (7.17), system (7.18) is RISS with respect to μ with state x and input $\text{col}\left(z, \hat{u}\right)$. In particular, for any given C^1 class $\mathcal{K}_\infty$ function $\kappa(\cdot)$, $\rho(x)$ can be chosen such that the solution of (7.18) satisfies the inequality (7.19) with a known class $\mathcal{K}_\infty$ function $\gamma_z(\cdot)$ satisfying the small gain condition (7.20), and a known C^1 class $\mathcal{K}_\infty$ function $\gamma_u(\cdot)$.*

Proof. By assumption (ii) of Proposition 7.7, there exist $b_M \geq b_m > 0$ such that $b_M \geq \psi(\mu) \geq b_m$ for all $\mu \in \Sigma$. Also, since $\phi(z, x, \mu)$ is a C^1 function satisfying $\phi(0, 0, \mu) = 0$ for all $\mu \in \Sigma \subset \mathcal{R}^{n_\mu}$, by Lemma 7.8, there exist smooth functions $F_1(x)$ with $F_1(0) = 0$ and $F_2(y)$ with $F_2(0) = 0$ such that

$$|\phi(z, x, \mu)| \leq F_1(x) + F_2(z), \quad \forall z \in \mathcal{R}^m, \ x \in \mathcal{R}, \ \text{and } \mu \in \Sigma.$$

Moreover, by Taylor's theorem, there exist smooth functions $\phi_0(\cdot) \geq 1$ and $\phi_1(\cdot) \geq 1$ such that $F_1(x) \leq |x|\phi_0(x)$ for all $x \in \mathcal{R}$ and $F_2(z) \leq \|z\|\phi_1(z)$ for all $z \in \mathcal{R}^m$. Thus,

$$|\phi(z, x, \mu)| \leq |x|\phi_0(x) + \|z\|\phi_1(z), \quad \forall z \in \mathcal{R}^m, \ x \in \mathcal{R}, \ \text{and } \mu \in \Sigma. \tag{7.23}$$

As a result, the function $V(x) = x^2$ satisfies

$$\begin{aligned}
&\frac{\partial V}{\partial x}[\phi(z, x, \mu) + \psi(\mu)(-x\rho(x) + \hat{u})] \\
&\quad = 2x\phi(z, x, \mu) - 2x^2\psi(\mu)\rho(x) + 2x\psi(\mu)\hat{u} \\
&\quad \leq 2x^2\phi_0(x) + 2|x|\|z\|\phi_1(z) - 2x^2 b_m \rho(x) + 2x\psi(\mu)\hat{u} \\
&\quad \leq 2x^2\phi_0(x) + x^2 + \|z\|^2\phi_1^2(z) - 2x^2 b_m \rho(x) + x^2 b_M^2 + \hat{u}^2 \\
&\quad = -x^2(-2\phi_0(x) - 1 + 2b_m\rho(x) - b_M^2) + \|z\|^2\phi_1^2(z) + \hat{u}^2
\end{aligned}$$

for all $x \in \mathcal{R}$, all $z \in L_\infty^m$, all $\hat{u} \in L_\infty$, and all $\mu \in \Sigma$.

Now, given any smooth function $\alpha_0 : \mathcal{R} \to [0, \infty)$, letting $\rho(x) \geq 0$ be any smooth function satisfying

$$\rho(x) \geq \frac{\alpha_0(x) + 2\phi_0(x) + 1 + b_M^2}{2b_m} \tag{7.24}$$

gives

$$\frac{\partial V}{\partial x}[\phi(z, x, \mu) + \psi(\mu)(-x\rho(x) + \hat{u})] \leq -x^2\alpha_0(x) + \|z\|^2\phi_1^2(z) + \hat{u}^2 \tag{7.25}$$

for all $x \in \mathcal{R}$, all $z \in L_\infty^m$, all $\hat{u} \in L_\infty$, and all $\mu \in \Sigma$. In particular, assume $\alpha_0(x)$ is even and nondecreasing in $[0, +\infty)$, and let $\alpha(|x|) = x^2\alpha_0(x)$ for all $x \in \mathcal{R}$. Then $\alpha(\cdot)$ is a class $\mathcal{K}_\infty$ function.

Since $\phi_1(z) \geq 1$ for all $z \in \mathcal{R}^m$, there exists a smooth nondecreasing function c : $[0, \infty) \to [0, \infty)$ satisfying $c(\|z\|) \geq \phi_1^2(z)$. Letting $\sigma(s) = s^2(1 + c(s))$, which is a class $\mathcal{K}_\infty$ function, gives

$$\frac{\partial V}{\partial x}[\phi(z, x, \mu) + \psi(\mu)(-x\rho(x) + \hat{u})] \leq -\alpha(|x|) + \sigma(\|\mathrm{col}(z, \hat{u})\|) \tag{7.26}$$

for all $x \in \mathcal{R}$, $z \in L_\infty^m$, $\hat{u} \in L_\infty$, and $\mu \in \Sigma$. Also letting $\underline{\alpha}(s) = \bar{\alpha}(s) = s^2$, which are class $\mathcal{K}_\infty$ functions, gives $\underline{\alpha}(\|x\|) \leq V(x) \leq \bar{\alpha}(\|x\|)$ for all $x \in \mathcal{R}$. Thus, by Theorem 2.17 as well as Remark 2.21, the closed-loop system (7.18) is RISS with respect to μ with state x and input col $(z, \hat{u})$.

To obtain an estimation of the form (7.19), let $\sigma_z(s) = s^2c(s)$, $\sigma_u(s) = s^2$, $\chi_z(s) = \alpha^{-1}(\delta\sigma_z(s))$, and $\chi_u(s) = \alpha^{-1}(\delta\sigma_u(s))$, where $\delta > 2$. Then the inequality $|x| \geq \max\{\chi_z(\|z\|), \chi_u(|\hat{u}|)\}$ implies

$$\frac{1}{\delta}\alpha(|x|) \geq \max\{\sigma_z(\|z\|), \sigma_u(|\hat{u}|)\},$$

which in turn implies

$$\frac{\partial V}{\partial x}[\phi(z, x, \mu) + \psi(\mu)(-x\rho(x) + \hat{u})] \leq -\frac{\delta - 2}{\delta}\alpha(|x|) \tag{7.27}$$

for all $x \in \mathcal{R}$, $z \in L_\infty^m$, $\hat{u} \in L_\infty$, and $\mu \in \Sigma$. By Theorem 2.16, an inequality of the form (7.19) holds, where $\gamma_z(s) = \underline{\alpha}^{-1} \circ \bar{\alpha} \circ \chi_z(s)$ and $\gamma_u(s) = \underline{\alpha}^{-1} \circ \bar{\alpha} \circ \chi_u(s)$. Since $\underline{\alpha}(s) = \bar{\alpha}(s) = s^2$, we have $\gamma_z(s) = \chi_z(s) = \alpha^{-1}(\delta\sigma_z(s)) = \alpha^{-1}(\delta s^2 c(s))$ and $\gamma_u(s) = \chi_u(s) = \alpha^{-1}(\delta\sigma_u(s)) = \alpha^{-1}(\delta s^2)$.

Clearly, $\gamma_u(\cdot)$ is a class $\mathcal{K}_\infty$ function. It remains to show that, the function $\alpha(\cdot)$, hence $\rho(\cdot)$, can be chosen to satisfy the small gain condition (7.20) and $\gamma_u(\cdot)$ is C^1. To this end, for the given C^1 class $\mathcal{K}_\infty$ function $\kappa(\cdot)$, let $\tilde{\alpha}(s) = \delta\sigma_z(\kappa(2s)) = \delta\kappa^2(2s)c(\kappa(2s))$, which is a class $\mathcal{K}_\infty$ function, and satisfies, for all $s \geq 0$,

$$\tilde{\alpha}^{-1}(\delta\sigma_z(s)) = \kappa^{-1}\left(\frac{s}{2}\right).$$

Since $\kappa(\cdot)$ is C^1 and $\kappa(0) = 0$, there exists a C^0 function $\hat{\alpha}(x)$ such that $\tilde{\alpha}(|x|) = x^2\hat{\alpha}(x)$ for all x. Letting $\alpha_0(x) \geq \hat{\alpha}(x)$ shows that $\alpha(s) \geq \tilde{\alpha}(s)$, hence,

$$\kappa \circ \gamma_z(s) = \kappa \circ \alpha^{-1}(\delta\sigma_z(s)) \leq \kappa \circ \tilde{\alpha}^{-1}(\delta\sigma_z(s)) = \frac{s}{2}.$$

Thus, the small gain condition (7.20) is satisfied. To show that $\gamma_u(\cdot)$ is C^1, note that $\gamma_u^{-1}(s) = s\sqrt{\alpha_0(s)}/\sqrt{\delta}$. Thus $\gamma_u^{-1}(s)$ is C^1 and its derivative with respect to s is greater than 0 for all $s \geq 0$. By the Inverse Function Theorem, $\gamma_u(\cdot)$ is also C^1. □

Remark 7.10. In summary, the function $\rho(x)$ can be obtained as follows:

(i) Obtain $\phi_0(x)$ and $\phi_1(z)$ from (7.23).

(ii) Obtain the function $\tilde{\alpha}(s) = \delta\kappa^2(2s)c(\kappa(2s))$, where $\delta > 2$ and $c(\cdot)$ is a nondecreasing smooth function such that $c(||z||) \geq \phi_1^2(z)$.

(iii) Let $\hat{\alpha}(s)$ be a C^0 function such that $\tilde{\alpha}(|x|) = x^2\hat{\alpha}(x)$, and let $\alpha_0(x)$ be a smooth and even function nondecreasing in $[0, \infty)$ such that $\alpha_0(x) \geq \hat{\alpha}(x)$ for all $x \in \mathcal{R}$.

(iv) Obtain $\rho(x)$ from (7.24). ∎

Lemma 7.11. *Consider the system*

$$\dot{x} = f(x, u, \mu(t)), \quad t \geq t_0, \tag{7.28}$$

in which $x \in \mathcal{R}^n$, $u \in \mathcal{R}^m$, $\mu : [t_0, \ \infty) \to \Sigma$ piecewise continuous with Σ a compact set of $\mathcal{R}^{n_\mu}$, and $f(0, 0, \mu) = 0$ for all $\mu \in \Sigma$. Suppose system (7.28) is RISS with respect to μ with x as state and u as input and has a known C^1 class $\mathcal{K}_\infty$ gain function $\kappa(\cdot)$. Then, for any square matrix $G(u)$ of dimension m with its entries being sufficiently smooth functions of u, the system

$$\dot{x} = f(x, G(u)u, \mu(t)), \quad t \geq t_0 \geq 0, \tag{7.29}$$

is also RISS with respect to μ with x as state and u as input and has a known C^1 class $\mathcal{K}_\infty$ gain function $\tilde{\gamma}(s) = \kappa(sc(s))$, where $c : [0, \ \infty) \to [0, \ \infty)$ is some C^1 nondecreasing function such that $c(||u||) \geq ||G(u)||$ for all $u \in \mathcal{R}^m$.

Proof. By the assumption, there exist some class $\mathcal{KL}$ function $\beta^{kl}(\cdot, \cdot)$ and some known C^1 class $\mathcal{K}_\infty$ function $\kappa(\cdot)$, independent of $\mu(\cdot)$, such that the solution of (7.28) exists and satisfies, for all $x(t_0) \in \mathcal{R}^n$, all $u \in L_\infty^m$, all $\mu \in \sum$ and all $t \geq t_0 \geq 0$,

$$\|x(t)\| \leq \max\left\{\beta^{kl}(\|x(t_0)\|, t - t_0), \kappa\left(\sup_{t_0 \leq \tau \leq t} \|u(\tau)\|\right)\right\}.$$

Let $\gamma(s) = sc(s)$. Then $\gamma(s)$ is a C^1 class $\mathcal{K}_\infty$ function satisfying $\gamma(||u||) \geq ||G(u)||||u||$. It is now possible to verify that the solution of (7.29) satisfies, for all $t \geq t_0 \geq 0$,

$$\begin{aligned}
\|x(t)\| &\leq \max\left\{\beta^{kl}(\|x(t_0)\|, t - t_0), \kappa\left(\sup_{t_0 \leq \tau \leq t} \|G(u(\tau))u(\tau)\|\right)\right\} \\
&\leq \max\left\{\beta^{kl}(\|x(t_0)\|, t - t_0), \kappa\left(\sup_{t_0 \leq \tau \leq t} \gamma(\|u(\tau)\|)\right)\right\} \\
&\leq \max\left\{\beta^{kl}(\|x(t_0)\|, t - t_0), \kappa\left(\gamma(\sup_{t_0 \leq \tau \leq t} (\|u(\tau)\|)\right)\right\} \\
&\leq \max\left\{\beta^{kl}(\|x(t_0)\|, t - t_0), \tilde{\gamma}\left(\sup_{t_0 \leq \tau \leq t} \|u(\tau)\|\right)\right\},
\end{aligned}$$

where $\tilde{\gamma}(s) = \kappa(\gamma(s)) = \kappa(sc(s))$ is a known C^1 class $\mathcal{K}_\infty$ function. The proof is thus completed. □

We are now ready to complete the proof of Theorem 7.6. For this purpose, we consider the following transformation:

$$\begin{aligned} \tilde{x}_1 &= x_1, \\ \tilde{x}_2 &= x_2 - \alpha_1(\tilde{x}_1), \\ &\vdots \\ \tilde{x}_{j+1} &= x_{j+1} - \alpha_j(\tilde{x}_j) \end{aligned} \tag{7.30}$$

for some integer $1 \leq j \leq r$, where $\alpha_j(\tilde{x}_j) = -\tilde{x}_j \rho_j(\tilde{x}_j)$, with $\rho_j(\tilde{x}_j) \geq 0$ some smooth scalar function. Then, for any $1 \leq j \leq r$, under the transformation (7.30), system (7.12) can be put into the following form:

$$\begin{aligned} \dot{\bar{z}}_j &= F_j(\bar{z}_j, \tilde{x}_{j+1}, \mu), \\ \dot{z}_{j+1} &= \tilde{Q}_{j+1}(\bar{z}_j, z_{j+1}, \tilde{x}_{j+1}, \mu), \\ \dot{\tilde{x}}_{j+1} &= \tilde{f}_{j+1}(\bar{z}_j, z_{j+1}, \tilde{x}_{j+1}, \mu) + b_{j+1}(\mu) x_{j+2}, \\ \dot{z}_i &= Q_i(z_0, \ldots, z_i, \tilde{x}_1, \tilde{x}_2 + \alpha_1(\tilde{x}_1), \ldots, \tilde{x}_{j+1} + \alpha_j(\tilde{x}_j), x_{j+2}, \ldots, x_i, \mu), \\ \dot{x}_i &= f_i(z_0, \ldots, z_i, \tilde{x}_1, \tilde{x}_2 + \alpha_1(\tilde{x}_1), \ldots, \tilde{x}_{j+1} + \alpha_j(\tilde{x}_j), x_{j+2}, \ldots, x_i, \mu) + b_i(\mu) x_{i+1}, \\ & i = j+2, \ldots, r, \end{aligned} \tag{7.31}$$

where $\bar{z}_j = \text{col}\left(z_0, z_1, \tilde{x}_1, \ldots, z_j, \tilde{x}_j\right)$ and the other functions are defined recursively as follows:

$$F_j\left(\bar{z}_j, \tilde{x}_{j+1}, \mu\right) = \begin{bmatrix} F_{j-1}\left(\bar{z}_{j-1}, \tilde{x}_j, \mu\right) \\ \tilde{Q}_j(\bar{z}_{j-1}, z_j, \tilde{x}_j, \mu) \\ \tilde{f}_j(\bar{z}_{j-1}, z_j, \tilde{x}_j, \mu) + b_j(\mu)\left(\alpha_j(\tilde{x}_j) + \tilde{x}_{j+1}\right) \end{bmatrix}, \quad 1 \leq j \leq r,$$

$$\begin{aligned} \tilde{Q}_j(\bar{z}_{j-1}, z_j, \tilde{x}_j, \mu) &= Q_j(z_0, \ldots, z_j, \tilde{x}_1, \tilde{x}_2 + \alpha_1(\tilde{x}_1), \ldots, \tilde{x}_j + \alpha_{j-1}(\tilde{x}_{j-1}), \mu), \\ \tilde{f}_j(\bar{z}_{j-1}, z_j, \tilde{x}_j, \mu) &= f_j(z_0, \ldots, z_j, \tilde{x}_1, \tilde{x}_2 + \alpha_1(\tilde{x}_1), \ldots, \tilde{x}_j + \alpha_{j-1}(\tilde{x}_{j-1}), \mu) \\ &\quad - \nabla\alpha_{j-1}(\tilde{x}_{j-1})\dot{\tilde{x}}_{j-1}, \end{aligned}$$

with $\nabla\alpha_j(\tilde{x}_j) = \frac{\partial \alpha_j(\tilde{x}_j)}{\partial \tilde{x}_j}$ for $1 \leq j \leq r$. It is clear that system (7.12) itself is also in the form (7.31) with $j = 0$ upon defining $\bar{z}_0 = z_0$, $F_0(\bar{z}_0, \tilde{x}_1, \mu) = f_0(z_0, x_1, \mu)$, and $\alpha_0(\tilde{x}_0) = 0$.

Lemma 7.12. *Under Assumptions 7.2 to 7.4, for any $0 \leq j \leq r$, there exist smooth functions $\rho_1(\cdot), \ldots, \rho_j(\cdot)$ such that, with $\alpha_1(\tilde{x}_1) = -\tilde{x}_1\rho_1(\tilde{x}_1), \ldots, \alpha_j(\tilde{x}_j) = -\tilde{x}_j\rho_j(\tilde{x}_j)$, system (7.31) satisfies the following property.*

Property 7.3. The subsystem $\dot{\bar{z}}_j = F_j(\bar{z}_j, \tilde{x}_{j+1}, \mu)$, $t \geq t_0 \geq 0$, is RISS with respect to μ with state $\bar{z}_j$ and input $\tilde{x}_{j+1}$, and has a known C^1 class $\mathcal{K}_\infty$ gain function $\bar{\kappa}_j(\cdot)$.

Proof. We will prove it by using mathematical induction. When $j = 0$, Property 7.3 is implied by Assumption 7.3 with the known C^1 class $\mathcal{K}_\infty$ gain function being given by $\bar{\kappa}_0(\cdot) = \kappa_0(\cdot)$.

Now assume that, for some integer $0 \leq J < r$, there exist smooth functions $\rho_1(\cdot), \ldots, \rho_J(\cdot)$ such that, with $\alpha_1(\tilde{x}_1) = -\tilde{x}_1\rho_1(\tilde{x}_1), \ldots, \alpha_J(\tilde{x}_J) = -\tilde{x}_J\rho_J(\tilde{x}_J)$, system (7.31) with $j = J$ satisfies Property 7.3.

First note that applying a coordinate transformation $\tilde{x}_{J+2} = x_{J+2} - \alpha_{J+1}(\tilde{x}_{J+1})$ to system (7.31) with $j = J$ yields a system of the same form as (7.31) with $j = J + 1$. Then we will show that system (7.31) with $j = J + 1$ satisfies Property 7.3. Consider the system composed of the following three equations:

$$\dot{\bar{z}}_J = F_J(\bar{z}_J, \tilde{x}_{J+1}, \mu), \tag{7.32}$$

$$\dot{z}_{J+1} = \tilde{Q}_{J+1}(\bar{z}_J, z_{J+1}, \tilde{x}_{J+1}, \mu), \tag{7.33}$$

$$\dot{\tilde{x}}_{J+1} = \tilde{f}_{J+1}(\bar{z}_J, z_{J+1}, \tilde{x}_{J+1}, \mu) + b_{J+1}(\mu)x_{J+2}. \tag{7.34}$$

By induction assumption, system (7.32) is RISS with respect to μ, with $\bar{z}_J$ as state and $\tilde{x}_{J+1}$ as input, and has a known class $\mathcal{K}_\infty$ gain function $\bar{\kappa}_J(\cdot)$. Consider system (7.33). By definition,

$$\begin{aligned} &\tilde{Q}_{J+1}(\bar{z}_J, z_{J+1}, \tilde{x}_{J+1}, \mu) \\ &\quad = Q_{J+1}(z_0, \ldots, z_{J+1}, \tilde{x}_1, \tilde{x}_2 + \alpha_1(\tilde{x}_1), \ldots, \tilde{x}_{J+1} + \alpha_J(\tilde{x}_J), \mu). \end{aligned} \tag{7.35}$$

Let $\bar{u}_{J+1} = \operatorname{col}(\bar{z}_J, \tilde{x}_{J+1})$. Then there exists a square matrix $G_{J+1}(\bar{u}_{J+1})$ of dimension $n_0 + \cdots + n_J + J + 1$ with its entries a smooth function of $\bar{u}_{J+1}$ such that

$$\operatorname{col}(z_0, z_1, \ldots, z_J, \tilde{x}_1, \tilde{x}_2 + \alpha_2(\tilde{x}_1), \ldots, \tilde{x}_{J+1} + \alpha_J(\tilde{x}_J)) = G_{J+1}(\bar{u}_{J+1})\bar{u}_{J+1}.$$

Therefore, (7.33) can be written as

$$\dot{z}_{J+1} = Q_{J+1}(z_{J+1}, G_{J+1}(\bar{u}_{J+1})\bar{u}_{J+1}, \mu). \tag{7.36}$$

By Lemma 7.11, Assumption 7.4 implies that system (7.36), hence (7.33), is RISS with respect to μ with z_{J+1} as state and $\operatorname{col}(\bar{z}_J, \tilde{x}_{J+1})$ as input, and has some known C^1 gain function $\tilde{\gamma}_{J+1}(s)$. Thus by Corollary 2.20, the system consisting of (7.32) and (7.33) is RISS with $\operatorname{col}(\bar{z}_J, z_{J+1})$ as state and $\tilde{x}_{J+1}$ as input and has some known C^1 gain function $\tilde{\kappa}_{J+1}(s)$. Finally, note that equations (7.32) and (7.33) can be viewed as the upper subsystem of (7.16), and equation (7.34) can be viewed as the lower subsystem of (7.16). Applying Proposition 7.7 to system (7.32) to (7.34) shows the existence of a smooth function $\rho_{J+1}(\cdot)$ such that the following system:

$$\begin{aligned} \dot{\bar{z}}_J &= F_J(\bar{z}_J, \tilde{x}_{J+1}, \mu), \\ \dot{z}_{J+1} &= \tilde{Q}_{J+1}(\bar{z}_J, z_{J+1}, \tilde{x}_{J+1}, \mu), \\ \dot{\tilde{x}}_{J+1} &= \tilde{f}_{J+1}(\bar{z}_J, z_{J+1}, \tilde{x}_{J+1}, \mu) + b_{J+1}(\mu)(\alpha_{J+1}(\tilde{x}_{J+1}) + \tilde{x}_{J+2}), \end{aligned} \tag{7.37}$$

where $\alpha_{J+1}(\tilde{x}_{J+1}) = -\tilde{x}_{J+1}\rho_{J+1}(\tilde{x}_{J+1})$ is RISS with respect to μ with $\bar{z}_{J+1}$ as state and $\tilde{x}_{J+2}$ as input and has some known C^1 class $\mathcal{K}_\infty$ gain function $\bar{\kappa}_{J+1}(\cdot)$. The induction is completed upon noting that the system $\dot{\bar{z}}_{J+1} = F_{J+1}(\bar{z}_{J+1}, \tilde{x}_{J+2}, \mu)$ is nothing but (7.37) in a compact form. □

When $j = r$, system (7.31) becomes $\dot{\bar{z}}_r = F_r(\bar{z}_r, \tilde{x}_{r+1}, \mu)$, it is RISS with respect to μ with state $\bar{z}_r$ and input $\tilde{x}_{r+1}$, and it has a known C^1 class $\mathcal{K}_\infty$ gain function $\bar{\kappa}_r(\cdot)$. In particular, when $\tilde{x}_{r+1} = 0$, that is, under the state feedback controller $u = \alpha_r(\tilde{x}_r)$, the equilibrium point $\bar{z}_r = 0$ of the closed-loop system $\dot{\bar{z}}_r = F_r(\bar{z}_r, 0, \mu)$ is globally asymptotically stable for all $\mu \in \Sigma$. As a result, the proof of Theorem 7.6 is completed.

It is not difficult to see that the controller that solves the global robust stabilization problem for the original system (7.12) is given by

$$\begin{aligned} u &= \alpha_r(\tilde{x}_r), \\ \tilde{x}_i &= x_i - \alpha_{i-1}(\tilde{x}_{i-1}), \quad i = r, \ldots, 2, \\ \tilde{x}_1 &= x_1. \end{aligned}$$

Remark 7.13. For $0 \le j < r$, let $\beta_j^{kl}(\cdot, \cdot)$ be some class $\mathcal{KL}$ function and $\kappa_j(\cdot)$ some known C^1 class $\mathcal{K}_\infty$ gain functions as described in (7.14) and (7.15) and let $\bar{\kappa}_j(\cdot)$ be the C^1 class $\mathcal{K}_\infty$ gain function of the subsystem $\dot{\bar{z}}_j = F_j(\bar{z}_j, \tilde{x}_{j+1}, \mu)$. Then the C^1 class $\mathcal{K}_\infty$ gain function $\bar{\kappa}_{j+1}(\cdot)$ of the subsystem $\dot{\bar{z}}_{j+1} = F_{j+1}(\bar{z}_{j+1}, \tilde{x}_{j+2}, \mu)$ can be obtained recursively as follows. Let $c_{j+1} : [0, \infty) \to [0, \infty)$ be any C^1 nondecreasing function such that $c_{j+1}(\|\bar{u}_{j+1}\|) \ge \|G(\bar{u}_{j+1})\|$ for all $\bar{u}_{j+1} \in \mathcal{R}^{n_0+\cdots+n_j+j+1}$. Then, Lemma 7.11 states that $\tilde{\gamma}_{j+1}(s) = \kappa_j(sc_{j+1}(s))$ is a C^1 class $\mathcal{K}_\infty$ gain function for the subsystem (7.33). Hence, the solution of the subsystem (7.33) exists and satisfies, for all $t \ge t_0 \ge 0$,

$$\begin{aligned} \|z_{j+1}(t)\| &\le \max\left\{\beta_{j+1}^{kl}(\|z_{j+1}(t_0)\|, t - t_0), \tilde{\gamma}_{j+1}\left(\sup_{t_0 \le \tau \le t} \|\mathrm{col}(\bar{z}_j(\tau), \tilde{x}_{j+1}(\tau))\|\right)\right\} \\ &\le \max\left\{\beta_{j+1}^{kl}(\|z_{j+1}(t_0)\|, t - t_0), \tilde{\gamma}_{j+1}\left(2\sup_{t_0 \le \tau \le t} \|\bar{z}_j(\tau)\|\right), \tilde{\gamma}_{j+1}\left(2\sup_{t_0 \le \tau \le t} \|\tilde{x}_{j+1}(\tau)\|\right)\right\}. \end{aligned}$$

Then, by Corollary 2.20, the gain function of the system consisting of (7.32) and (7.33) is any C^1 class $\mathcal{K}_\infty$ function $\tilde{\kappa}_{j+1}(\cdot)$ satisfying

$$\tilde{\kappa}_{j+1}(s) \ge \max\left\{2\bar{\kappa}_j(s), 2\tilde{\gamma}_{j+1}(2s), 2\tilde{\gamma}_{j+1} \circ (2\bar{\kappa}_j(s))\right\}.$$

Finally, by Proposition 7.7, there exists a smooth function $\rho_{j+1}(\tilde{x}_{j+1})$ such that the subsystem

$$\dot{\tilde{x}}_{j+1} = \tilde{f}_{j+1}(\bar{z}_j, z_{j+1}, \tilde{x}_{j+1}, \mu) + b_{j+1}(\mu)(\alpha_{j+1}(\tilde{x}_{j+1}) + \tilde{x}_{j+2}), \tag{7.38}$$

where $\alpha_{j+1}(\tilde{x}_{j+1}) = -\tilde{x}_{j+1}\rho_{j+1}(\tilde{x}_{j+1})$ is RISS with respect to μ with $\tilde{x}_{j+1}$ as state and $\mathrm{col}(\tilde{u}_{j+1}, \tilde{x}_{j+2})$, where $\tilde{u}_{j+1} = (\bar{z}_j, z_{j+1})$, as input, and has some known C^1 gain function. In particular, the solution of system (7.38) exists and satisfies, for all $t \ge t_0 \ge 0$, for some class $\mathcal{KL}$ function $\tilde{\beta}_{j+1}^{kl}(\cdot, \cdot)$, known class $\mathcal{K}_\infty$ function $\gamma_{\tilde{u}_{j+1}}(\cdot)$, and known C^1 class $\mathcal{K}_\infty$ function $\gamma_{\tilde{x}_{j+2}}(\cdot)$,

$$\begin{aligned} |\tilde{x}_{j+1}(t)| \le \max\Bigg\{&\tilde{\beta}_{j+1}^{kl}(|\tilde{x}_{j+1}(t_0)|, t - t_0), \\ &\gamma_{\tilde{u}_{j+1}}\left(\sup_{t_0 \le \tau \le t} \|\tilde{u}_{j+1}(\tau)\|\right), \gamma_{\tilde{x}_{j+2}}\left(\sup_{t_0 \le \tau \le t} \|\tilde{x}_{j+2}(\tau)\|\right)\Bigg\} \end{aligned}$$

for all $\tilde{x}_{j+1}(t_0) \in \mathcal{R}$. Moreover, by Proposition 7.7, the gain function $\bar{\kappa}_{j+1}(\cdot)$ is given by any C^1 class $\mathcal{K}_\infty$ function satisfying

$$\bar{\kappa}_{j+1}(s) \ge \max\{2\tilde{\kappa}_{j+1} \circ \gamma_{\tilde{x}_{j+2}}(s), 2\gamma_{\tilde{x}_{j+2}}(s)\}. \quad \blacksquare \tag{7.39}$$

Remark 7.14. As pointed out before, system (7.13) is a special case of (7.12) when $n_1 = \cdots = n_r = 0$. Under the transformation (7.30), for $0 \leq j \leq r$, system (7.13) can be put into the following form:

$$\begin{aligned}
\dot{\bar{z}}_j &= F_j(\bar{z}_j, \tilde{x}_{j+1}, \mu),\\
\dot{\tilde{x}}_{j+1} &= \tilde{f}_{j+1}(\bar{z}_j, \tilde{x}_{j+1}, \mu) + b_{j+1}(\mu)x_{j+2},\\
\dot{x}_i &= f_i(z, \tilde{x}_1, \tilde{x}_2 + \alpha_1(\tilde{x}_1), \ldots, \tilde{x}_{j+1} + \alpha_j(\tilde{x}_j), x_{j+2}, \ldots, x_i, \mu) + b_i(\mu)x_{i+1},\\
i &= j+2, \ldots, r,
\end{aligned} \tag{7.40}$$

where $\bar{z}_j = \operatorname{col}\left(z, \tilde{x}_1, \ldots, \tilde{x}_j\right)$ and the other functions are defined recursively as follows:

$$F_j\left(\bar{z}_j, \tilde{x}_{j+1}, \mu\right) = \begin{bmatrix} F_{j-1}\left(\bar{z}_{j-1}, \tilde{x}_j, \mu\right) \\ \tilde{f}_j(\bar{z}_{j-1}, \tilde{x}_j, \mu) + b_j(\mu)\left(\alpha_j(\tilde{x}_j) + \tilde{x}_{j+1}\right) \end{bmatrix}, \quad 1 \leq j \leq r,$$

$$\tilde{f}_j(\bar{z}_{j-1}, \tilde{x}_j, \mu) = f_j(z, \tilde{x}_1, \tilde{x}_2 + \alpha_1(\tilde{x}_1), \ldots, \tilde{x}_j + \alpha_{j-1}(\tilde{x}_{j-1}), \mu) - \nabla\alpha_{j-1}(\tilde{x}_{j-1})\dot{\tilde{x}}_{j-1}$$

with $\nabla\alpha_j(\tilde{x}_j) = \frac{\partial\alpha_j(\tilde{x}_j)}{\partial\tilde{x}_j}$ for $j = 1, \ldots, r$. ■

Since system (7.13) satisfies Assumption 7.4 automatically, we have the following result on the solvability of the global robust stabilization of the lower triangular system (7.13) as follows.

Corollary 7.15. *Under Assumptions 7.2 and 7.3, for any $0 \leq j \leq r$, there exist smooth functions $\alpha_i(\cdot)$, $i = 0, \ldots, j$, such that system (7.40) satisfies the following property.*

Property 7.4. The subsystem $\dot{\bar{z}}_j = F_j(\bar{z}_j, \tilde{x}_{j+1}, \mu)$ is RISS with respect to μ with state $\bar{z}_j$ and input $\tilde{x}_{j+1}$ and has a known C^1 class $\mathcal{K}_\infty$ gain function $\bar{\kappa}_j(\cdot)$.

As a result, there exists a smooth state feedback controller $u = k(x_1, \ldots, x_r)$ that solves the global robust stabilization problem of system (7.13).

Remark 7.16. For system (7.40), for any $0 \leq j < r$, a C^1 class $\mathcal{K}_\infty$ gain function $\bar{\kappa}_{j+1}(\cdot)$ of the subsystem $\dot{\bar{z}}_j = F_j(\bar{z}_j, \tilde{x}_{j+1}, \mu)$ can be more easily obtained from a given C^1 gain function $\bar{\kappa}_j(\cdot)$ as follows. In fact, applying Proposition 7.7 to the system consisting of the subsystem $\dot{\bar{z}}_j = F_j(\bar{z}_j, \tilde{x}_{j+1}, \mu)$ and (7.38) immediately concludes that the gain function $\bar{\kappa}_{j+1}(\cdot)$ is given by any C^1 class $\mathcal{K}_\infty$ function satisfying

$$\bar{\kappa}_{j+1}(s) \geq \max\{2\bar{\kappa}_j \circ \gamma_{\tilde{x}_{j+2}}(s), 2\gamma_{\tilde{x}_{j+2}}(s)\}, \tag{7.41}$$

where $\gamma_{\tilde{x}_{j+2}}(s)$ is as defined in Remark 7.13. ■

7.3 Global Robust Output Regulation for Output Feedback Systems

Consider the class of nonlinear systems described by

$$\begin{aligned}
\dot{x} &= \bar{F}(w)x + \bar{G}(y, w)y + g(w)u,\\
\dot{y} &= \bar{H}(w)x + \bar{K}(y, w)y,
\end{aligned} \tag{7.42}$$

where $\text{col}(x, y) \in \mathcal{R}^n$ is the state, $y \in \mathcal{R}$ the output, $u \in \mathcal{R}$ the input, $w \in \mathcal{R}^{n_w}$ is a vector of uncertain parameter, and all the functions are sufficiently smooth.

Systems described by (7.42) are called nonlinear systems in output feedback form. The problem of global robust stabilization of such systems by output feedback control has been well studied in the literature [86]. In this section, we will further study the robust output regulation problem for a modified version of (7.42) as follows:

$$\begin{aligned} \dot{x} &= \bar{F}(w)x + \bar{G}(y, v, w)y + g(w)u + \bar{D}_1(v, w), \\ \dot{y} &= \bar{H}(w)x + \bar{K}(y, v, w)y + \bar{D}_2(v, w), \\ e &= y - q(v, w), \end{aligned} \tag{7.43}$$

where v is the exogenous signal generated by $\dot{v} = A_1 v$ and $\bar{D}_1(v, w)$, $\bar{D}_2(v, w)$, and $q(v, w)$ are sufficiently smooth functions satisfying $\bar{D}_1(0, w) = 0$, $\bar{D}_2(0, w) = 0$, and $q(0, w) = 0$ for all $w \in \mathcal{R}^{n_w}$.

The first step towards solving the robust output regulation problem for system (7.43) is to convert the system into the lower triangular form through a suitable dynamic extension and coordinate transformation. For this purpose, let us first make the following assumption.

Assumption 7.5. System (7.43) has a uniform relative degree $r \geq 2$; that is, for all $w \in \mathcal{R}^{n_w}$,

$$\bar{H}(w)g(w) = \bar{H}(w)\bar{F}(w)g(w) = \cdots = \bar{H}(w)\bar{F}^{r-3}(w)g(w) = 0$$

and

$$\bar{H}(w)\bar{F}^{r-2}(w)g(w) \neq 0.$$

Now define the following dynamic extension:

$$\dot{\xi} = F_e \xi + G_e u, \tag{7.44}$$

where $\xi = \text{col}(\xi_1, \ldots, \xi_{r-1})$ with $\xi_i \in \mathcal{R}$ for $i = 1, \ldots, r-1$,

$$F_e = \begin{bmatrix} -\lambda_1 & 1 & 0 & \cdots & 0 \\ 0 & -\lambda_2 & 1 & \cdots & 0 \\ \vdots & \vdots & \vdots & \cdots & \vdots \\ 0 & 0 & 0 & \cdots & -\lambda_{r-1} \end{bmatrix}, \quad G_e = \begin{bmatrix} 0 \\ \vdots \\ 0 \\ 1 \end{bmatrix},$$

with $\lambda_i, i = 1, \ldots, r-1$, being positive numbers.

Next we perform on the extended system (7.43) and (7.44) the following coordinate transformation:

$$z = x - D(w)\xi - h(w)y, \tag{7.45}$$

which turns the extended system (7.43) and (7.44) into the following:

$$\begin{aligned} \dot{z} &= (\bar{F}(w) - h(w)\bar{H}(w))z \\ &\quad + [(\bar{F}(w) - h(w)\bar{H}(w))h(w) + \bar{G}(y, v, w) - h(w)\bar{K}(y, v, w)]y \\ &\quad + (\bar{F}(w)D(w) - D(w)F_e - h(w)\bar{H}(w)D(w))\xi \\ &\quad + (g(w) - D(w)G_e)u + \bar{D}_1(v, w) - h(w)\bar{D}_2(v, w), \\ \dot{y} &= \bar{H}(w)z + (\bar{H}(w)h(w) + \bar{K}(y, v, w))y + \bar{H}(w)D(w)\xi + \bar{D}_2(v, w), \\ \dot{\xi} &= F_e\xi + G_e u. \end{aligned} \tag{7.46}$$

Clearly, in order to render system (7.46) a lower triangular form, it suffices to choose $D(w)$ and $h(w)$ such that, for some scalar function $b(w)$,

$$\begin{aligned} \bar{F}(w)D(w) - D(w)F_e &= h(w)\bar{H}(w)D(w), \\ g(w) &= D(w)G_e, \\ \bar{H}(w)D(w)\xi &= b(w)\xi_1, \end{aligned}$$

or, equivalently, for some scalar function $b(w)$,

$$\begin{aligned} \bar{F}(w)D(w) - D(w)F_e &= h(w)[b(w), 0, \ldots, 0], \\ g(w) &= D(w)G_e, \\ \bar{H}(w)D(w) &= [b(w), 0, \ldots, 0]. \end{aligned} \tag{7.47}$$

Let us first obtain $D(w)$. For this purpose, assume

$$D(w) = [d_1(w), d_2(w), \ldots, d_{r-1}(w)]. \tag{7.48}$$

Substituting (7.48) into the first equation of (7.47) gives

$$d_{i-1}(w) = (\lambda_i I + \bar{F}(w))d_i(w), \quad i = r-1, \ldots, 2, \tag{7.49}$$

with $h(w)$ and $b(w)$ satisfying

$$(\lambda_1 I + \bar{F})d_1(w) = h(w)b(w). \tag{7.50}$$

Substituting (7.48) into the second equation of (7.47) gives

$$d_{r-1}(w) = g(w). \tag{7.51}$$

Substituting (7.51) into (7.49) gives

$$\begin{aligned} d_{r-1}(w) &= g(w), \\ d_{r-2}(w) &= (\bar{F} + \lambda_{r-1}I)g(w), \\ &\vdots \\ d_1(w) &= (\bar{F} + \lambda_2 I)\cdots(\bar{F} + \lambda_{r-1}I)g(w). \end{aligned} \tag{7.52}$$

It is noted that, when $r = 2$, the last equation of (7.52) should be understood as $d_1(w) = g(w)$.

It is now possible to verify, using Assumption 7.5, that $D(w)$ as defined in (7.52) indeed satisfies the third equation of (7.47) by letting

$$b(w) = \bar{H}(w)\bar{F}(w)^{r-2}g(w).$$

Finally, substituting $d_1(w)$ into (7.50) gives

$$h(w) = \frac{d(w)}{b(w)}, \tag{7.53}$$

where

$$d(w) = (\bar{F} + \lambda_1 I)(\bar{F} + \lambda_2 I)\cdots(\bar{F} + \lambda_{r-1}I)g(w).$$

With $D(w)$, $h(w)$, and $b(w)$ defined as above, the extended system together with the exosystem takes the following form:

$$\begin{aligned}
\dot{z} &= F(w)\,z + G(y, v, w)\,y + D_1(v, w),\\
\dot{y} &= H(w)\,z + K(y, v, w)\,y + b(w)\,\xi_1 + D_2(v, w),\\
\dot{\xi} &= F_e\xi + G_e u,\\
\dot{v} &= A_1 v,\\
e &= y - q(v, w),
\end{aligned} \tag{7.54}$$

where

$$\begin{aligned}
F(w) &= \bar{F}(w) - \frac{d(w)}{b(w)}\bar{H}(w),\\
G(y, v, w) &= \left(\bar{F}(w) - \frac{d(w)}{b(w)}\bar{H}(w)\right)\frac{d(w)}{b(w)} + \bar{G}(y, v, w) - \frac{d(w)}{b(w)}\bar{K}(y, v, w),\\
H(w) &= \bar{H}(w),\\
K(y, v, w) &= \bar{H}(w)\frac{d(w)}{b(w)} + \bar{K}(y, v, w),\\
D_1(v, w) &= \bar{D}_1(v, w) - \frac{d(w)}{b(w)}\bar{D}_2(v, w),\\
D_2(v, w) &= \bar{D}_2(v, w).
\end{aligned}$$

Finally, we will establish a property regarding the matrix $F(w)$ as follows.

Lemma 7.17. *Assume $\lambda_i > 0$ for $i = 1, \dots, r-1$. Then the eigenvalues of the matrix $F(w)$ have negative real parts for all $w \in \mathcal{R}^{n_w}$ if and only if the following assumption holds.*

Assumption 7.6. For all $w \in \mathcal{R}^{n_w}$, the linear system

$$\dot{x} = \bar{F}(w)\,x + g(w)\,u, \quad \dot{y} = \bar{H}(w)\,x \tag{7.55}$$

with y as output is a minimum phase system.

Proof. The numerator polynomial of the transfer function from u to y of (7.55) is given by

$$\det\begin{bmatrix} sI_{n-1} - \bar{F} & 0 & -g(w)\\ -\bar{H} & s & 0\\ 0 & 1 & 0 \end{bmatrix}. \tag{7.56}$$

On the other hand, the numerator polynomial of the transfer function from u to y of the following system:

$$\begin{aligned}
\dot{x} &= \bar{F}(w)\,x + g(w)\,u,\\
\dot{y} &= \bar{H}(w)\,x,\\
\dot{\xi} &= F_e\xi + G_e u
\end{aligned} \tag{7.57}$$

is given by

$$\det\begin{bmatrix} sI_{n-1}-\bar{F} & 0 & 0 & -g(w) \\ -\bar{H} & s & 0 & 0 \\ 0 & 0 & sI_{r-1}-F_e & -G_e \\ 0 & 1 & 0 & 0 \end{bmatrix}$$

$$= \det\begin{bmatrix} sI_{n-1}-\bar{F} & 0 & -g(w) & 0 \\ -\bar{H} & s & 0 & 0 \\ 0 & 1 & 0 & 0 \\ 0 & 0 & -G_e & sI_{r-1}-F_e \end{bmatrix} \tag{7.58}$$

$$= \det\begin{bmatrix} sI_{n-1}-\bar{F} & 0 & -g(w) \\ -\bar{H} & s & 0 \\ 0 & 1 & 0 \end{bmatrix}\det(sI_{r-1}-F_e). \tag{7.59}$$

Thus, under assumption $\lambda_i > 0$, $i = 1, \ldots, r-1$, system (7.57) with y as output is minimum phase if and only if Assumption 7.6 holds.

Now by a mere inspection (refer to Remark 2.46), it can easily be found that the zero dynamics of the following system:

$$\begin{aligned} \dot{z} &= F(w)\,z, \\ \dot{y} &= H(w)\,z + b(w)\,\xi_1, \\ \dot{\xi} &= F_e\xi + G_e u, \end{aligned} \tag{7.60}$$

with y as the output, is given by

$$\dot{z} = F(w)z.$$

The proof follows from the fact that the zero dynamics of systems (7.57) and (7.60) with y as the output are the same (modulo the coordinate transformation (7.45)), and, from part (iii) of Remark 2.45, that the eigenvalues of the matrix F coincide with the roots of the numerator polynomial of the transfer function from u to y of system (7.57). $\square$

We are now ready to consider the robust output regulation problem of system (7.54). We need two more assumptions.

Assumption 7.7. For all $w \in \mathcal{R}^{n_w}$, $b(w) > 0$.

Assumption 7.8. There exists a sufficiently smooth function $\mathbf{z}(v, w)$ with $\mathbf{z}(0, 0) = 0$ satisfying, for all $v \in \mathcal{R}^q$ and all $w \in \mathcal{R}^{n_w}$,

$$\frac{\partial \mathbf{z}(v, w)}{\partial v} A_1 v = F(w)\mathbf{z}(v, w) + G(q(v, w), v, w)q(v, w) + D_1(v, w). \tag{7.61}$$

Under Assumptions 7.7 and 7.8, let

$$\begin{aligned} \mathbf{y}(v, w) &= q(v, w), \\ \Xi_1(v, w) &= \frac{1}{b(w)}\Big(\frac{\partial q(v, w)}{\partial v} A_1 v - H(w)\mathbf{z}(v, w) \\ &\qquad - K(q(v, w), v, w)q(v, w) - D_2(v, w)\Big) \end{aligned}$$

and

$$\Xi_i(v,w) = \frac{\partial \Xi_{i-1}(v,w)}{\partial v} A_1 v + \lambda_{i-1} \Xi_{i-1}(v,w), \; i = 2, \ldots, r-1,$$
$$\mathbf{u}(v,w) = \frac{\partial \Xi_{r-1}(v,w)}{\partial v} A_1 v + \lambda_{r-1} \Xi_{r-1}(v,w). \tag{7.62}$$

Then it can be verified that the regulator equations associated with system (7.54) have a solution given by col $(\mathbf{z}(v,w), \mathbf{y}(v,w), \Xi(v,w))$ and $\mathbf{u}(v,w)$, where $\Xi(v,w) = \mathrm{col}(\Xi_1(v,w), \ldots, \Xi_{r-1}(v,w))$.

Remark 7.18. Equation (7.61) is a type of center manifold equation studied in Section 4.4. By Lemma 4.13, if none of the eigenvalues of the matrix $F(w)$ coincide with any λ given by $\{ \lambda \mid \lambda = l_1\lambda_1 + \cdots + l_q\lambda_q, l_1 + \cdots + l_q = l, \; l = 1, 2, \ldots, \; l_1, \ldots, l_q = 0, 1, \ldots, l \}$, where $\lambda_1, \ldots, \lambda_q$ are eigenvalues of the matrix A_1, then (7.61) has a formal power series solution of the form

$$\mathbf{z}(v,w) = \sum_{l \geq 1} Z_l(w) v^{[l]}, \tag{7.63}$$

where, for all $l = 1, 2, \ldots$, $Z_l(w)$ satisfies the Sylvester equation of the form

$$Z_l A^{[l]}(w) = F(w) Z_l(w) + G_l(w), \tag{7.64}$$

where $G_l(w)$ is such that $G(q(v,w), v, w) q(v,w) + D_1(v,w) = \sum_{l \geq 1} G_l(w) v^{[l]}$. In particular, when $q(v,w)$ and $D_1(v,w)$ are polynomials in v and $G(y,v,w)$ is a polynomial in v and y, then for some integer k, $G(q(v,w), v, w) q(v,w) + D_1(v,w)$ is a degree k polynomial in v. In this case, equation (7.61) has a unique globally defined solution which is a polynomial of degree k in v. ■

Next, we will convert the robust output regulation problem for system (7.54) into a robust stabilization problem for an augmented system. For this purpose, we will follow the procedure detailed in Section 6.2 to obtain the steady-state generator of (7.54) and a corresponding internal model.

Lemma 7.19. *Assume that there exist pairwise coprime polynomials $\pi_1(v,w), \ldots, \pi_I(v,w)$, with $r_1, \ldots, r_I$ being the degrees of their minimal zeroing polynomials $P_1(s), \ldots, P_I(s)$ and sufficiently smooth function $\Gamma_1 : \mathcal{R}^{r_1 + \cdots + r_I} \to \mathcal{R}$ vanishing at the origin such that, for all $v \in \mathcal{R}^q$ and all $w \in \mathcal{R}^{n_w}$,*

$$\Xi_1(v,w) = \Gamma_1\Big(\pi_1(v,w), \dot{\pi}_1(v,w), \ldots, \pi_1^{(r_1-1)}(v,w), \ldots, \pi_I(v,w), \dot{\pi}_I(v,w), \ldots, \pi_I^{(r_I-1)}(v,w)\Big) \tag{7.65}$$

and

$$\textit{for } i = 1, \ldots, I, \textit{ the pair } (\Psi_i, \Phi_i) \textit{ is observable}, \tag{7.66}$$

where $\Psi = (\Psi_1, \ldots, \Psi_l)$ is the gradient of Γ_1 at the origin with $\Psi_i \in \mathcal{R}^{1\times r_i}$, and Φ_i is the companion matrix of $P_i(s)$. Then system (7.54) has a linearly observable steady-state generator $\{\theta, \alpha, \beta\}$ with output $g_o(z, y, \xi_1, \ldots, \xi_{r-1}, u) = \mathrm{col}(\xi_1, \ldots, \xi_{r-1}, u)$.

Proof. By Lemma 6.17, system (7.54) has a linearly observable steady-state generator $\{\theta, \alpha, \beta_1\}$ with output ξ_1. Specifically, let

$$\theta(v, w) = T \begin{bmatrix} \pi_1(v, w) \\ \dot{\pi}_1(v, w) \\ \vdots \\ \pi_1^{(r_1-1)}(v, w) \\ \vdots \\ \pi_l(v, w) \\ \dot{\pi}_l(v, w) \\ \vdots \\ \pi_l^{(r_l-1)}(v, w) \end{bmatrix}, \quad \alpha(\theta) = T\Phi T^{-1}\theta, \quad \beta_1(\theta) = \Gamma_1(T^{-1}\theta),$$

where $\Phi = \mathrm{diag}(\Phi_1, \ldots, \Phi_l)$ and T is any nonsingular matrix with the appropriate dimension. Then, $\dot{\theta} = \alpha(\theta) = T\Phi T^{-1}\theta$ and $\Xi_1(v, w) = \beta_1(\theta(v, w)) = \Gamma_1(T^{-1}\theta(v, w))$. Now, utilizing the relation (7.62) gives a linearly observable steady-state generator with output $g_o(z, y, \xi_1, \ldots, \xi_{r-1}, u) = \mathrm{col}(\xi_1, \ldots, \xi_{r-1}, u)$ as follows. Let $\beta(\theta(v, w)) = \mathrm{col}(\beta_1(\theta(v, w)), \ldots, \beta_r(\theta(v, w)))$ where

$$\beta_i(\theta(v, w)) = \dot{\beta}_{i-1}(\theta(v, w)) + \lambda_{i-1}\beta_{i-1}(\theta(v, w)), \ i = 2, \ldots, r.$$

Then, clearly,

$$g_o\left(\mathbf{z}(v, w), \mathbf{y}(v, w), \Xi_1(v, w), \ldots, \Xi_{r-1}(v, w), \mathbf{u}(v, w)\right) = \beta(\theta(v, w)).$$

Therefore, $\{\theta, \alpha, \beta\}$ is a steady-state generator with output $g_o(z, y, \xi_1, \ldots, \xi_{r-1}, u) = \mathrm{col}(\xi_1, \ldots, \xi_{r-1}, u)$. Moreover, since the pair (β_1, α) is linearly observable, so is the pair (β, α). □

Note that in synthesizing the steady-state generator with output $\mathrm{col}(\xi_1, \ldots, \xi_{r-1}, u)$, we have taken advantage of the fact that the functions $\Xi_i(v, w), i = 1, \ldots, r-1$, and $\mathbf{u}(v, w)$ rely on the same set of polynomials. Therefore, the dimension of the steady-state generator with output $\mathrm{col}(\xi_1, \ldots, \xi_{r-1}, u)$ is the same as that of the steady-state generator with output ξ_1. As a result, the dimension of the steady-state generator with output $\mathrm{col}(\xi_1, \ldots, \xi_{r-1}, u)$ is much smaller than what would have been obtained by the general approach given in Lemma 6.17.

Taking advantage of the lower dimensional steady-state generator obtained here, we can also obtain a lower dimensional internal model. Pick any matrices

$$M \in \mathcal{R}^{(r_1+\cdots+r_l)\times(r_1+\cdots+r_l)} \quad \text{and} \quad N \in \mathcal{R}^{(r_1+\cdots+r_l)\times 1}$$

such that (M, N) is controllable and M Hurwitz. Then there exists a unique nonsingular matrix T satisfying the Sylvester equation

$$T\Phi - MT = N\Psi$$

since the pair (Ψ, Φ) is observable. Let

$$\dot{\eta} = M\eta + N(\xi_1 - \beta_1(\eta) + \Psi T^{-1}\eta) \stackrel{def}{=} \gamma(\eta, \xi_1). \tag{7.67}$$

Then

$$\begin{aligned}\gamma(\theta, \Xi_1) &= M\theta + N(\Xi_1 - \beta_1(\theta) + \Psi T^{-1}\theta) \\ &= M\theta + N\Psi T^{-1}\theta \\ &= T\Phi T^{-1}\theta = \alpha(\theta).\end{aligned}$$

Thus, (7.67) is an internal model of system (7.54) with output $g_o(z, y, \xi_1, \dots, \xi_{r-1}, u) = \text{col}(\xi_1, \dots, \xi_{r-1}, u)$. It will be seen later, in Theorem 7.21, that this particular internal model will facilitate the solution of the robust stabilization problem of the augmented system composed of the given plant and the internal model.

Remark 7.20. It is known from Remark 6.22 that in the special case where $\Xi_1(v, w)$ is a polynomial in v, the function Γ_1 is linear, and therefore $\beta_1(\eta) = \Psi T^{-1}\eta$. The internal model (7.67) becomes $\dot{\eta} = M\eta + N\xi_1$. ■

Now attaching the internal model (7.67) to system (7.54) yields the augmented system with the state variables $(z, y, \xi_1, \dots, \xi_{r-1}, \eta)$. Performing on the augmented system the following coordinate and input transformation:

$$\begin{aligned}\bar{z} &= z - \mathbf{z}(v, w), \\ e &= y - q(v, w), \\ \bar{\xi}_i &= \xi_i - \beta_i(\eta), \qquad i = 1, \dots, r-1, \\ \bar{\eta} &= \eta - \theta(v, w), \\ \bar{u} &= u - \beta_r(\eta)\end{aligned} \tag{7.68}$$

defines the augmented system in new coordinates and input as follows:

$$\dot{\bar{z}} = F(w)\bar{z} + \tilde{G}(e, v, w)e, \tag{7.69}$$

$$\dot{e} = H(w)\bar{z} + \tilde{K}(e, v, w)e + b(w)(\beta_1(\bar{\eta} + \theta) - \beta_1(\theta)) + b(w)\bar{\xi}_1, \tag{7.70}$$

$$\dot{\bar{\xi}}_i = -\frac{\partial \beta_i(\eta)}{\partial \eta} N\bar{\xi}_1 - \lambda_i \bar{\xi}_i + \bar{\xi}_{i+1}, \; i = 1, \dots, r-1, \tag{7.71}$$

$$\dot{\bar{\eta}} = (M + N\Psi T^{-1})\bar{\eta} + N\bar{\xi}_1, \tag{7.72}$$

where

$$\begin{aligned}\tilde{G}(e, v, w)e &= G(q + e, v, w)(q + e) - G(q, v, w)q, \\ \tilde{K}(e, v, w)e &= K(q + e, v, w)(q + e) - K(q, v, w)q,\end{aligned}$$

and $\bar{\xi}_r = \bar{u}$. It is noted that, in deriving equation (7.71), we have used the following identity:

$$\frac{\partial \beta_i(\eta)}{\partial \eta} T\Phi T^{-1}\eta + \lambda_i \beta_i(\eta) - \beta_{i+1}(\eta) = 0, \; i = 1, \dots, r-1.$$

By Corollary 7.4, all we need to do is globally stabilize the transformed augmented system consisting of (7.69) to (7.72). However, this system is not in the familiar lower triangular form (7.13) yet. Therefore, let us perform on the subsystem (7.72) another coordinate transformation as follows:

$$\tilde{\eta} = \bar{\eta} - Nb^{-1}(w)e, \tag{7.73}$$

which yields

$$\begin{aligned}\dot{\tilde{\eta}} &= \dot{\bar{\eta}} - Nb^{-1}(w)\dot{e} \\ &= (M + N\Psi T^{-1})(\tilde{\eta} + Nb^{-1}(w)e) + N\bar{\xi}_1 \\ &\quad - Nb^{-1}(w)(H(w)\bar{z} + \tilde{K}(e, v, w)e + b(w)(\beta_1(\bar{\eta} + \theta) - \beta_1(\theta)) + b(w)\bar{\xi}_1).\end{aligned} \tag{7.74}$$

Introducing the notation $\beta_1^{[2]}(x)$ to denote the nonlinear part of $\beta_1(x)$, that is,

$$\beta_1^{[2]}(x) = \beta_1(x) - \Psi T^{-1}x, \tag{7.75}$$

gives

$$\beta_1(\bar{\eta} + \theta) - \beta_1(\theta) = \beta_1^{[2]}(\tilde{\eta} + Nb^{-1}(w)e + \theta(v, w)) - \beta_1^{[2]}(\theta(v, w)) + \Psi T^{-1}\bar{\eta}. \tag{7.76}$$

Substituting (7.76) into (7.74) gives

$$\dot{\tilde{\eta}} = M\tilde{\eta} - N(\beta_1^{[2]}(\tilde{\eta} + Nb^{-1}(w)e + \theta(v, w)) - \beta_1^{[2]}(\theta(v, w))) + \bar{\phi}(\bar{z}, e, v, w), \tag{7.77}$$

where

$$\bar{\phi}(\bar{z}, e, v, w) = MNb^{-1}(w)e - b^{-1}(w)N(H(w)\bar{z} + \tilde{K}(e, v, w)e)$$

with $\bar{\phi}(0, 0, v, w) = 0$.

Denoting $Z = \text{col}(\bar{z}, \tilde{\eta})$, $x = \text{col}(x_1, \ldots, x_r) = \text{col}(e, \bar{\xi}_1, \ldots, \bar{\xi}_{r-1})$, $x_{r+1} = \bar{u} = \bar{\xi}_r$, $\mu = \text{col}(v, w)$, and $b_1(\mu) = b(w)$, $b_i(\mu) = 1, i = 2, \ldots, r$, puts equations (7.77) and (7.69) to (7.71) into the following form:

$$\begin{aligned}\dot{Z} &= f_0(Z, x_1, \mu), \\ \dot{x}_i &= f_i(Z, x_1, \ldots, x_i, \mu) + b_i(\mu)x_{i+1}, \quad i = 1, \ldots, r,\end{aligned} \tag{7.78}$$

where

$$f_0(Z, x_1, \mu) = \begin{bmatrix} F(w)\bar{z} + \tilde{G}(e, v, w)e \\ M\tilde{\eta} - N\left(\beta_1^{[2]}(\tilde{\eta} + Nb^{-1}(w)e + \theta) - \beta_1^{[2]}(\theta)\right) + \bar{\phi}(\bar{z}, e, v, w) \end{bmatrix},$$

$$f_1(Z, x_1, \mu) = H(w)\bar{z} + \tilde{K}(e, v, w)e + b(w)\big(\beta_1(\tilde{\eta} + b^{-1}(w)Ne + \theta) - \beta_1(\theta)\big),$$

and, for $i = 1, 2, \ldots, r - 1$,

$$f_{i+1}(Z, x_1, \ldots, x_{i+1}, \mu) = -\frac{\partial\beta_i(\eta)}{\partial\eta}N\bar{\xi}_1 - \lambda_i\bar{\xi}_i.$$

System (7.78) is in the form of the lower triangular systems described in (7.13). By appealing to Corollary 7.15, we can obtain the solvability conditions for the global robust stabilization problem for (7.78), and hence the solvability conditions for the global robust output regulation problem for system (7.54) as follows.

Theorem 7.21. *Under Assumptions 7.6 to 7.8, assume that*

(i) *the function $\Xi_1(v, w)$ satisfies conditions (7.65) and (7.66), and*

(ii) *system (7.77) is RISS with respect to μ viewing $\tilde{\eta}$ as state and* $\mathrm{col}(\bar{z}, e)$ *as input with a known C^1 gain function.*

Then, the global robust output regulation problem for system (7.54) is solvable.

Proof. By Corollary 7.15, it suffices to show that under condition (ii), the subsystem $\dot{Z} = f_0(Z, x_1, \mu)$ of (7.78) is RISS with respect to μ, viewing Z as state and x_1 as input with a C^1 gain function. This can be done by utilizing Corollary 2.20 as follows.

First, let us show that the following system:

$$\dot{\bar{z}} = F(w)\,\bar{z} + \tilde{G}(e, v, w)\,e \tag{7.79}$$

is RISS with respect to μ with a C^1 gain function, viewing $\bar{z}$ as state and e as input. By Lemma 7.17, the matrix $F(w)$ is Hurwitz for all $w \in \mathcal{R}^{n_w}$. Therefore, there exists a symmetric positive definite matrix $Q(w)$ continuously depending on w, such that

$$Q(w)F(w) + F^T(w)Q(w) = -I.$$

Let $V_{\bar{z}}(\bar{z}, w) = \bar{z}^T Q(w)\bar{z}$. Clearly, for all $w \in W$, $V_{\bar{z}}(\bar{z}, w)$ satisfies

$$\underline{a}\|\bar{z}\|^2 \leq V_{\bar{z}}(\bar{z}, w) \leq \bar{a}\|\bar{z}\|^2$$

for suitable $\underline{a} > 0$ and $\bar{a} > 0$, and its derivative along (7.79) satisfies

$$\frac{\partial V_{\bar{z}}(\bar{z}, w)}{\partial \bar{z}}(F(w)\,\bar{z} + \tilde{G}(e, v, w)\,e) \leq -\|\bar{z}\|^2 + 2\|z\|\|Q(w)\|\|\tilde{G}(e, v, w)\,e\|.$$

Pick any $0 < \epsilon < 1$ and let $\chi(\cdot)$ be a C^1 class $\mathcal{K}$ function satisfying, for all $v \in V$ and all $w \in W$,

$$\chi(\|e\|) \geq \frac{2}{1-\epsilon}\|Q(w)\|\|\tilde{G}(e, v, w)\,e\|.$$

Then

$$\|\bar{z}\| \geq \chi(\|e\|) \Rightarrow \frac{\partial V_{\bar{z}}(\bar{z}, w)}{\partial \bar{z}}(F(w)\,\bar{z} + \tilde{G}(e, v, w)\,e) \leq -\epsilon\|\bar{z}\|^2.$$

Thus, by Theorem 2.16, system (7.79) is RISS with respect to μ with state $\bar{z}$ and input e and with a C^1 gain function $\gamma_1^e(s) = \underline{a}^{-1} \circ \bar{a} \circ \chi(s) = (\bar{a}/\underline{a})\chi(s)$; in particular, for all $t \geq t_0 \geq 0$,

$$\|\bar{z}(t)\| \leq \max\left\{\beta_1^{kl}(\|\bar{z}(t_0)\|, t - t_0), \gamma_1^e\left(\sup_{t_0 \leq \tau \leq t}\|e(\tau)\|\right)\right\} \tag{7.80}$$

for some class $\mathcal{KL}$ function β_1^{kl}.

Next, note that condition (ii) guarantees the existence of some class $\mathcal{KL}$ function β_2^{kl} and two known C^1 class $\mathcal{K}$ functions $\gamma_2^{\bar{z}}$ and γ_2^e such that the solution of system (7.77) satisfies, for all $t \geq t_0 \geq 0$,

$$\|\tilde{\eta}(t)\| \leq \max\left\{\beta_2^{kl}(\|\tilde{\eta}(t_0)\|, t - t_0), \gamma_2^{\bar{z}}\left(\sup_{t_0 \leq \tau \leq t} \|\bar{z}(\tau)\|\right), \gamma_2^e\left(\sup_{t_0 \leq \tau \leq t} \|e(\tau)\|\right)\right\}. \quad (7.81)$$

Since the subsystem $\dot{Z} = f_0(Z, x_1, \mu)$ consists of (7.77) and (7.79), applying Corollary 2.20 to the subsystem $\dot{Z} = f_0(Z, x_1, \mu)$ shows that this subsystem is RISS with respect to μ, viewing Z as state and x_1 as input with a C^1 gain function, which is any C^1 class $\mathcal{K}_\infty$ function $\gamma(\cdot)$ satisfying

$$\gamma(s) \geq \max\left\{2\gamma_1^e(s), 2\gamma_2^e(s), 2\gamma_2^{\bar{z}} \circ \gamma_1^e(s)\right\}, \quad s \geq 0. \quad (7.82)$$

Since γ_2^e, $\gamma_2^{\bar{z}}$, and γ_1^e are C^1 functions, it is always possible to choose a C^1 class $\mathcal{K}_\infty$ function γ satisfying (7.82). The proof is completed. □

Remark 7.22. The controller that solves the robust stabilization problem for the lower triangular system (7.78) takes the following form:

$$\begin{aligned}
\bar{u} &= \alpha_r(\bar{x}_r), \\
\bar{x}_{i+1} &= x_{i+1} - \alpha_i(\bar{x}_i), \; i = 1, \ldots, r-1, \\
\bar{x}_1 &= x_1,
\end{aligned}$$

where the smooth functions α_i, $i = 1, \ldots, r$, can be obtained by the algorithm described in Remark 7.10. By Corollary 7.4, the controller that solves the robust output regulation problem of system (7.54) is

$$\begin{aligned}
u &= \alpha_r(\bar{x}_r) + \beta_r(\eta), \\
\bar{x}_{i+1} &= \xi_i - \beta_i(\eta) - \alpha_i(\bar{x}_i), \; i = 1, \ldots, r-1, \\
\bar{x}_1 &= e, \\
\dot{\eta} &= M\eta + N(\xi_1 - \beta_1(\eta) + \Psi T^{-1}\eta).
\end{aligned}$$

Finally, the controller that solves the global robust output regulation problem of system (7.43) is given by

$$\begin{aligned}
u &= \alpha_r(\bar{x}_r) + \beta_r(\eta), \\
\bar{x}_{i+1} &= \xi_i - \beta_i(\eta) - \alpha_i(\bar{x}_i), \; i = 1, \ldots, r-1, \\
\bar{x}_1 &= e, \\
\dot{\eta} &= M\eta + N(\xi_1 - \beta_1(\eta) + \Psi T^{-1}\eta), \\
\dot{\xi} &= F_e\xi + G_e u,
\end{aligned}$$

which only relies on the error output e of system (7.43). ■

Remark 7.23. Since M is Hurwitz, there exists a symmetric positive definite matrix P such that

$$PM + M^T P = -I. \quad (7.83)$$

To guarantee condition (ii) in Theorem 7.21, that is, the RISS property of (7.77), it suffices to suppose that there exists a positive number $r_0 < 1$ satisfying

$$-2\tilde{\eta}^T PN\left(\beta_1^{[21]}(\tilde{\eta}+d)-\beta_1^{[21]}(d)\right) \leq (1-r_0)\tilde{\eta}^T\tilde{\eta} \tag{7.84}$$

for all $\tilde{\eta}, d$. This assumption is to restrict the growth of the nonlinear part of the function $\beta_1(\cdot)$. Indeed, rewrite (7.77) as follows:

$$\dot{\tilde{\eta}} = M\tilde{\eta} - N\left(\beta_1^{[21]}(\tilde{\eta}+d(e,v,w)) - \beta_1^{[21]}(d(e,v,w))\right) + \phi(\bar{z},e,v,w), \tag{7.85}$$

where

$$\begin{aligned} d(e,v,w) &= Nb^{-1}(w)e + \theta(v,w), \\ \phi(\bar{z},e,v,w) &= -N\left(\beta_1^{[21]}(d(e,v,w)) - \beta_1^{[21]}(\theta(v,w))\right) + \bar{\phi}(\bar{z},e,v,w). \end{aligned}$$

Let $V(\tilde{\eta}) = \frac{2}{r_0}\tilde{\eta}^T P\tilde{\eta}$. Then $\frac{2}{r_0}\lambda_{min}\|\tilde{\eta}\|^2 \leq V(\tilde{\eta}) \leq \frac{2}{r_0}\lambda_{max}\|\tilde{\eta}\|^2$, where λ_{max} (λ_{min}) is the maximal (minimal) eigenvalue of P. And the derivative of $V(\tilde{\eta})$ along system (7.85) satisfies

$$\begin{aligned} &\frac{\partial V(\tilde{\eta})}{\partial\tilde{\eta}}\left(M\tilde{\eta} - N\left(\beta_1^{[21]}(\tilde{\eta}+d(e,v,w)) - \beta_1^{[21]}(d(e,v,w))\right) + \phi(\bar{z},e,v,w)\right) \\ &= \frac{2}{r_0}\left[2\tilde{\eta}^T PM\tilde{\eta} - 2\tilde{\eta}^T PN\left(\beta^{[21]}(\tilde{\eta}+d) - \beta^{[21]}(d)\right) + 2\tilde{\eta}^T P\phi(\bar{z},e,v,w)\right] \\ &= \frac{2}{r_0}\left[-\tilde{\eta}^T\tilde{\eta} - 2\tilde{\eta}^T PN\left(\beta^{[21]}(\tilde{\eta}+d) - \beta^{[21]}(d)\right) + 2\tilde{\eta}^T P\phi(\bar{z},e,v,w)\right] \\ &\leq \frac{2}{r_0}\left[-r_0\tilde{\eta}^T\tilde{\eta} + 2\tilde{\eta}^T P\phi(\bar{z},e,v,w)\right] \\ &\leq \frac{2}{r_0}\left[-\frac{r_0}{2}||\tilde{\eta}||^2 + \frac{2}{r_0}||P\phi(\bar{z},e,v,w)||^2\right] \\ &\leq -||\tilde{\eta}||^2 + \left\|\frac{2}{r_0}P\phi(\bar{z},e,v,w)\right\|^2. \end{aligned} \tag{7.86}$$

Noting that the function $\phi(\bar{z},e,v,w)$ is C^1 satisfying $\phi(0,0,v,w) = 0$ and that $(v,w) \in V \times W$, with $V \times W$ a compact set, we have

$$\left\|\frac{2}{r_0}P\phi(\bar{z},e,v,w)\right\| \leq \|\text{col}(\bar{z},e)\|a_1(\bar{z},e)$$

for some smooth function $a_1(\bar{z},e) \geq 1$. And there exists a smooth nondecreasing function $a_2(\cdot)$ satisfying

$$a_2(\|\text{col}(\bar{z},e)\|) \geq a_1(\bar{z},e).$$

As a result, we have

$$\frac{dV(\tilde{\eta})}{dt} \leq -\|\tilde{\eta}\|^2 + a^2(\|\mathrm{col}(\bar{z}, e)\|)$$

for some smooth class $\mathcal{K}_\infty$ function $a(s) = sa_2(s)$. Thus, for any $0 < \epsilon < 1$,

$$\|\tilde{\eta}\| \geq \frac{a(\|\mathrm{col}(\bar{z}, e)\|)}{1-\epsilon} \quad \Rightarrow \quad \frac{dV(\tilde{\eta})}{dt} \leq -\epsilon\|\tilde{\eta}\|^2.$$

Thus, by Theorem 2.16, choosing $\underline{\alpha}(s) = \frac{2}{r_0}\lambda_{min}s^2$, $\bar{\alpha}(s) = \frac{2}{r_0}\lambda_{max}s^2$, $\chi(s) = \frac{1}{1-\epsilon}a(s)$ shows that the condition (ii) holds for a known C^1 gain function

$$\kappa(s) \geq \underline{\alpha}^{-1}(\bar{\alpha}(\chi(s))) = \sqrt{\frac{\lambda_{max}}{\lambda_{min}}}\frac{a(s)}{1-\epsilon}. \quad \blacksquare$$

Remark 7.24. The inequality (7.84) is satisfied in at least two meaningful cases. First, (7.84) holds for some $0 < r_0 < 1$ if $|\beta_1^{[2]}(\tilde{\eta} + d) - \beta_1^{[2]}(d)| < \frac{1}{2\|PN\|}\|\tilde{\eta}\|$. Thus, (7.84) holds if $\beta^{[2]}$ is globally Lipschitz, that is, $|\beta_1^{[2]}(\tilde{\eta} + d) - \beta_1^{[2]}(d)| \leq L\|\tilde{\eta}\|$ for some positive number L, and the Lipschitz constant L satisfies $L < \frac{1}{2\|PN\|}$. Second, when the solution of the regulator equations is a trigonometric polynomial in t, condition (i) of Theorem 7.21 is automatically satisfied and the function $\beta_1(\cdot)$ is linear. In this case, condition (7.84), and hence condition (ii) of Theorem 7.21, is also automatically satisfied. Thus we obtain the following corollary of Theorem 7.21. $\blacksquare$

Corollary 7.25. *Under Assumptions 7.6 to 7.8, assume the solution of the regulator equations of (7.54) is a polynomial or a trigonometric polynomial in t. Then the global robust output regulation problem for system (7.54) is solvable.*

Example 7.26. Consider the following system:

$$\begin{aligned} \dot{x} &= x - 2y - 2\sin^2(0.1wy) + 10u + (v_1^2 + 20wv_1 + 20v_2), \\ \dot{y} &= x - 2y - 0.1v_1 y - \sin^2(0.1wy) + (10wv_1 + 10v_2), \\ e &= y - 10v_1, \end{aligned} \tag{7.87}$$

and the exosystem

$$\dot{v}_1 = v_2, \qquad \dot{v}_2 = -v_1.$$

It is assumed that $v(t) \in V = \{v_1^2 + v_2^2 \leq 1\}$ and $-1 \leq w \leq 1$.

This system is in the form (7.43) with

$$\begin{aligned} &\bar{F}(w) = 1, \quad \bar{G}(y, v, w)y = -2y - 2\sin^2(0.1wy), \\ &g(w) = 10, \quad \bar{D}_1(w) = v_1^2 + 20wv_1 + 20v_2, \\ &\bar{H}(w) = 1, \quad \bar{K}(y, v, w)y = -2y - 0.1v_1 y - \sin^2(0.1wy), \quad \bar{D}_2(w) = 10wv_1 + 10v_2. \end{aligned}$$

It can be verified that the system has a uniform relative degree $r = 2$. Using (7.53) and (7.52) gives $D(w) = g(w) = 10$ and $h(w) = 2$. Thus, applying the coordinate transformation $z = x - 10\xi_1 - 2y$, where $\dot{\xi}_1 = -\xi_1 + u$, gives the following extended system:

$$\begin{aligned}
\dot{z} &= -z + 0.2v_1 y + v_1^2,\\
\dot{y} &= z - 0.1v_1 y - \sin^2(0.1wy) + 10\xi_1 + (10wv_1 + 10v_2),\\
\dot{\xi}_1 &= -\xi_1 + u,\\
e &= y - 10v_1.
\end{aligned} \tag{7.88}$$

This system is clearly in the form (7.54) with

$$\begin{aligned}
&F(w) = -1, \quad G(y, v, w)y = 0.2v_2 y, \quad D_1(w) = v_1^2, \quad b(w) = 10,\\
&H(w) = 1, \quad K(y, v, w)y = -0.1v_1 y - \sin^2(0.1wy), \quad D_2(w) = 10wv_1 + 10v_2,\\
&F_e = -1, \quad G_e = 1.
\end{aligned}$$

In order to solve the global robust output regulation problem for this system, let us first verify, by inspection, that the solution of the regulator equations exists globally and is given by

$$\begin{aligned}
\mathbf{z}(v, w) &= v_1^2,\\
\mathbf{y}(v, w) &= 10v_1,\\
\Xi_1(v, w) &= -wv_1 + 0.1\sin^2(wv_1),\\
\mathbf{u}(v, w) &= \dot{\Xi}_1(v, w) + \Xi_1(v, w)\\
&= -wv_2 + 0.2\sin(wv_1)\cos(wv_1)wv_2 - wv_1 + 0.1\sin^2(wv_1).
\end{aligned}$$

Let $\pi_1(v, w) = wv_1$. Then the minimal zeroing polynomial of $\pi_1(v, w)$ is $P_1(\lambda) = \lambda^2 + 1$ and

$$\Xi_1(v, w) = \Gamma_1(\pi_1, \dot{\pi}_1) = -\pi_1(v, w) + 0.1\sin^2\pi_1(v, w).$$

Thus, the system has a steady-state generator $\{\theta, \alpha, \beta\}$ with output $\mathrm{col}(\xi_1, u)$, where

$$\theta = T\begin{bmatrix} \pi_1(v, w) \\ \dot{\pi}_1(v, w) \end{bmatrix} = T\begin{bmatrix} wv_1 \\ wv_2 \end{bmatrix}, \quad \Phi = \begin{bmatrix} 0 & 1 \\ -1 & 0 \end{bmatrix},$$

and $T \in \mathcal{R}^{2\times 2}$ is any nonsingular matrix. Since $\Psi = \begin{bmatrix} -1 & 0 \end{bmatrix}$, the pair (Ψ, Φ) is observable. Thus the generator is linearly observable.

Choose

$$M = \begin{bmatrix} -1 & 0 \\ 0 & -2 \end{bmatrix}, \quad N = \begin{bmatrix} 0.2 \\ 0.4 \end{bmatrix},$$

which makes a controllable pair. For this pair of matrices, the solution of the Sylvester equation $MT + N\Psi = T\Phi$ is given by

$$T = \begin{bmatrix} -0.1 & 0.1 \\ -0.16 & 0.08 \end{bmatrix},$$

which is nonsingular with

$$T^{-1} = \begin{bmatrix} 10 & -12.5 \\ 20 & -12.5 \end{bmatrix}.$$

Under this design,

$$\theta = \begin{bmatrix} -0.1wv_1 + 0.1wv_2 \\ -0.16wv_1 + 0.8wv_2 \end{bmatrix} = \begin{bmatrix} \theta_1 \\ \theta_2 \end{bmatrix}$$

and

$$\begin{aligned} \beta_1(\theta) &= \Gamma_1(T^{-1}\theta) \\ &= -10\theta_1 + 12.5\theta_2 + 0.1\sin^2(10\theta_1 - 12.5\theta_2), \\ \beta_2(\theta) &= \dot{\beta}_1(\theta) + \beta_1(\theta) \\ &= -20\theta_1 + 12.5\theta_2 + 0.2\sin(10\theta_1 - 12.5\theta_2)\cos(10\theta_1 - 12.5\theta_2)(20\theta_1 - 12.5\theta_2) \\ &\quad - 10\theta_1 + 12.5\theta_2 + 0.1\sin^2(10\theta_1 - 12.5\theta_2). \end{aligned}$$

Using the internal model (7.67) and the coordinate transformation (7.68) gives the following augmented system:

$$\begin{aligned} \dot{\bar{z}} &= -\bar{z} + 0.2v_2 e, \\ \dot{\bar{\eta}} &= \left(M + N\Psi T^{-1}\right)\bar{\eta} + N\bar{\xi}_1, \\ \dot{e} &= \bar{z} - \sin^2(0.1we + wv_1) + \sin^2(wv_1) - 0.1v_1 e + 10\beta_1(\bar{\eta} + \theta) - 10\beta_1(\theta) + 10\bar{\xi}_1, \\ \dot{\bar{\xi}}_1 &= -\left(1 + \frac{\partial \beta_1(\eta)}{\partial \eta} N\right)\bar{\xi}_1 + \bar{u}. \end{aligned} \tag{7.89}$$

A further coordinate transformation of the form (7.73) puts (7.89) into the lower triangular system of the form (7.78) with $r = 2$, $Z = \mathrm{col}(\bar{z}, \tilde{\eta})$, and $x = \mathrm{col}(x_1, x_2) = \mathrm{col}(e, \bar{\xi}_1)$:

$$\begin{aligned} \dot{Z} &= \begin{bmatrix} \dot{\bar{z}} \\ \dot{\tilde{\eta}} \end{bmatrix} = \begin{bmatrix} -\bar{z} + 0.2v_2 x_1 \\ M\tilde{\eta} - N(\beta_1{}^{[2]}(\tilde{\eta} + d) - \beta_1{}^{[2]}(d)) + \phi(\bar{z}, e, \mu) \end{bmatrix} = f_0(Z, x_1, \mu), \\ \dot{x}_1 &= \bar{z} - \sin^2(0.1wx_1 + wv_1) + \sin^2(wv_1) - 0.1v_1 x_1 + 10\beta_1(\tilde{\eta} + \theta) - 10\beta_1(\theta) + 10x_2, \\ \dot{x}_2 &= -\left(1 + \frac{\partial \beta_1(\eta)}{\partial \eta} N\right)x_2 + \bar{u}, \end{aligned}$$

where

$$\begin{aligned} d &= 0.1Ne + \theta, \\ \beta_1^{[2]}(\theta) &= 0.1\sin^2(10\theta_1 - 12.5\theta_2), \\ \phi(\bar{z}, e, v, w) &= -N\left(\beta_1^{[2]}(0.1Ne + \theta) - \beta_1^{[2]}(\theta)\right) + 0.1MNe \\ &\quad - 0.1N\left(\bar{z} - \sin^2(0.1we + wv_1) + \sin^2(wv_1) - 0.1v_1 e\right). \end{aligned}$$

To verify condition (ii) of Theorem 7.21, we resort to Remark 7.23. Solving the Lyapunov equation (7.83) gives

$$P = \begin{bmatrix} 0.5 & 0 \\ 0 & 0.25 \end{bmatrix}.$$

Simple calculation gives

$$\begin{aligned}
&-2\tilde{\eta}^T \begin{bmatrix} 0.5 & 0 \\ 0 & 0.25 \end{bmatrix} \begin{bmatrix} 0.2 \\ 0.4 \end{bmatrix} (\beta_1{}^{[2]}(\tilde{\eta} + d) - \beta_1{}^{[2]}(d)) \\
&\quad = 0.02(\tilde{\eta}_1 + \tilde{\eta}_2)\left(-\sin^2(10(\tilde{\eta}_1 + d_1) - 12.5(\tilde{\eta}_2 + d_2)) + \sin^2(10d_1 - 12.5d_2)\right) \\
&\quad \leq 0.02\,|(\tilde{\eta}_1 + \tilde{\eta}_2)(10\tilde{\eta}_1 - 12.5\tilde{\eta}_2)| < 0.28\|\tilde{\eta}_1\|^2.
\end{aligned}$$

Thus, the inequality (7.84) holds for $0 < r_0 < 0.72$. Therefore, condition (ii) of Theorem 7.21 also holds.

Thus, by Theorem 7.21, the global output regulation problem for system (7.87) is solvable. Finally, by Remark 7.22, an output feedback controller can be synthesized and is given as follows:

$$\begin{aligned}
u &= \beta_2(\eta) - 1190\bar{x}_2, \\
\bar{x}_2 &= \xi_1 - \beta_1(\eta) + 17.3\bar{x}_1, \\
\bar{x}_1 &= e, \\
\dot{\eta} &= M\eta + N(\xi_1 - \beta_1(\eta) + \Psi T^{-1}\eta), \\
\dot{\xi}_1 &= -\xi_1 + u. \quad \blacksquare
\end{aligned}$$

7.4 Global Robust Output Regulation for Nonlinear Systems in Lower Triangular Form

In this section, we will consider the global robust output regulation problem for the class of the lower triangular systems described in Section 7.2. When taking into account the effect of the exogenous signals v, system (7.13) can be modified into the following form:

$$\begin{aligned}
\dot{z} &= f(z, x_1, v, w), \\
\dot{x}_1 &= f_1(z, x_1, v, w) + b_1(v, w)x_2, \\
&\vdots \\
\dot{x}_r &= f_r(z, x_1, \ldots, x_r, v, w) + b_r(v, w)u, \\
\dot{v} &= A_1 v, \\
e &= x_1 - q_d(v, w),
\end{aligned} \tag{7.90}$$

where $z \in \mathcal{R}^m$, $x_i \in \mathcal{R}$, $i = 1, \ldots, r$, $u, y \in \mathcal{R}$, $v \in \mathcal{R}^q$, $w \in \mathcal{R}^{n_w}$, and the functions $f, f_i, b_i, i = 1, \ldots, r$, and q_d, are sufficiently smooth functions satisfying $f(0, 0, 0, w) = 0$, $f_i(0, \ldots, 0, w) = 0$, $i = 1, \ldots, r$, and $q_d(0, w) = 0$, for all $w \in \mathcal{R}^{n_w}$.

Again, all the eigenvalues of the matrix A_1 are simple with zero real part.

At the outset, let us make the following assumptions.

Assumption 7.9. For $i = 1, \ldots, r$, $b_i(v, w) > 0$ for all $v \in \mathcal{R}^q$ and $w \in \mathcal{R}^{n_w}$.

Assumption 7.10. There exists a sufficiently smooth function $\mathbf{z}(v, w)$ with $\mathbf{z}(0, 0) = 0$ satisfying the following equation for all $v \in \mathcal{R}^q$ and $w \in \mathcal{R}^{n_w}$:

$$\frac{\partial \mathbf{z}(v, w)}{\partial v} A_1 v = f(\mathbf{z}(v, w), q_d(v, w), v, w). \tag{7.91}$$

Remark 7.27. Under Assumptions 7.9 and 7.10, the solution of the regulator equations of system (7.90) exists globally and can be obtained as follows:

$$\begin{aligned}
\mathbf{x}_1(v, w) &= q_d(v, w), \\
\mathbf{x}_i(v, w) &= \frac{1}{b_{i-1}(v, w)} \Big(\frac{\partial \mathbf{x}_{i-1}(v, w)}{\partial v} A_1 v \\
&\quad - f_{i-1}(\mathbf{z}(v, w), \mathbf{x}_1(v, w), \ldots, \mathbf{x}_{i-1}(v, w), v, w) \Big), \quad i = 2, \ldots, r, \\
\mathbf{u}(v, w) &= \frac{1}{b_r(v, w)} \left(\frac{\partial \mathbf{x}_r(v, w)}{\partial v} A_1 v - f_r(\mathbf{z}(v, w), \mathbf{x}_1(v, w), \ldots, \mathbf{x}_r(v, w), v, w) \right).
\end{aligned}$$

The solution of the regulator equations will be denoted by $\mathbf{z}(v, w)$, $\mathbf{x}(v, w)$, $\mathbf{u}(v, w)$ with $\mathbf{x}(v, w) = \operatorname{col}(\mathbf{x}_1(v, w), \ldots, \mathbf{x}_r(v, w))$. Also, for convenience, we define $\mathbf{x}_{r+1}(v, w) = \mathbf{u}(v, w)$. ∎

As before, we need to convert the global robust output regulation problem of system (7.90) into the global robust stabilization problem of an augmented system. For this purpose, we will assume that the solution of the regulator equations satisfies the following assumption.

Assumption 7.11. For $i = 1, \ldots, r$, there exist pairwise coprime polynomials $\pi_i^1(v, w)$, $\ldots, \pi_i^{l_i}(v, w)$ with $r_i^1, \ldots, r_i^{l_i}$ being the degrees of their minimal zeroing polynomials $P_i^1(s), \ldots, P_i^{l_i}(s)$, and sufficiently smooth function $\Gamma_i : \mathcal{R}^{r_i^1 + \cdots + r_i^{l_i}} \to \mathcal{R}$ vanishing at the origin such that, for all trajectories $v(t)$ of the exosystem, and $w \in \mathcal{R}^{n_w}$,

$$\begin{aligned}
\mathbf{x}_{i+1}(v, w) = \Gamma_i \Bigg(& \pi_i^1(v, w), \dot{\pi}_i^1(v, w), \ldots, \frac{d^{(r_i^1 - 1)} \pi_i^1(v, w)}{dt^{(r_i^1 - 1)}}, \ldots, \\
& \pi_i^{l_i}(v, w), \dot{\pi}_i^{l_i}(v, w), \ldots, \frac{d^{(r_i^{l_i} - 1)} \pi_i^{l_i}(v, w)}{dt^{(r_i^{l_i} - 1)}} \Bigg)
\end{aligned} \tag{7.92}$$

and

$$\text{the pair } (\Psi_i, \Phi_i) \text{ is observable,} \tag{7.93}$$

where Ψ_i is the gradient of Γ_i at the origin, and $\Phi_i = \text{block diag}\left(\Phi_i^1, \ldots, \Phi_i^{l_i}\right)$ with Φ_i^j, $j = 1, \ldots, l_i$, being the companion matrix of the polynomial $P_i^j(s)$.

By Lemma 6.17, under Assumption 7.11, system (7.90) has a linearly observable steady-state generator $\{\theta_i, \alpha_i, \beta_i\}$ with output x_{i+1}, $i = 1, 2, \ldots, r$. To be more specific, let

$$\theta_i(v, w) = T_i \begin{bmatrix} \pi_i^1(v, w) \\ \dot{\pi}_i^1(v, w) \\ \vdots \\ \dfrac{d^{(r_i^1-1)}\pi_i^1(v, w)}{dt^{(r_i^1-1)}} \\ \vdots \\ \pi_i^{l_i}(v, w) \\ \dot{\pi}_i^{l_i}(v, w) \\ \vdots \\ \dfrac{d^{(r_i^{l_i}-1)}\pi_i^{l_i}(v, w)}{dt^{(r_i^{l_i}-1)}} \end{bmatrix}, \quad \alpha_i(\theta_i) = T_i\Phi_i T_i^{-1}\theta_i, \quad \beta_i(\theta_i) = \Gamma_i(T_i^{-1}\theta_i),$$

where T_i is any nonsingular matrix with the appropriate dimension. Then, $\dot{\theta}_i = \alpha_i(\theta_i) = T_i\Phi_i T_i^{-1}\theta_i$, and $\mathbf{x}_{i+1}(v, w) = \beta_i(\theta_i(v, w)) = \Gamma_i(T_i^{-1}\theta_i(v, w))$.

Further, by Proposition 6.21, the following system:

$$\dot{\eta}_i = M_i\eta_i + N_i(x_{i+1} - \beta_i(\eta_i) + \Psi_i T_i^{-1}\eta_i), \quad i = 1, \ldots, r, \tag{7.94}$$

is an internal model of (7.90) with output x_{i+1}, where the pair (M_i, N_i) is controllable with M_i Hurwitz, and T_i satisfies the Sylvester equation $T_i\Phi_i - M_iT_i = N_i\Psi_i$. Clearly, putting the r systems given by (7.94) with $i = 1, \ldots, r$ gives an internal model of system (7.90) with output $g_o(z, x_1, \ldots, x_r, u) = \mathrm{col}(x_2, \ldots, x_r, u)$.

Next, define the coordinate and input transformation according to (6.9), which becomes

$$\begin{aligned}
\bar{z} &= z - \mathbf{z}(v, w), \\
\bar{\eta}_i &= \eta_i - \theta_i(v, w), \quad i = 1, \ldots, r, \\
\bar{x}_1 &= x_1 - \mathbf{x}_1(v, w) = e, \\
\bar{x}_{i+1} &= x_{i+1} - \beta_i(\eta_i), \quad i = 1, \ldots, r-1, \\
\bar{u} &= u - \beta_r(\eta_r).
\end{aligned}$$

This transformation converts the augmented system composed of the original plant (7.90) and the internal model (7.94) into the following form:

$$\begin{aligned}
\dot{\bar{z}} &= \hat{f}_0(\bar{z}, \bar{x}_1, v, w), \\
\dot{\bar{\eta}}_i &= \left(M_i + N_i\Psi_i T_i^{-1}\right)\bar{\eta}_i + N_i\bar{x}_{i+1}, \quad i = 1, \ldots, r, \\
\dot{\bar{x}}_i &= \hat{f}_i(\bar{z}, \bar{\eta}_1, \ldots, \bar{\eta}_i, \bar{x}_1, \ldots, \bar{x}_i, v, w) + b_i(v, w)\bar{x}_{i+1}, \quad i = 1, \ldots, r,
\end{aligned} \tag{7.95}$$

where $\bar{x}_{r+1} = \bar{u}$ and

$$\begin{aligned}
\hat{f}_0(\bar{z}, \bar{x}_1, v, w) &= f(\bar{z} + \mathbf{z}(v, w), \bar{x}_1 + \mathbf{x}_1(v, w), v, w) \\
&\quad - f(\mathbf{z}(v, w), \mathbf{x}_1(v, w), v, w), \\
\hat{f}_1(\bar{z}, \bar{\eta}_1, \bar{x}_1, v, w) &= f_1(\bar{z} + \mathbf{z}(v, w), \bar{x}_1 + \mathbf{x}_1(v, w), v, w) \\
&\quad + b_1(v, w)\beta_1(\eta_1) - \frac{\partial \mathbf{x}_1(v, w)}{\partial v} A_1 v, \\
\hat{f}_i(\bar{z}, \bar{\eta}_1, \dots, \bar{\eta}_i, \bar{x}_1, \dots, \bar{x}_i, v, w) &= f_i\Big(\bar{z} + \mathbf{z}(v, w), \bar{x}_1 + \mathbf{x}_1(v, w), \\
&\quad \bar{x}_2 + \beta_1(\bar{\eta}_1 + \theta_1(v, w)), \dots, \\
&\quad \bar{x}_i + \beta_{i-1}(\bar{\eta}_{i-1} + \theta_{i-1}(v, w)), v, w\Big) \\
&\quad + b_i(v, w)\beta_i(\eta_i) - \frac{\partial \beta_{i-1}(\eta_{i-1})}{\partial \eta_{i-1}} \dot{\eta}_{i-1}, \\
&\quad i = 2, \dots, r.
\end{aligned} \tag{7.96}$$

By Corollary 7.4, the global robust output regulation problem for system (7.90) will be solved if we can make the equilibrium point of system (7.95) at $(\bar{z}, \bar{x}, \bar{\eta}) = (0, 0, 0)$ globally asymptotically stable for all trajectories $v(t) \in V$ of the exosystem, and all $w \in W$. An inspection of the structure of (7.95) reveals that (7.95) is in the lower triangular form (7.12) if we identify z_0 with z and z_i with $\bar{\eta}_i$, $i = 1, \dots, r$. However, since $M_i + N_i \Psi_i T_i^{-1} = T_i^{-1} \Phi_i T_i$ and all the eigenvalues of the matrix Φ_i have zero real part, the subsystems described by the second equation of (7.95) does not satisfy Assumption 7.4. Therefore, Theorem 7.6 cannot be directly applied to system (7.95). To circumvent this difficulty, similar to what has been done in Section 7.3, we further perform on (7.95) another coordinate transformation:

$$z_i = \bar{\eta}_i - b_i^{-1}(v, w) N_i \bar{x}_i, \ i = 1, \dots, r, \tag{7.97}$$

which yields

$$\begin{aligned}
\dot{z}_i &= \dot{\bar{\eta}}_i - \frac{\partial b_i^{-1}(v, w)}{\partial v} A_1 v N_i \bar{x}_i - b_i^{-1}(v, w) N_i \dot{\bar{x}}_i \\
&= (M_i + N_i \Psi_i T_i^{-1})\bar{\eta}_i + N_i \bar{x}_{i+1} - \frac{\partial b_i^{-1}(v, w)}{\partial v} A_1 v N_i \bar{x}_i \\
&\quad - b_i^{-1}(v, w) N_i \left(\hat{f}_i(\bar{z}, \bar{\eta}_1, \dots, \bar{\eta}_i, \bar{x}_1, \dots, \bar{x}_i, v, w) + b_i(v, w)\bar{x}_{i+1}\right) \\
&= (M_i + N_i \Psi_i T_i^{-1})\bar{\eta}_i - N_i \beta_i(\bar{\eta}_i + \theta_i) - \frac{\partial b_i^{-1}(v, w)}{\partial v} A_1 v N_i \bar{x}_i \\
&\quad - b_i^{-1}(v, w) N_i \left(\hat{f}_i(\bar{z}, \bar{\eta}_1, \dots, \bar{\eta}_i, \bar{x}_1, \dots, \bar{x}_i, v, w) - b_i(v, w)\beta_i(\bar{\eta}_i + \theta_i)\right).
\end{aligned}$$

Using the identity $\beta_i(\bar{\eta}_i + \theta_i) = \Psi_i T_i^{-1}(\bar{\eta}_i + \theta_i) + \beta_i^{[2]}(\bar{\eta}_i + \theta_i)$ in the above equation gives

$$\begin{aligned}
\dot{z}_i &= M_i \bar{\eta}_i - N_i \Psi_i T_i^{-1} \theta_i(v, w) - N_i \beta_i^{[2]}(\bar{\eta}_i + \theta_i) - \frac{\partial b_i^{-1}(v, w)}{\partial v} A_1 v N_i \bar{x}_i \\
&\quad - b_i^{-1}(v, w) N_i \left(\hat{f}_i(\bar{z}, \bar{\eta}_1, \dots, \bar{\eta}_i, \bar{x}_1, \dots, \bar{x}_i, v, w) - b_i(v, w)\beta_i(\bar{\eta}_i + \theta_i)\right).
\end{aligned}$$

Substituting (7.97) into the above equations gives

$$
\begin{aligned}
\dot{z}_i &= M_i z_i - N_i \beta_i^{[2]}(z_i + b_i^{-1}(v,w)N_i\bar{x}_i + \theta_i) \\
&\quad + \left(b_i^{-1}(v,w)M_i - \frac{\partial b_i^{-1}(v,w)}{\partial v} A_1 v \right) N_i \bar{x}_i \\
&\quad - b_i^{-1}(v,w)N_i\Big(\hat{f}_i(\bar{z}, \bar{\eta}_1, \ldots, \bar{\eta}_i, \bar{x}_1, \ldots, \bar{x}_i, v, w) \\
&\qquad - b_i(v,w)(\beta_i(\bar{\eta}_i + \theta_i) - \Psi_i T_i^{-1}\theta_i(v,w))\Big).
\end{aligned}
$$

Let $z_0 = \bar{z}$ and $\mu = \mathrm{col}(v, w)$. Then, in terms of the coordinate $\mathrm{col}(z_0, z_1, \ldots, z_r, \bar{x}_1, \ldots, \bar{x}_r)$, equation (7.95) can be put into the standard lower triangular form (7.12) as follows:

$$
\begin{aligned}
\dot{z}_0 &= \bar{f}_0(z_0, \bar{x}_1, \mu(t)), \\
\dot{z}_i &= Q_i(z_0, z_1, \ldots, z_i, \bar{x}_1, \ldots, \bar{x}_i, \mu(t)), \quad i = 1, \ldots, r, \\
\dot{\bar{x}}_i &= \bar{f}_i(z_0, z_1, \ldots, z_i, \bar{x}_1, \ldots, \bar{x}_i, \mu(t)) + b_i(\mu(t))\bar{x}_{i+1}, \quad i = 1, \ldots, r,
\end{aligned} \tag{7.98}
$$

where, for $i = 1, \ldots, r$,

$$
\begin{aligned}
\bar{f}_0(z_0, \bar{x}_1, \mu) &= \hat{f}_0(\bar{z}_0, \bar{x}_1, v, w), \\
Q_i(z_0, z_1, \ldots, z_i, \bar{x}_1, \ldots, \bar{x}_i, \mu) &= M_i z_i - N_i \beta_i^{[2]}(z_i + b_i^{-1}(\mu)N_i\bar{x}_i + \theta_i) \\
&\quad + \hat{\gamma}_i(z_0, z_1, \ldots, z_{i-1}, \bar{x}_1, \ldots, \bar{x}_i, \mu), \\
\hat{\gamma}_i(z_0, z_1, \ldots, z_{i-1}, \bar{x}_1, \ldots, \bar{x}_i, \mu) &= \left(b_i^{-1}(\mu)M_i - \frac{\partial b_i^{-1}(\mu)}{\partial v} A_1 v \right) N_i \bar{x}_i \\
&\quad - b_i^{-1}(\mu)N_i\Big(\hat{f}_i(\bar{z}_0, \bar{\eta}_1, \ldots, \bar{\eta}_i, \bar{x}_1, \ldots, \bar{x}_i, \mu) \\
&\qquad - b_i(\mu)\beta_i(\bar{\eta}_i + \theta_i) + b_i(\mu)\Psi_i T_i^{-1}\theta_i(\mu)\Big), \\
\bar{f}_i(z_0, z_1, \ldots, z_i, \bar{x}_1, \ldots, \bar{x}_i, \mu) &= \hat{f}_i\Big(\bar{z}_0, z_1 + b_1^{-1}(\mu)N_1\bar{x}_1, \ldots, \\
&\qquad z_i + b_i^{-1}(\mu)N_i\bar{x}_i, \bar{x}_1, \ldots, \bar{x}_i, \mu\Big).
\end{aligned}
$$

The functions $\bar{f}_0$, Q_i, $\hat{\gamma}_i$, $\bar{f}_i$ are all sufficiently smooth in their arguments.

It is important to note that $\hat{\gamma}_i(z_0, z_1, \ldots, z_{i-1}, \bar{x}_1, \ldots, \bar{x}_i, \mu)$ does not depend on the variable z_i since, from (7.96),

$$
\hat{f}_1(\bar{z}_0, \bar{\eta}_1, \bar{x}_1, \mu) - b_i(\mu)\beta_1(\bar{\eta}_1 + \theta_1) = f_1(\bar{z}_0 + \mathbf{z}(\mu), \bar{x}_1 + \mathbf{x}_1(\mu), \mu) - \frac{\partial \mathbf{x}_1(v,w)}{\partial v} A_1 v,
$$

which does not depend on η_1, and for $i = 2, \ldots, r$,

$$
\begin{aligned}
&\hat{f}_i(\bar{z}_0, \bar{\eta}_1, \ldots, \bar{\eta}_i, \bar{x}_1, \ldots, \bar{x}_i, \mu) - b_1(\mu)\beta_i(\bar{\eta}_i + \theta_i) \\
&\quad = f_i(\bar{z}_0 + \mathbf{z}(\mu), \bar{x}_1 + \mathbf{x}_1(\mu), \bar{x}_2 + \beta_1(\bar{\eta}_1 + \theta_1(\mu)), \ldots, \bar{x}_i + \beta_{i-1}(\bar{\eta}_{i-1} + \theta_{i-1}(\mu)), \mu) \\
&\quad\quad - \frac{\partial \beta_{i-1}(\eta_{i-1})}{\partial \eta_{i-1}} \dot{\eta}_{i-1},
\end{aligned}
$$

which does not depend on η_i.

It can be seen that, under the coordinate transformation (7.97), the transformed augmented system (7.98) is still in the lower triangular form (7.12) with the dynamics of the internal model as the dynamic uncertainty. Moreover, the linear approximation of the function $Q_i(0, 0, \ldots, 0, z_i, 0, \ldots, 0, \mu)$ is given by $M_i z_i$, with M_i a Hurwitz matrix. Therefore, as will be seen later in Remark 7.31, in many interesting cases, the subsystems described by the second equation of (7.98) do satisfy Assumption 7.4. Thus, appealing to Theorem 7.6 immediately gives the following solvability conditions of the global robust stabilization problem of system (7.98):

Proposition 7.28. *Suppose system (7.98) satisfies the following two conditions.*

(i) $\dot{z}_0 = \bar{f}_0(z_0, \bar{x}_1, \mu)$ *is RISS with respect to* μ, *with* z_0 *as state and* $\bar{x}_1$ *as input and has a known* C^1 *gain function* $\kappa_0(\cdot)$.

(ii) *For all* $i = 1, \ldots, r$, $\dot{z}_i = Q_i(z_0, z_1, \ldots, z_i, \bar{x}_1, \ldots, \bar{x}_i, \mu)$ *is RISS with respect to* μ, *with* z_i *as state and* $\operatorname{col}(z_0, z_1, \ldots, z_{i-1}, \bar{x}_1, \ldots, \bar{x}_i)$ *as input, and has a known* C^1 *gain function* $\kappa_i(\cdot)$.

Then, there exists a smooth feedback control $\bar{u} = k(\bar{x}_1, \ldots, \bar{x}_r)$ *with* $k(0, \ldots, 0) = 0$ *such that the equilibrium point of the closed-loop system at the origin is globally asymptotically stable for all* $\mu \in V \times W$.

Combining Proposition 7.28 and Corollary 7.4 gives the solvability condition of the global robust output regulation problem for the original system (7.90) as follows.

Theorem 7.29. *Suppose system (7.90) satisfies Assumptions 7.9 to 7.11, and the same conditions (i) and (ii) of Proposition 7.28. Then the global robust output regulation problem can be solved by a dynamic state feedback controller of the form*

$$
\begin{aligned}
u &= \beta_r(\eta_r) + k(e, x_2 - \beta_1(\eta_1), \ldots, x_r - \beta_{r-1}(\eta_{r-1})), \\
\dot{\eta}_i &= M_i \eta_i + N_i(x_{i+1} - \beta_i(\eta_i) + \Psi_i T_i^{-1} \eta_i), \ i = 1, \ldots, r. \qquad (7.99)
\end{aligned}
$$

Remark 7.30. The three Assumptions 7.9 to 7.11 of Theorem 7.29 are mainly made for the existence of the regulator equations and the appropriate nonlinear internal model. Similar assumptions have to be made even for the solvability of the (local) robust output regulation problem. Conditions (i) and (ii) of Theorem 7.29 are made so that the augmented system can be globally robustly stabilized. ∎

Similar to Remark 7.23, we can identify two nontrivial cases where condition (ii) of Theorem 7.29 is satisfied as follows.

Remark 7.31. When the solution of the regulator equations, $\mathbf{x}_2(v, w), \ldots, \mathbf{x}_r(v, w), \mathbf{u}(v, w)$, are polynomial, the equation governing z_i, $i = 1, \ldots, r$, takes the special form as follows:

$$Q_i(z_0, z_1, \ldots, z_i, \bar{x}_1, \ldots, \bar{x}_i, \mu) = M_i z_i + \hat{\gamma}_i(z_0, z_1, \ldots, z_{i-1}, \bar{x}_1, \ldots, \bar{x}_i, \mu).$$

Thus, for this special case, condition (ii) of Theorem 7.29 automatically holds. In the current case, condition (ii) of Theorem 7.29 has to be verified. The way that we have already used in Remark 7.23 can be used directly to verify condition (ii) here and, for convenience, is repeated here. For $i = 1, \ldots, r$, denote $d_i = b_i^{-1}(v, w)N_i\bar{x}_i + \theta_i$. Then the second equation of (7.98) can be written as follows:

$$\dot{z}_i = M_i z_i - N_i \left(\beta_i^{[2]}(z_i + d_i) - \beta_i^{[2]}(d_i)\right) + \bar{\gamma}_i(z_0, z_1, \ldots, z_{i-1}, \bar{x}_1, \ldots, \bar{x}_i, \mu), \tag{7.100}$$

where

$$\bar{\gamma}_i(z_0, z_1, \ldots, z_{i-1}, \bar{x}_1, \ldots, \bar{x}_i, \mu) = \hat{\gamma}_i(z_0, z_1, \ldots, z_{i-1}, \bar{x}_1, \ldots, \bar{x}_i, \mu) - N_i\beta_i^{[2]}(d_i).$$

As in Remark 7.23, let P_i be a symmetric positive definite matrix such that

$$P_i M_i + M_i^T P_i = -I. \tag{7.101}$$

We will show that condition (ii) is verified if there exists a positive number $R_i < 1$ satisfying

$$-2z_i^T P_i N_i \left(\beta_i^{[2]}(z_i + d_i) - \beta_i^{[2]}(d_i)\right) \leq (1 - R_i)\|z_i\|^2 \tag{7.102}$$

for all z_i, d_i. In fact, let $V_{z_i}(z_i) = \frac{2}{R_i} z_i^T P_i z_i$. Then $\frac{2}{R_i}\lambda_{min}\|z_i\|^2 \leq V_{z_i}(z_i) \leq \frac{2}{R_i}\lambda_{max}\|z_i\|^2$, where λ_{max} (λ_{min}) is the maximal (minimal) eigenvalue of P_i. Further, in exactly the same way as deriving inequality (7.86), we can show that the derivative of $V_{z_i}(z_i)$ along system (7.100) satisfies

$$\frac{dV_{z_i}(z_i)}{dt} \leq -\|z_i\|^2 + \left\| \frac{2}{R_i} P_i \bar{\gamma}_i(z_0, z_1, \ldots, z_{i-1}, \bar{x}_1, \ldots, \bar{x}_i, \mu) \right\|^2. \tag{7.103}$$

Noting that function $\bar{\gamma}_i$ is C^1 satisfying $\bar{\gamma}_i(0, \ldots, 0, \mu) = 0$ and $\mu \in \Sigma$, with Σ a compact set, we have

$$\begin{aligned}&\left\| \frac{2}{R_i} P_i \bar{\gamma}_i(z_0, z_1, \ldots, z_{i-1}, \bar{x}_1, \ldots, \bar{x}_i, \mu)) \right\| \\ &\quad \leq \|\mathrm{col}(z_0, z_1, \ldots, z_{i-1}, \bar{x}_1, \ldots, \bar{x}_i)\| a_{i1}(z_0, z_1, \ldots, z_{i-1}, \bar{x}_1, \ldots, \bar{x}_i)\end{aligned}$$

for some smooth function $a_{i1}(z_0, z_1, \ldots, z_{i-1}, \bar{x}_1, \ldots, \bar{x}_i) \geq 1$. And there exists a smooth nondecreasing function $a_{i2}(\cdot)$ satisfying

$$a_{i2}(\|\mathrm{col}(z_0, z_1, \ldots, z_{i-1}, \bar{x}_1, \ldots, \bar{x}_i)\|) \geq a_{i1}(z_0, z_1, \ldots, z_{i-1}, \bar{x}_1, \ldots, \bar{x}_i).$$

As a result, we have

$$\frac{dV_{z_i}(z_i)}{dt} \leq -\|z_i\|^2 + a_i^2\,(\|\mathrm{col}(z_0, z_1, \ldots, z_{i-1}, \bar{x}_1, \ldots, \bar{x}_i)\|)$$

for some smooth class $\mathcal{K}_\infty$ function $a_i(s) = sa_{i2}(s)$. Thus, for any $0 < \epsilon_i < 1$,

$$\|z_i\| \geq \frac{a_i(\|\text{col}(z_0, z_1, \dots, z_{i-1}, \bar{x}_1, \dots, \bar{x}_i)\|)}{1-\epsilon_i} \quad \Rightarrow \quad \frac{dV_{z_i}(z_i)}{dt} \leq -\epsilon_i \|z_i\|^2.$$

Thus, by Theorem 2.16, choosing $\chi_i(s) = \frac{1}{1-\epsilon_i} a_i(s)$ gives that the condition (ii) holds for the known C^1 gain function

$$\kappa_i(s) \geq \sqrt{\frac{\lambda_{max}}{\lambda_{min}}} \chi_i(s) = \sqrt{\frac{\lambda_{max}}{\lambda_{min}}} \frac{a_i(s)}{(1-\epsilon_i)}.$$

Again, it can be seen that (7.102) holds for some $0 < R_i < 1$ when

$$\left|\beta_i^{[2]}(z_i + d) - \beta_i^{[2]}(d)\right| < \frac{1}{2\|P_i N_i\|}\|z_i\|.$$

Thus, (7.102) holds if $\beta_i^{[2]}$ is globally Lipschitz, that is

$$\left|\beta_i^{[2]}(z_i + d) - \beta_i^{[2]}(d)\right| \leq L_i\|z_i\|$$

for some positive number L_i, and if the Lipschitz constant L_i satisfies $L_i < \frac{1}{2\|P_i N_i\|}$ for all $v \in V$ and $w \in W$. ∎

Example 7.32. Consider the following lower triangular system:

$$\begin{aligned}
\dot{z} &= -5z^3 + w_1 z^2 e,\\
\dot{x}_1 &= z + 0.1w_1 e + 12.5x_2,\\
\dot{x}_2 &= -0.2v_1 + (0.1z - 0.8w_2 v_2)x_1 + \sin^2(w_2 v_1 x_2) + u,\\
y &= x_1,\\
e &= y - 10v_1,
\end{aligned} \tag{7.104}$$

with the exosystem

$$\dot{v}_1 = -0.5v_2, \qquad \dot{v}_2 = 0.5v_1. \tag{7.105}$$

These equations formulate the control problem of designing a state feedback regulator to have the output y of system (7.104) asymptotically track a sinusoidal signal of frequency 0.5 with arbitrarily large amplitude in the presence of two uncertain parameters w_1, w_2. Denote $u \equiv x_3$. It can be verified that this system satisfies Assumptions 7.9 to 7.11. In particular, the regulator equations associated with this system have a globally defined solution as follows:

$$\begin{aligned}
\mathbf{z}(v, w) &= 0,\\
\mathbf{x}_1(v, w) &= 10v_1,\\
\mathbf{x}_2(v, w) &= -0.4v_2,\\
\mathbf{u}(v, w) &= 8w_2 v_1 v_2 - \sin^2(0.4w_2 v_1 v_2).
\end{aligned} \tag{7.106}$$

Let $g_o(x,u) = \text{col}(x_2, u)$, $\pi_1^1(v,w) = -0.4v_2$, and $\pi_2^1(v,w) = 8w_2v_1v_2$. Then, the minimal zeroing polynomials of $\pi_1^1(v,w)$ and $\pi_2^1(v,w)$ are $4\lambda^2+1$ and λ^2+1, respectively. Assumption 7.11 is satisfied with

$$\begin{aligned}\Gamma_1(\pi_1^1(v,w), \dot{\pi}_1^1(v,w)) &= \pi_1^1(v,w),\\ \Gamma_2(\pi_2^1(v,w), \dot{\pi}_2^1(v,w)) &= \pi_2^1(v,w) - \sin^2(0.05\pi_2^1(v,w)),\end{aligned}$$

and the corresponding gradients and companion matrices are

$$\Psi_1 = \Psi_2 = [1\ 0], \text{ and } \Phi_1 = \begin{bmatrix} 0 & 1 \\ -0.25 & 0 \end{bmatrix}, \Phi_2 = \begin{bmatrix} 0 & 1 \\ -1 & 0 \end{bmatrix}.$$

For each $i = 1, 2$, the steady-state generator with output x_{i+1} is given by

$$\begin{aligned}\theta_i(v,w) &= T_i \text{col}(\pi_i^1(v,w), \dot{\pi}_i^1(v,w)),\\ \alpha_i(\theta_i) &= T_i \Phi_i T_i^{-1}\theta_i,\\ \beta_i(\theta_i) &= \Gamma_i(T_i^{-1}\theta_i),\end{aligned}$$

where T_i is any nonsingular matrix. To design an internal model, let

$$M_1 = M_2 = \begin{bmatrix} -2 & 0 \\ 0 & -1 \end{bmatrix}, \quad N_1 = N_2 = \begin{bmatrix} 2 \\ 1 \end{bmatrix}.$$

Solving the pertinent Sylvester equation gives

$$T_1 = \begin{bmatrix} 0.9412 & -0.4706 \\ 0.8 & -0.8 \end{bmatrix} \text{ and } T_2 = \begin{bmatrix} 0.8 & -0.4 \\ 0.5 & -0.5 \end{bmatrix}.$$

Thus,

$$\begin{aligned}\beta_1(\theta_1) &= \begin{bmatrix} 2.125 & -1.25 \end{bmatrix}\theta_1,\\ \beta_2(\theta_2) &= \begin{bmatrix} 2.5 & -2 \end{bmatrix}\theta_2 - \sin^2\left(\begin{bmatrix} 0.125 & -0.1 \end{bmatrix}\theta_2\right).\end{aligned}$$

Then

$$\begin{aligned}\beta_1^{[2]}(\theta_1) &= 0,\\ \beta_2^{[2]}(\theta_2) &= -\sin^2\left(\begin{bmatrix} 0.125 & -0.1 \end{bmatrix}\theta_2\right).\end{aligned} \tag{7.107}$$

Thus, the internal model is as follows:

$$\begin{aligned}\dot{\eta}_1 &= M_1\eta_1 + N_1 x_2,\\ \dot{\eta}_2 &= M_2\eta_2 + N_2\left(u + \sin^2\left(\begin{bmatrix} 0.125 & -0.1 \end{bmatrix}\theta_2\right)\right).\end{aligned} \tag{7.108}$$

Using the canonical coordinate and input transformation

$$\begin{aligned}\bar{\eta}_i &= \eta_i - \theta_i(v,w), \quad i = 1, 2,\\ \bar{z} &= z - \mathbf{z}(v,w),\\ \bar{x}_1 &= x_1 - \mathbf{x}_1(v,w) = e,\\ \bar{x}_2 &= x_2 - \beta_1(\eta_1),\\ \bar{u} &= u - \beta_2(\eta_2),\end{aligned}$$

and $z_0 = \bar{z}$, $z_1 = \bar{\eta}_1 - 0.08N_1\bar{x}_1$, $z_2 = \bar{\eta}_2 - N_2\bar{x}_2$ puts the augmented system (7.104) and (7.108) into the following form:

$$
\begin{aligned}
\dot{z}_0 &= -5z_0^3 + w_1 z_0{}^2\bar{x}_1,\\
\dot{z}_1 &= M_1 z_1 + 0.08M_1N_1\bar{x}_1 - 0.08N_1\left(z_0 + 0.1w_1\bar{x}_1\right),\\
\dot{\bar{x}}_1 &= z_0 + 0.1w_1\bar{x}_1 + 12.5\beta_1(z_1 + 0.08N_1\bar{x}_1) + 12.5\bar{x}_2,\\
\dot{z}_2 &= M_2z_2 - N_2\left(\beta_2^{[2]}(z_2 + N_2\bar{x}_2 + \theta_2) - \beta_2^{[2]}(N_2\bar{x}_2 + \theta_2)\right) + M_2N_2\bar{x}_2\\
&\quad - N_2F_2(z_0, z_1, \bar{x}_1, \bar{x}_2, v, w) - N_2\left(\beta_2^{[2]}(N_2\bar{x}_2 + \theta_2) - \beta_2^{[2]}(\theta_2)\right),\\
\dot{\bar{x}}_2 &= F_2(z_0, z_1, \bar{x}_1, \bar{x}_2, v, w) + \beta_2\left(z_2 + N_2\bar{x}_2 + \theta_2\right) - \beta_2(\theta_2) + \bar{u},
\end{aligned}
\tag{7.109}
$$

where

$$
\begin{aligned}
&F_2(z_0, z_1, \bar{x}_1, \bar{x}_2, v, w)\\
&\quad = 0.1z_0(\bar{x}_1 + 10v_1) + \sin^2\left(w_2v_1(\bar{x}_2 + \beta_1(\bar{\eta}_1) - 0.4v_2)\right)\\
&\quad\quad - \sin^2(0.4w_2v_1v_2) - 0.8w_2v_2\bar{x}_1 - \frac{\partial\beta_1(\bar{\eta}_1)}{\partial\bar{\eta}_1}\dot{\bar{\eta}}_1.
\end{aligned}
$$

Next, we will verify that all the solvability conditions given in Theorem 7.29 are satisfied. To be specific, we assume that $v(t) \in V = \{v_1^2 + v_2^2 \le 1\}$ and $-1 \le w_i \le 1$, $i = 1, 2$. Let us first verify the condition (i) of Proposition 7.28. Let $V_{z_0} = \frac{1}{2}z_0^2$. Then $\underline{a}z_0^2 \le V_{z_0}(z_0) \le \bar{a}z_0^2$ with $\underline{a} = \bar{a} = \frac{1}{2}$. And the derivative of $V_{z_0}(z_0)$ along the first equation of (7.109) is

$$
\frac{dV_{z_0}(z_0)}{dt} = z_0(-5z_0^3 + w_1z_0{}^2\bar{x}_1) \le -5z_0^4 + |z_0|^3|\bar{x}_1|.
$$

Thus, for any $0 < \epsilon_0 < 1$,

$$
|z_0| \ge \frac{0.2|x_1|}{1-\epsilon_0} \quad\Rightarrow\quad \frac{dV_{z_0}(z_0)}{dt} \le -5\epsilon_0 z_0^4.
$$

Thus, choosing $\chi_{z_0}(s) = \frac{0.2}{1-\epsilon_0}s$ gives a C^1 gain function

$$
\kappa_0(s) = 0.21s \ge \underline{a}^{-1}\circ\bar{a}\circ\chi_{z_0}(s) = \frac{0.2}{1-\epsilon_0}s.
$$

To verify condition (ii), we will resort to Remark 7.31. First, solving the Lyapunov equation (7.101) gives $P_1 = P_2 = \begin{bmatrix} 0.25 & 0 \\ 0 & 0.5 \end{bmatrix}$. When $i = 1$, it is satisfied since $\beta_1^{[2]}(\cdot) \equiv 0$. In particular, let $V_{z_1}(z_1) = 2z_1{}^TP_1z_1$. Then, from (7.103), the derivative of $V_{z_1}(z_1)$ along the trajectories of the second equation of system (7.109) is

$$
\begin{aligned}
\frac{dV_{z_1}(z_1)}{dt} &\le -\|z_1\|^2 + \|2P_1\left(0.08M_1N_1\bar{x}_1 - 0.08N_1\left(z_0 + 0.1w_1\bar{x}_1\right)\right)\|^2\\
&\le -\|z_1\|^2 + \|[0.16\ 0.08]|\bar{x}_1| + [0.08\ 0.08]\left(|z_0| + 0.1|\bar{x}_1|\right)\|^2\\
&\le -\|z_1\|^2 + \|[0.168\ 0.088]|\bar{x}_1| + [0.08\ 0.08]|z_0|\|^2\\
&\le -\|z_1\|^2 + (0.221\|\mathrm{col}(z_0, \bar{x}_1)\|)^2.
\end{aligned}
$$

Thus, when $i = 1$, condition (ii) is satisfied with $\kappa_1(s) = 0.313s \ge \frac{\sqrt{2}}{(1-\epsilon_1)} \times 0.221s$ for some $0 < \epsilon_1 < 1$.

When $i = 2$, the inequality (7.102) becomes

$$-2z_2^T \begin{bmatrix} 0.25 & 0 \\ 0 & 0.5 \end{bmatrix} \begin{bmatrix} 2 \\ 1 \end{bmatrix} \left(\beta_2^{[2]}(z_2 + d_2) - \beta_2^{[2]}(d_2) \right) \leq (1 - R_2)\|z_2\|^2, \tag{7.110}$$

where $d_2 = N_2\bar{x}_2 + \theta_2$. Letting $z_2 = \mathrm{col}(z_{21}, z_{22})$ leads (7.110) to

$$(z_{21} + z_{22}) \left(\sin^2([0.125 \ \ -0.1](z_2 + d_2)) - \sin^2([0.125 \ \ -0.1]d_2) \right) \leq (1 - R_2)(z_{21}^2 + z_{22}^2). \tag{7.111}$$

Simple manipulation shows that (7.111), hence the inequality (7.102), holds for $0 < R_2 < 0.773$, and we choose $R_2 = 0.77$. In particular, let $V_{z_2}(z_2) = \frac{2}{R_2} z_2{}^T P_2 z_2$ and note that

$$\begin{aligned}
&\|F_2(z_0, z_1, \bar{x}_1, \bar{x}_2, v, w)\| \\
&\leq \Big| 0.1z_0(\bar{x}_1 + 10v_1) + \sin^2\big(w_2v_1(\bar{x}_2 + \beta_1(\bar{\eta}_1) - 0.4v_2)\big) \\
&\quad - \sin^2(0.4w_2v_1v_2) - 0.8w_2v_2\bar{x}_1 - \frac{\partial\beta_1(\bar{\eta}_1)}{\partial\bar{\eta}_1}\dot{\bar{\eta}}_1 \Big| \\
&\leq 0.1|z_0\bar{x}_1| + |z_0| + |\bar{x}_2| + 2.47\|z_1\| + 0.24|\bar{x}_1| \\
&\quad + 0.8|\bar{x}_1| + 3.29\|z_1\| + 0.14|\bar{x}_1| + 3|\bar{x}_2| \\
&\leq 0.1|z_0\bar{x}_1| + |z_0| + 5.76\|z_1\| + 1.18|\bar{x}_1| + 4|\bar{x}_2|.
\end{aligned}$$

Then, from (7.103), the derivative of $V_{z_2}(z_2)$ along the trajectories of the fourth equation of system (7.109) is

$$\begin{aligned}
&\frac{dV_{z_2}(z_2)}{dt} \\
&\leq -\|z_2\|^2 \\
&\quad + \left\| \frac{2}{R_2} P_2 \left[M_2N_2\bar{x}_2 - N_2F_2(z_0, z_1, \bar{x}_1, \bar{x}_2, v, w) - N_2 \left(\beta_2^{[2]}(N_2\bar{x}_2 + \theta_2) - \beta_2^{[2]}(\theta_2) \right) \right] \right\|^2 \\
&\leq -\|z_2\|^2 \\
&\quad + \left[2.91|\bar{x}_2| + 1.84\big(0.1|z_0\bar{x}_1| + |z_0| + 5.76\|z_1\| + 1.18|\bar{x}_1| + 4|\bar{x}_2|\big) + 0.276|\bar{x}_2| \right]^2 \\
&\leq -\|z_2\|^2 + [0.184|z_0\bar{x}_1| + 1.84|z_0| + 10.6\|z_1\| + 2.18|\bar{x}_1| + 10.6|\bar{x}_2|]^2 \\
&\leq -\|z_2\|^2 + \left(0.092\|\mathrm{col}(z_0, z_1, \bar{x}_1, \bar{x}_2)\|^2 + 15.3\|\mathrm{col}(z_0, z_1, \bar{x}_1, \bar{x}_2)\|\right)^2.
\end{aligned}$$

Thus, when $i = 2$, condition (ii) is satisfied with

$$\kappa_2(s) = 0.14s^2 + 21.7s \geq \sqrt{2}\frac{0.092s^2 + 15.3s}{1 - \epsilon_2}$$

for some $0 < \epsilon_2 < 1$.

By Proposition 7.28, the global robust stabilization problem of system (7.109) is solvable. In fact, using the procedure described in the proof of Proposition 7.28 shows that the following controller:

$$\begin{aligned} \bar{u} &= -(0.3\tilde{x}_2^2 + 500)^2 \tilde{x}_2, \\ \tilde{x}_2 &= \bar{x}_2 + 6.47\bar{x}_1 \end{aligned}$$

globally robustly stabilizes system (7.109) for all $v(t) \in V = \{v_1^2 + v_2^2 \leq 1\}$ and all $-1 \leq w_i \leq 1$, $i = 1, 2$. Further, by Corollary 7.4, the overall controller for solving the global robust output regulation problem for system (7.104) is given by the composition of the internal model (7.108) and

$$\begin{aligned} u &= -(0.3\tilde{x}_2^2 + 500)^2 \tilde{x}_2 + \beta_2(\eta_2), \\ \tilde{x}_2 &= x_2 + 6.47\bar{x}_1 - \beta_1(\eta_1). \quad \blacksquare \end{aligned}$$

Chapter 8

Output Regulation for Singular Nonlinear Systems

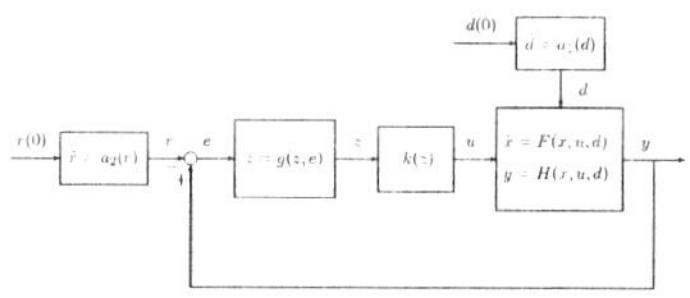

Singular systems are dynamical systems whose behaviors are governed by both differential equations and algebraic equations. Such systems arise in electrical networks, power systems, large-scale systems, and so on. In this chapter, we study the output regulation problem for singular nonlinear systems. In Section 8.1, we give a formulation of the output regulation problem for singular nonlinear systems. In Section 8.2, we review some basic results on singular linear systems that will be invoked in subsequent sections. Section 8.3 starts from a generalized version of the Center Manifold Theorem that applies to singular nonlinear systems and then presents the solvability conditions of the output regulation problem by both state feedback control and *singular* output feedback control. In Section 8.4, we further give the solvability conditions of the output regulation problem by *normal* output feedback control. Section 8.5 studies the approximation of the output regulation problem for singular systems. Finally, in Section 8.6, we turn to the study of the robust output regulation problem for uncertain singular systems.

8.1 Problem Formulation

Consider the plant described by

$$\begin{aligned} S\dot{x}(t) &= f(x(t), u(t), v(t)), \ x(0) = x_0, \\ e(t) &= h(x(t), v(t)), \ t \geq 0, \end{aligned} \tag{8.1}$$

and an exosystem described by

$$\dot{v}(t) = a(v(t)), \quad v(0) = v_0, \tag{8.2}$$

where $x(t) \in \mathcal{R}^n$ is the plant state, $u(t) \in \mathcal{R}^m$ the plant input, $e(t) \in \mathcal{R}^p$ the plant output representing the tracking error, $v(t) \in \mathcal{R}^q$ the exogenous signal representing the disturbance and/or the reference input, and $S \in \mathcal{R}^{n \times n}$ a constant matrix. When S is an identity matrix, (8.1) is called a normal system, and when S is singular, (8.1) is called a singular system. Throughout this chapter, we assume that S is singular and denote rank $S = n_s$.

We will focus on two classes of control laws, namely,

1. **Static State Feedback Control Laws:**

$$u(t) = k(x(t), v(t)). \tag{8.3}$$

2. **Dynamic Output Feedback Control Laws:**

$$\begin{aligned} u(t) &= k(z(t), e(t)), \\ S_z\dot{z}(t) &= g(z(t), e(t)), \end{aligned} \tag{8.4}$$

where $z(t)$ is the compensator state vector of dimension n_z to be specified later, and $S_z \in \mathcal{R}^{n_z \times n_z}$ is a constant matrix.

Equation (8.4) is said to be a *normal* controller if S_z is an identity matrix. The closed-loop system composed of plant (8.1), (8.2), and control law (8.3) or (8.4) can be put into the following form:

$$\begin{aligned} S_c\dot{x}_c(t) &= f_c(x_c(t), v(t)), \; x_c(0) = x_{c0}, \\ \dot{v}(t) &= a(v(t)), \; v(0) = v_0, \\ e(t) &= h_c(x_c(t), v(t)), \end{aligned} \tag{8.5}$$

where for the state feedback case, $x_c = x$, $S_c = S$, $f_c(x, v) = f(x, k(x, v), v)$, and $h_c(x_c, v) = h(x, v)$, and for the output feedback case, $x_c = \text{col}(x, z)$ and

$$\begin{aligned} f_c(x_c, v) &= \begin{bmatrix} f(x, k(z, h(x, v)), v) \\ g(z, h(x, v)) \end{bmatrix}, \quad S_c = \begin{bmatrix} S & 0 \\ 0 & S_z \end{bmatrix}, \\ h_c(x_c, v) &= h(x, v). \end{aligned} \tag{8.6}$$

Again, all functions involved in this setup are assumed to be sufficiently smooth and defined globally on the appropriate Euclidean spaces, with the value zero at the respective origins. As in Chapter 3, the results will be stated locally in terms of an open neighborhood V of the origin in $\mathcal{R}^q$, and we implicitly permit V to be made smaller to accommodate subsequent local arguments. We denote the dimension of x_c by n_c with the understanding that $n_c = n$ for the static state feedback case and $n_c = n + n_z$ for the output feedback case.

Remark 8.1. Unlike the normal systems studied in the previous chapters, the input u does not appear on the right-hand side of the second equation of (8.1). This is because, as will be seen later, we usually need to resort to a dynamic control law of the form (8.4) to control a singular system. Using the simplified output equation can avoid inconsistent feedback composition of the plant and the control law. ∎

Before stating the objective of control, let us first introduce some notation and terminologies. Let $S, A \in \mathcal{R}^{n \times n}$, $B \in \mathcal{R}^{n \times m}$, and $C \in \mathcal{R}^{p \times n}$. Let $\sigma(S, A) = \{\lambda \mid \lambda \in \mathcal{C}, \; \det(\lambda S - A) = 0\}$, $\mathcal{C}_- = \{\lambda \mid \lambda \in \mathcal{C}, \; \text{Re}(\lambda) < 0\}$, and $\bar{\mathcal{C}}_- = \{\lambda \mid \lambda \in \mathcal{C}, \; \text{Re}(\lambda) \le 0\}$. A complex number λ is said to be the eigenvalue of (S, A) if $\lambda \in \sigma(S, A)$. (S, A) is said to be *stable* if $\sigma(S, A) \subset \mathcal{C}_-$; (S, A, B) is *stabilizable* if there exists $K \in \mathcal{R}^{m \times n}$ such that $(S, A + BK)$ is stable. (S, C, A) is *detectable* if there exists an $L \in \mathcal{R}^{n\times p}$ such that $(S, A - LC)$ is stable. (S, A) is said to be *standard* if $\deg\det(\lambda S - A) = \text{rank } S$. (S, A) is said to be *strongly stable* if it is both stable and standard. (S, A, B) is *strongly stabilizable* if there exists a matrix $K \in \mathcal{R}^{m\times n}$ such that $(S, A + BK)$ is strongly stable. (S, C, A) is *strongly detectable* if there exists a matrix $L \in \mathcal{R}^{n\times m}$ such that $(S, A + LC)$ is strongly stable. (A, B) is said to be *normalizable* if there exists a matrix $K \in \mathcal{R}^{m\times n}$ such that $A + BK$ is nonsingular.

Our objective is to find a controller (static state feedback or dynamic output feedback) such that the closed-loop system (8.5) has the following two properties.

Property 8.1. The pair (S_c, A_c) is strongly stable where

$$A_c = \frac{\partial f_c(0, 0)}{\partial x_c}. \tag{8.7}$$

Property 8.2. The trajectory starting from any sufficiently small initial state $\text{col}(x_{c0}, v_0)$ satisfies

$$\lim_{t\to\infty} e(t) = \lim_{t\to\infty} h_c(x_c(t), v(t)) = 0. \tag{8.8}$$

Remark 8.2. Property 8.1 is slightly stronger than the stability of (S_c, A_c). The additional condition $\deg(\det(\lambda S_c - A_c)) = \text{rank}(S_c)$ guarantees, as will be seen later from the proof of Lemma 8.9, that the closed-loop system (8.5) will induce a stable center manifold passing the origin of $\mathcal{R}^{n_c+q}$ that is crucial for the fulfillment of Property 8.2. Moreover, it is well known that the response of a strongly stable singular linear system is impulse free, a desirable property by all practical engineering systems. We will see in Remark 8.10 that this nice property will also be retained by nonlinear systems with Property 8.1. Thus, Property 8.1 will guarantee that the trajectories of the closed-loop system exist and are bounded for all $t \geq 0$ and for all sufficiently small initial states. ∎

Many of our results will rely on the properties of the linear approximation of the plant and the exosystem. Therefore, we introduce the following familiar notation:

$$A = \frac{\partial f}{\partial x}(0, 0, 0), \quad B = \frac{\partial f}{\partial u}(0, 0, 0), \quad E = \frac{\partial f}{\partial v}(0, 0, 0),$$

$$C = \frac{\partial h}{\partial x}(0, 0), \qquad F = \tfrac{\partial h}{\partial v}(0, 0), \qquad A_1 = \frac{\partial a}{\partial v}(0).$$

As a result, the linear approximation of the plant and the exosystem at the origin can be described by

$$\begin{aligned} S\dot{x} &= Ax + Bu + Ev, \\ \dot{v} &= A_1 v, \\ e &= Cx + Fv, \end{aligned} \tag{8.9}$$

where $S, A \in \mathcal{R}^{n\times n}$, $B \in \mathcal{R}^{n\times m}$, $E \in \mathcal{R}^{n\times q}$, $C \in \mathcal{R}^{p\times n}$, $F \in \mathcal{R}^{p\times q}$, and $A_1 \in \mathcal{R}^{q\times q}$.

Now we are ready to list the following assumptions.

Assumption 8.1. The triple (S, A, B) is strongly stabilizable.

Assumption 8.2. The triple

$$\left(\begin{bmatrix} S & 0 \\ 0 & I_q \end{bmatrix}, [C \quad F], \begin{bmatrix} A & E \\ 0 & A_1 \end{bmatrix}\right)$$

is strongly detectable.

Remark 8.3. Assumptions 8.1 and 8.2 are made to ensure the fulfillment of Property 8.1 by state feedback and/or output feedback control. We note that, in the special case where $S = I$, Assumptions 8.1 and 8.2 reduce to Assumptions 3.2 and 3.3 made for the solvability of the output regulation problem of the *normal* systems. ∎

8.2 Preliminaries of Singular Linear Systems

In this section, we will introduce some properties of a singular linear system of the form (8.9). These properties will be used in the subsequent sections. Let us first note that there exist two nonsingular matrices $T_1, T_2 \in \mathcal{R}^{n\times n}$ such that

$$T_1 S T_2 = \begin{bmatrix} I_{n_s} & 0 \\ 0 & 0 \end{bmatrix} = \bar{S}.$$

Let

$$T_1 A T_2 = \bar{A} = \begin{bmatrix} \bar{A}_{11} & \bar{A}_{12} \\ \bar{A}_{21} & \bar{A}_{22} \end{bmatrix}, \quad T_1 B = \bar{B} = \begin{bmatrix} \bar{B}_1 \\ \bar{B}_2 \end{bmatrix}, \quad T_1 E = \bar{E} = \begin{bmatrix} \bar{E}_1 \\ \bar{E}_2 \end{bmatrix},$$

$$C T_2 = \bar{C} = \begin{bmatrix} \bar{C}_1 & \bar{C}_2 \end{bmatrix}, \quad T_2^{-1} x = \bar{x} = \begin{bmatrix} \bar{x}_1 \\ \bar{x}_2 \end{bmatrix},$$

where $\bar{A}_{11} \in \mathcal{R}^{n_s\times n_s}$, $\bar{B}_1 \in \mathcal{R}^{n_s\times m}$, $\bar{E}_1 \in \mathcal{R}^{n_s\times q}$, $\bar{C}_1 \in \mathcal{R}^{p\times n_s}$, $\bar{x}_1 \in \mathcal{R}^{n_s}$, and all other matrices have appropriate dimensions. Then the coordinate transformation $\bar{x} = T_2^{-1}x$ on (8.9) leads to a singular system of the form

$$\begin{aligned} \dot{\bar{x}}_1 &= \bar{A}_{11}\bar{x}_1 + \bar{A}_{12}\bar{x}_2 + \bar{B}_1 u + \bar{E}_1 v, \\ 0 &= \bar{A}_{21}\bar{x}_1 + \bar{A}_{22}\bar{x}_2 + \bar{B}_2 u + \bar{E}_2 v, \\ e &= \bar{C}\bar{x} + F v. \end{aligned} \tag{8.10}$$

From $\det(T_1(\lambda S - A)T_2) = \det(\lambda\bar{S} - \bar{A})$, it is clear that (S, A) is standard or strongly stable if and only if $(\bar{S}, \bar{A})$ is standard or strongly stable. Moreover, system (8.10) will retain the strong stabilizability and detectability properties of (8.9) as shown below.

Lemma 8.4.

(i) *$(\bar{S}, \bar{A})$ is standard if and only if $\bar{A}_{22}$ is nonsingular and is strongly stable if and only if $\bar{A}_{22}$ is nonsingular and $\bar{A}_{11} - \bar{A}_{12}\bar{A}_{22}^{-1}\bar{A}_{21}$ is Hurwitz.*

(ii) *$\left(\bar{S}, \bar{A}, \bar{B}\right)$ is strongly stabilizable if and only if (S, A, B) is, and $\left(\bar{S}, \bar{C}, \bar{A}\right)$ is strongly detectable if and only if (S, C, A) is.*

(iii)

$$\left(\begin{bmatrix} \bar{S} & 0 \\ 0 & I_q \end{bmatrix}, \begin{bmatrix} \bar{C} & F \end{bmatrix}, \begin{bmatrix} \bar{A} & \bar{E} \\ 0 & A_1 \end{bmatrix}\right)$$

is strongly detectable if and only if

$$\left(\begin{bmatrix} S & 0 \\ 0 & I_q \end{bmatrix}, \begin{bmatrix} C & F \end{bmatrix}, \begin{bmatrix} A & E \\ 0 & A_1 \end{bmatrix}\right)$$

is.

Proof. (i) From

$$\det\left(\lambda\bar{S} - \bar{A}\right) = \det\begin{bmatrix} \lambda I_{n_s} - \bar{A}_{11} & -\bar{A}_{12} \\ -\bar{A}_{21} & -\bar{A}_{22} \end{bmatrix},$$

$(\bar{S}, \bar{A})$ is standard if and only if $\bar{A}_{22}$ is nonsingular. On the other hand, if $\bar{A}_{22}$ is nonsingular, then

$$\begin{aligned} \det\left(\lambda\bar{S} - \bar{A}\right) &= \det\begin{bmatrix} \lambda I_{n_s} - \bar{A}_{11} & -\bar{A}_{12} \\ -\bar{A}_{21} & -\bar{A}_{22} \end{bmatrix} \\ &= \det(-\bar{A}_{22})\det(\lambda I_{n_s} - (\bar{A}_{11} - \bar{A}_{12}\bar{A}_{22}^{-1}\bar{A}_{21})). \end{aligned}$$

Thus, $(\bar{S}, \bar{A})$ is standard and is strongly stable if, additionally, $\bar{A}_{11} - \bar{A}_{12}\bar{A}_{22}^{-1}\bar{A}_{21}$ is Hurwitz.

(ii) The proof follows from

$$\begin{aligned} \det\left(T_1(\lambda S - (A + BK))\, T_2\right) &= \det\left(\lambda\bar{S} - \left(\bar{A} + \bar{B}\bar{K}\right)\right) \text{ where } \bar{K} = KT_2, \\ \det\left(T_1(\lambda S - (A + LC))\, T_2\right) &= \det\left(\lambda\bar{S} - \left(\bar{A} + \bar{L}\bar{C}\right)\right) \text{ where } \bar{L} = T_1 L. \end{aligned}$$

(iii) The proof follows from

$$\begin{aligned} &\begin{bmatrix} T_1 & 0 \\ 0 & I_q \end{bmatrix}\left(\lambda\begin{bmatrix} S & 0 \\ 0 & I_q \end{bmatrix} - \left(\begin{bmatrix} A & E \\ 0 & A_1 \end{bmatrix} - \begin{bmatrix} L_1 \\ L_2 \end{bmatrix}\begin{bmatrix} C & F \end{bmatrix}\right)\right)\begin{bmatrix} T_2 & 0 \\ 0 & I_q \end{bmatrix} \\ &\quad = \lambda\begin{bmatrix} \bar{S} & 0 \\ 0 & I_q \end{bmatrix} - \left(\begin{bmatrix} \bar{A} & \bar{E} \\ 0 & A_1 \end{bmatrix} - \begin{bmatrix} \bar{L}_1 \\ L_2 \end{bmatrix}\begin{bmatrix} \bar{C} & F \end{bmatrix}\right), \end{aligned}$$

where $\bar{L}_1 = T_1 L_1$. □

If (S, A) is standard, we can always define a reduced-order normal system from (8.10) as follows. Let

$$\bar{x}_2 = -\bar{A}_{22}^{-1}\left(\bar{A}_{21}\bar{x}_1 + \bar{B}_2 u + \bar{E}_2 v\right). \tag{8.11}$$

Substituting (8.11) into the first and third equations of (8.10) gives a reduced-order normal system as follows:

$$\begin{aligned} \dot{\bar{x}}_1 &= A_r \bar{x}_1 + B_r u + E_r v, \\ e &= C_r \bar{x}_1 + D_r u + F_r v, \end{aligned} \tag{8.12}$$

where

$$\begin{aligned} A_r &= \bar{A}_{11} - \bar{A}_{12}\bar{A}_{22}^{-1}\bar{A}_{21}, \\ B_r &= \bar{B}_1 - \bar{A}_{12}\bar{A}_{22}^{-1}\bar{B}_2, \\ E_r &= \bar{E}_1 - \bar{A}_{12}\bar{A}_{22}^{-1}\bar{E}_2, \\ C_r &= \bar{C}_1 - \bar{C}_2\bar{A}_{22}^{-1}\bar{A}_{21}, \\ D_r &= -\bar{C}_2\bar{A}_{22}^{-1}\bar{B}_2, \\ F_r &= F - \bar{C}_2\bar{A}_{22}^{-1}\bar{E}_2. \end{aligned} \tag{8.13}$$

This normal system has the following property.

Lemma 8.5. *Assume* (S, A) *is standard. Then,*

(i) (A_r, B_r) *is stabilizable if* (S, A, B) *is strongly stabilizable, and* (C_r, A_r) *is detectable if* (S, C, A) *is strongly detectable,*

(ii) *The pair*

$$\left(\begin{bmatrix} C_r & F_r \end{bmatrix}, \begin{bmatrix} A_r & E_r \\ 0 & A_1 \end{bmatrix} \right)$$

is detectable if the triple

$$\left(\begin{bmatrix} S & 0 \\ 0 & I_q \end{bmatrix}, \begin{bmatrix} C & F \end{bmatrix}, \begin{bmatrix} A & E \\ 0 & A_1 \end{bmatrix} \right)$$

is strongly detectable.

(iii) *For all* $\lambda \in \mathcal{C}$,

$$\begin{aligned} \operatorname{rank} \begin{bmatrix} A - \lambda S & B \\ C & 0 \end{bmatrix} &= \operatorname{rank} \begin{bmatrix} \bar{A} - \lambda \bar{S} & \bar{B} \\ \bar{C} & 0 \end{bmatrix} \\ &= \operatorname{rank} \begin{bmatrix} A_r - \lambda I_{n_s} & B_r \\ C_r & D_r \end{bmatrix} + \operatorname{rank}(\bar{A}_{22}). \end{aligned}$$

Proof. (i) By part (ii) of Lemma 8.4 if (S, A, B) is strongly stabilizable, so is $(\bar{S}, \bar{A}, \bar{B})$. Thus, there exists a matrix $K \in \mathcal{R}^{m \times n}$ such that $(\bar{S}, \bar{A} + \bar{B}K)$ is strongly stable.

Denote $K = \begin{bmatrix} K_1 & K_2 \end{bmatrix}$ with $K_1 \in \mathcal{R}^{m \times n_s}$. Then

$$\begin{aligned} &\det\left(\lambda \bar{S} - (\bar{A} + \bar{B}K)\right) \\ &= \det\left(\begin{bmatrix} \lambda I_{n_s} - (\bar{A}_{11} + \bar{B}_1 K_1) & -(\bar{A}_{12} + \bar{B}_1 K_2) \\ -(\bar{A}_{21} + \bar{B}_2 K_1) & -(\bar{A}_{22} + \bar{B}_2 K_2) \end{bmatrix} \right). \end{aligned}$$

Using the fact that $\bar{A}_{22} + \bar{B}_2 K_2$ is nonsingular since $\left(\bar{S}, \bar{A} + \bar{B}K\right)$ is standard gives further

$$\begin{aligned}
&\det\left(\begin{bmatrix} \lambda I_{n_s} - \left(\bar{A}_{11} + \bar{B}_1 K_1\right) & -(\bar{A}_{12} + \bar{B}_1 K_2) \\ -\left(\bar{A}_{21} + \bar{B}_2 K_1\right) & -\left(\bar{A}_{22} + \bar{B}_2 K_2\right) \end{bmatrix}\right) \\
&\quad = \det\left(-\bar{A}_{22} - \bar{B}_2 K_2\right) \det\left(\lambda I_{n_s} - \left(\bar{A}_{11} + \bar{B}_1 K_1\right)\right. \\
&\qquad\qquad \left. + \left(\bar{A}_{12} + \bar{B}_1 K_2\right)\left(\bar{A}_{22} + \bar{B}_2 K_2\right)^{-1}\left(\bar{A}_{21} + \bar{B}_2 K_1\right)\right).
\end{aligned}$$

Noting that $\bar{A}_{22}$ is nonsingular since (S, A) is standard and using the following matrix identity:

$$\left(\bar{A}_{22} + \bar{B}_2 K_2\right)^{-1} = \bar{A}_{22}^{-1} - \bar{A}_{22}^{-1}\bar{B}_2\left(K_2\bar{A}_{22}^{-1}\bar{B}_2 + I\right)^{-1} K_2\bar{A}_{22}^{-1}$$

gives

$$\begin{aligned}
&\left(\bar{A}_{11} + \bar{B}_1 K_1\right) - \left(\bar{A}_{12} + \bar{B}_1 K_2\right)\left(\bar{A}_{22} + \bar{B}_2 K_2\right)^{-1}\left(\bar{A}_{21} + \bar{B}_2 K_1\right) \\
&\quad = \left(\bar{A}_{11} - \bar{A}_{12}\bar{A}_{22}^{-1}\bar{A}_{21}\right) + \left(\bar{B}_1 - \bar{A}_{12}\bar{A}_{22}^{-1}\bar{B}_2\right) K_r = A_r + B_r K_r,
\end{aligned}$$

where

$$K_r = K_1 - \left(K_2\bar{A}_{22}^{-1}\bar{B}_2 + I\right)^{-1} K_2\bar{A}_{22}^{-1}\left(\bar{A}_{21} + \bar{B}_2 K_1\right).$$

Therefore,

$$\det\left(\lambda\bar{S} - \left(\bar{A} + \bar{B}K\right)\right) = \det\left(-\bar{A}_{22} - \bar{B}_2 K_2\right) \times \det\left(\lambda I_{n_s} - (A_r + B_r K_r)\right),$$

which shows that the pair (A_r, B_r) is stabilizable. The detectability of (C_r, A_r) follows from the fact that (S, C, A) is detectable if and only if (S, A^T, C^T) is stabilizable, and (C_r, A_r) is detectable if and only if (A_r^T, C_r^T) is stabilizable.

(ii) Recall from part (iii) of Lemma 8.4 that

$$\left(\begin{bmatrix} \bar{S} & 0 \\ 0 & I_q \end{bmatrix}, \begin{bmatrix} \bar{C} & F \end{bmatrix}, \begin{bmatrix} \bar{A} & \bar{E} \\ 0 & A_1 \end{bmatrix}\right)$$

is strongly detectable if and only if

$$\left(\begin{bmatrix} S & 0 \\ 0 & I_q \end{bmatrix}, \begin{bmatrix} C & F \end{bmatrix}, \begin{bmatrix} A & E \\ 0 & A_1 \end{bmatrix}\right)$$

is. Let

$$M_1 = \begin{bmatrix} I_{n_s} & -\bar{A}_{12}\bar{A}_{22}^{-1} & 0 \\ 0 & 0 & I_q \\ 0 & I_{n-n_s} & 0 \end{bmatrix},$$

$$M_2 = \begin{bmatrix} I_{n_s} & 0 & 0 \\ -\bar{A}_{22}^{-1}\bar{A}_{21} & -\bar{A}_{22}^{-1}\bar{E}_2 & I_{n-n_s} \\ 0 & I_q & 0 \end{bmatrix}.$$

These two matrices are clearly nonsingular. A straightforward calculation shows

$$M_1\begin{bmatrix} \bar{S} & 0 \\ 0 & I_q \end{bmatrix} M_2 = \begin{bmatrix} I_{n_s} & 0 & 0 \\ 0 & I_q & 0 \\ 0 & 0 & 0 \end{bmatrix},$$

$$\begin{aligned} M_1\begin{bmatrix} \bar{A} & \bar{E} \\ 0 & A_1 \end{bmatrix} M_2 &= M_1 \begin{bmatrix} \bar{A}_{11} & \bar{A}_{12} & \bar{E}_1 \\ \bar{A}_{21} & \bar{A}_{22} & \bar{E}_2 \\ 0 & 0 & A_1 \end{bmatrix} M_2 \\ &= \begin{bmatrix} A_r & 0 & E_r \\ 0 & 0 & A_1 \\ \bar{A}_{21} & \bar{A}_{22} & \bar{E}_2 \end{bmatrix} M_2 \\ &= \begin{bmatrix} A_r & E_r & 0 \\ 0 & A_1 & 0 \\ 0 & 0 & \bar{A}_{22} \end{bmatrix}, \end{aligned}$$

$$\begin{aligned} \begin{bmatrix} \bar{C} & F \end{bmatrix} M_2 &= \begin{bmatrix} \bar{C}_1 & \bar{C}_2 & F \end{bmatrix} M_2 \\ &= \begin{bmatrix} C_r & F_r & \bar{C}_2 \end{bmatrix}. \end{aligned}$$

Thus the triple

$$\left(\begin{bmatrix} I_{n_s} & 0 & 0 \\ 0 & I_q & 0 \\ 0 & 0 & 0 \end{bmatrix}, \begin{bmatrix} C_r & F_r & \bar{C}_2 \end{bmatrix}, \begin{bmatrix} A_r & E_r & 0 \\ 0 & A_1 & 0 \\ 0 & 0 & \bar{A}_{22} \end{bmatrix} \right)$$

is strongly detectable, which in turn implies the detectability of

$$\left(\begin{bmatrix} C_r & F_r \end{bmatrix}, \begin{bmatrix} A_r & E_r \\ 0 & A_1 \end{bmatrix} \right).$$

(iii) The proof of the first equality follows directly from

$$\begin{bmatrix} T_1 & 0 \\ 0 & I_p \end{bmatrix} \begin{bmatrix} A - \lambda S & B \\ C & 0 \end{bmatrix} \begin{bmatrix} T_2 & 0 \\ 0 & I_m \end{bmatrix} = \begin{bmatrix} \bar{A} - \lambda \bar{S} & \bar{B} \\ \bar{C} & 0 \end{bmatrix}.$$

To show the second equality, let

$$N_1 = \begin{bmatrix} I_{n_s} & -\bar{A}_{12}\bar{A}_{22}^{-1} & 0 \\ 0 & 0 & I_p \\ 0 & I_{n-n_s} & 0 \end{bmatrix}, \quad N_2 = \begin{bmatrix} I_{n_s} & 0 & 0 \\ -\bar{A}_{22}^{-1}\bar{A}_{21} & -\bar{A}_{22}^{-1}\bar{B}_2 & I_{n-n_s} \\ 0 & I_m & 0 \end{bmatrix}.$$

Then it can be verified that

$$\begin{aligned} N_1 \begin{bmatrix} \bar{A} - \lambda \bar{S} & \bar{B} \\ \bar{C} & 0 \end{bmatrix} N_2 &= N_1 \begin{bmatrix} \bar{A}_{11} - \lambda I_{n_s} & \bar{A}_{12} & \bar{B}_1 \\ \bar{A}_{21} & \bar{A}_{22} & \bar{B}_2 \\ \bar{C}_1 & \bar{C}_2 & 0 \end{bmatrix} N_2 \\ &= \begin{bmatrix} A_r - \lambda I_{n_s} & 0 & B_r \\ \bar{C}_1 & \bar{C}_2 & 0 \\ \bar{A}_{21} & \bar{A}_{22} & \bar{B}_2 \end{bmatrix} N_2 \\ &= \begin{bmatrix} A_r - \lambda I_{n_s} & B_r & 0 \\ C_r & D_r & \bar{C}_2 \\ 0 & 0 & \bar{A}_{22} \end{bmatrix}. \quad \square \end{aligned}$$

Lemma 8.6. *Assume that (S, A) is standard. If a linear output feedback control law of the form*

$$u = K_z z, \qquad \dot{z} = G_1 z + G_2 e \tag{8.14}$$

stabilizes the reduced normal system (8.12), then it also strongly stabilizes the original singular system (8.9).

Proof. Let the closed-loop system composed of (8.12) and (8.14) be denoted by $\dot{x}_{cr} = A_{cr}x_{cr} + B_{cr}v$ with $x_{cr} = \mathrm{col}(\bar{x}_1, z)$, and the closed-loop system composed of (8.9) and (8.14) by $S_c\dot{x}_c = A_c x_c + B_c v$ with $x_c = \mathrm{col}(x, z)$. Then,

$$A_{cr} = \begin{bmatrix} A_r & B_r K_z \\ G_2 C_r & G_1 + G_2 D_r K_z \end{bmatrix}$$

and

$$S_c = \begin{bmatrix} S & 0 \\ 0 & I_{n_z} \end{bmatrix}, \quad A_c = \begin{bmatrix} A & BK_z \\ G_2 C & G_1 \end{bmatrix}.$$

Let

$$N_1 = \begin{bmatrix} I_{n_s} & -\bar{A}_{12}\bar{A}_{22}^{-1} & 0 \\ 0 & 0 & I_{n_z} \\ 0 & I_{n-n_s} & 0 \end{bmatrix},$$

$$N_2 = \begin{bmatrix} I_{n_s} & 0 & 0 \\ -\bar{A}_{22}^{-1}\bar{A}_{21} & -\bar{A}_{22}^{-1}\bar{B}_2 K_z & I_{n-n_s} \\ 0 & I_{n_z} & 0 \end{bmatrix}.$$

A simple calculation shows

$$\begin{aligned} N_1\begin{bmatrix} T_1 & 0 \\ 0 & I_{n_z} \end{bmatrix} A_c \begin{bmatrix} T_2 & 0 \\ 0 & I_{n_z} \end{bmatrix} N_2 &= N_1 \begin{bmatrix} \bar{A}_{11} & \bar{A}_{12} & \bar{B}_1 K_z \\ \bar{A}_{21} & \bar{A}_{22} & \bar{B}_2 K_z \\ G_2\bar{C}_1 & G_2\bar{C}_2 & G_1 \end{bmatrix} N_2 \\ &= \begin{bmatrix} A_r & 0 & B_r K_z \\ G_2\bar{C}_1 & G_2\bar{C}_2 & G_1 \\ \bar{A}_{21} & \bar{A}_{22} & \bar{B}_2 K_z \end{bmatrix} N_2 \\ &= \begin{bmatrix} A_r & B_r K_z & 0 \\ G_2 C_r & G_1 + G_2 D_r K_z & G_2\bar{C}_2 \\ 0 & 0 & \bar{A}_{22} \end{bmatrix}, \end{aligned}$$

$$N_1\begin{bmatrix} T_1 & 0 \\ 0 & I_{n_z} \end{bmatrix} S_c \begin{bmatrix} T_2 & 0 \\ 0 & I_{n_z} \end{bmatrix} N_2 = \begin{bmatrix} I_{n_s+n_z} & 0 \\ 0 & 0 \end{bmatrix},$$

from which we can verify that the stability of A_{cr} and the nonsingularity of $\bar{A}_{22}$ imply the strong stability of (S_c, A_c). □

When (S, A) is not standard, it is possible to employ an output feedback control to yield a new system that is standard and retains some desirable structural properties of the original system as shown by the following lemma.

Lemma 8.7. *Consider a singular linear system of the form (8.9). Assume (S, A) is not standard but (S, A, B) is strongly stabilizable and (S, C, A) is strongly detectable. Then, there exists a linear output feedback control*

$$u = K_e e + \bar{u} \tag{8.15}$$

such that the following system:

$$\begin{aligned} S\dot{x} &= (A + BK_eC)x + B\bar{u} + (E + BK_eF)v = \tilde{A}x + B\bar{u} + \tilde{E}v, \\ e &= Cx + Fv \end{aligned} \tag{8.16}$$

satisfies the following.

(i) $(S, \tilde{A})$ *is standard.*

(ii) $(S, \tilde{A}, B)$ *is strongly stabilizable and* $(S, C, \tilde{A})$ *is strongly detectable.*

(iii)

$$\left(\begin{bmatrix} S & 0 \\ 0 & I_q \end{bmatrix}, \begin{bmatrix} C & F \end{bmatrix}, \begin{bmatrix} \tilde{A} & \tilde{E} \\ 0 & A_1 \end{bmatrix} \right)$$

is strongly detectable if

$$\left(\begin{bmatrix} S & 0 \\ 0 & I_q \end{bmatrix}, \begin{bmatrix} C & F \end{bmatrix}, \begin{bmatrix} A & E \\ 0 & A_1 \end{bmatrix} \right)$$

is.

Proof. (i) Using the same transformation matrices T_1, T_2 as those used in Lemma 8.4, we can convert a system of the form (8.16) into the form given by (8.10), in particular,

$$T_1 \left(\lambda S - \tilde{A} \right) T_2 = \begin{bmatrix} \lambda I_{n_s} - \bar{A}_{11} - \bar{B}_1 K_e \bar{C}_1 & -\bar{A}_{12} - \bar{B}_1 K_e \bar{C}_2 \\ -\bar{A}_{21} - \bar{B}_2 K_e \bar{C}_1 & -\bar{A}_{22} - \bar{B}_2 K_e \bar{C}_2 \end{bmatrix}.$$

Thus,

$$\begin{aligned} &\deg \left(\det \left(\lambda S - \tilde{A} \right) \right) \\ &\quad = \deg \left(\det \left(\begin{bmatrix} \lambda I_{n_s} - \bar{A}_{11} - \bar{B}_1 K_e \bar{C}_1 & -\bar{A}_{12} - \bar{B}_1 K_e \bar{C}_2 \\ -\bar{A}_{21} - \bar{B}_2 K_e \bar{C}_1 & -\bar{A}_{22} - \bar{B}_2 K_e \bar{C}_2 \end{bmatrix} \right) \right). \end{aligned}$$

By part (i) of Lemma 8.4, $(S, \tilde{A})$ is standard if and only if there exists a matrix $K_e \in \mathcal{R}^{m \times p}$ such that $\det(\bar{A}_{22} + \bar{B}_2 K_e \bar{C}_2) \neq 0$.

Since (S, A, B) is strongly stabilizable, there exists a matrix K such that

$$\begin{aligned} &\deg (\det (\lambda S - (A + BK))) \\ &\quad = \deg \left(\det \left(\begin{bmatrix} \lambda I_{n_s} - \bar{A}_{11} - \bar{B}_1 \bar{K}_1 & -\bar{A}_{12} - \bar{B}_1 \bar{K}_2 \\ -\bar{A}_{21} - \bar{B}_2 \bar{K}_1 & -\bar{A}_{22} - \bar{B}_2 \bar{K}_2 \end{bmatrix} \right) \right) \\ &\quad = n_s, \end{aligned}$$

where $\left[\bar{K}_1, \bar{K}_2 \right] = KT_2$. By part (i) of Lemma 8.4, $\det(\bar{A}_{22} + \bar{B}_2 \bar{K}_2) \neq 0$; that is, the pair $(\bar{A}_{22}, \bar{B}_2)$ is normalizable.

Similarly, since (S, C, A) is strongly detectable, there exists a matrix L such that

$$\begin{aligned}
&\deg(\det(\lambda S - (A - LC))) \\
&= \deg\left(\det\left(\begin{bmatrix} \lambda I_{n_s} - \bar{A}_{11} + \bar{L}_1\bar{C}_1 & -\bar{A}_{12} + \bar{L}_1\bar{C}_2 \\ -\bar{A}_{21} + \bar{L}_2\bar{C}_1 & -\bar{A}_{22} + \bar{L}_2\bar{C}_2 \end{bmatrix}\right)\right) \\
&= n_s,
\end{aligned}$$

where

$$\begin{bmatrix} \bar{L}_1 \\ \bar{L}_2 \end{bmatrix} = T_1 L.$$

Hence, $\det(\bar{A}_{22} - \bar{L}_2\bar{C}_2) \neq 0$; that is, the pair $(\bar{A}_{22}^T, \bar{C}_2^T)$ is normalizable.

We now show that the normalizability of $(\bar{A}_{22}, \bar{B}_2)$ and the normalizability of $(\bar{A}_{22}^T, \bar{C}_2^T)$ guarantee the existence of a matrix $K_e \in \mathcal{R}^{m\times p}$ such that $\det(\bar{A}_{22} + \bar{B}_2 K_e \bar{C}_2) \neq 0$. For this purpose, denote rank $\bar{A}_{22} = n_a$. If $n_a = n - n_s$, $\bar{A}_{22}$ is nonsingular and it suffices to let $K_e = 0$ to solve the problem. Otherwise, suppose $n_a < n - n_s$. Then there exist two nonsingular matrices $P, Q \in \mathcal{R}^{(n-n_s)\times(n-n_s)}$ such that

$$P\bar{A}_{22}Q = \begin{bmatrix} I_{n_a} & 0 \\ 0 & 0 \end{bmatrix}, \; P\bar{B}_2 = \begin{bmatrix} \bar{B}_{21} \\ \bar{B}_{22} \end{bmatrix}, \; \bar{C}_2 Q = \left[\bar{C}_{21}, \bar{C}_{22}\right],$$

where $\bar{B}_{21} \in \mathcal{R}^{n_a\times m}$, $\bar{C}_{21} \in \mathcal{R}^{p\times n_a}$. We now claim that the gain $K_e = \bar{B}_{22}^T\bar{C}_{22}^T$ solves the problem. In fact,

$$\begin{aligned}
P(\bar{A}_{22} + \bar{B}_2\bar{B}_{22}^T\bar{C}_{22}^T\bar{C}_2)Q &= \begin{bmatrix} I_{n_a} & 0 \\ 0 & 0 \end{bmatrix} + \begin{bmatrix} \bar{B}_{21} \\ \bar{B}_{22} \end{bmatrix} \bar{B}_{22}^T\bar{C}_{22}^T\left[\bar{C}_{21}, \bar{C}_{22}\right] \\
&= \begin{bmatrix} I_{n_a} + \bar{B}_{21}\bar{B}_{22}^T\bar{C}_{22}^T\bar{C}_{21} & \bar{B}_{21}\bar{B}_{22}^T\bar{C}_{22}^T\bar{C}_{22} \\ \bar{B}_{22}\bar{B}_{22}^T\bar{C}_{22}^T\bar{C}_{21} & \bar{B}_{22}\bar{B}_{22}^T\bar{C}_{22}^T\bar{C}_{22} \end{bmatrix} \\
&= \begin{bmatrix} I_{n_a} & \bar{B}_{21}\bar{B}_{22}^T \\ 0 & \bar{B}_{22}\bar{B}_{22}^T \end{bmatrix} \begin{bmatrix} I_{n_a} & 0 \\ \bar{C}_{22}^T\bar{C}_{21} & \bar{C}_{22}^T\bar{C}_{22} \end{bmatrix}.
\end{aligned}$$

It follows from the above decomposition that the matrix $(\bar{A}_{22} + \bar{B}_2\bar{B}_{22}^T\bar{C}_{22}^T\bar{C}_2)$ is nonsingular if and only if $\bar{B}_{22}$ has full row rank and $\bar{C}_{22}$ has full column rank. Now let $\bar{K}_2$ be such that $\bar{A}_{22} + \bar{B}_2\bar{K}_2$ is nonsingular and denote $\bar{K}_2 Q = [\bar{K}_{21} \;\; \bar{K}_{22}]$. Then

$$\begin{aligned}
P(\bar{A}_{22} + \bar{B}_2\bar{K}_2)Q &= \begin{bmatrix} I_{n_a} & 0 \\ 0 & 0 \end{bmatrix} + \begin{bmatrix} \bar{B}_{21}\bar{K}_{21} & \bar{B}_{21}\bar{K}_{22} \\ \bar{B}_{22}\bar{K}_{21} & \bar{B}_{22}\bar{K}_{22} \end{bmatrix} \\
&= \begin{bmatrix} I_{n_a} & \bar{B}_{21} \\ 0 & \bar{B}_{22} \end{bmatrix} \begin{bmatrix} I_{n_a} & 0 \\ \bar{K}_{21} & \bar{K}_{22} \end{bmatrix}.
\end{aligned} \tag{8.17}$$

Thus the nonsingularity of the matrix $\bar{A}_{22} + \bar{B}_2\bar{K}_2$ implies that $\bar{B}_{22}$ has full row rank. Similarly, we can show that the nonsingularity of the matrix $\bar{A}_{22}^T - \bar{L}_2\bar{C}_2^T$ implies that $\bar{C}_{22}$ has full column rank.

(ii) To prove part (ii), one only needs to note that, for any matrices $K \in \mathcal{R}^{m\times n}$ and $L \in \mathcal{R}^{n\times p}$,

$$(S, A + BK) = \left(S, \tilde{A} + B(K - K_e C)\right)$$

and

$$(S, A - LC) = \left(S, \tilde{A} - (L + BK_e)C\right).$$

(iii) The proof follows from the fact that, for any $L \in \mathcal{R}^{(n+q)\times p}$,

$$\begin{aligned}&\left(\begin{bmatrix} S & 0 \\ 0 & I_q \end{bmatrix}, \begin{bmatrix} A & E \\ 0 & A_1 \end{bmatrix} - L\begin{bmatrix} C & F \end{bmatrix}\right)\\ &= \left(\begin{bmatrix} S & 0 \\ 0 & I_q \end{bmatrix}, \begin{bmatrix} \tilde{A} & \tilde{E} \\ 0 & A_1 \end{bmatrix} - \begin{bmatrix} L_1 + BK_e \\ L_2 \end{bmatrix}\begin{bmatrix} C & F \end{bmatrix}\right),\end{aligned}$$

where $L^T = [L_1^T, L_2^T]$. □

Remark 8.8. If there exist matrices L_1 and L_2 such that

$$\begin{aligned}&\deg\left(\det\left(\lambda\begin{bmatrix} S & 0 \\ 0 & I_q \end{bmatrix} - \left(\begin{bmatrix} A & E \\ 0 & A_1 \end{bmatrix} - \begin{bmatrix} L_1 \\ L_2 \end{bmatrix}\begin{bmatrix} C & F \end{bmatrix}\right)\right)\right)\\ &= \deg\left(\det\left(\begin{bmatrix} \lambda I_{n_s} - \bar{A}_{11} + \bar{L}_{11}\bar{C}_1 & -\bar{A}_{12} + \bar{L}_{11}\bar{C}_2 & -\bar{E}_1 + \bar{L}_{11}F \\ -\bar{A}_{21} + \bar{L}_{12}\bar{C}_1 & -\bar{A}_{22} + \bar{L}_{12}\bar{C}_2 & -\bar{E}_2 + \bar{L}_{12}F \\ L_2\bar{C}_1 & L_2\bar{C}_2 & \lambda I_q - A_1 + L_2F \end{bmatrix}\right)\right)\\ &= n_s + q,\end{aligned}$$

where

$$\begin{bmatrix} \bar{L}_{11} \\ \bar{L}_{12} \end{bmatrix} = T_1 L_1,$$

then, necessarily, $\det\left(\bar{A}_{22} - \bar{L}_{12}\bar{C}_2\right) \neq 0$; that is, the pair $\left(\bar{A}_{22}^T, \bar{C}_2^T\right)$ is normalizable. Thus part (i) still holds if we replace the strong detectability of (S, C, A) by that of

$$\left(\begin{bmatrix} S & 0 \\ 0 & I_q \end{bmatrix}, \begin{bmatrix} C & F \end{bmatrix}, \begin{bmatrix} A & E \\ 0 & A_1 \end{bmatrix}\right). \quad \blacksquare$$

8.3 Output Regulation by State Feedback and Singular Output Feedback

It is known that the Center Manifold Theorem plays a key role in solving the output regulation problem for *normal* nonlinear systems. In this section, we will establish a generalized version of the Center Manifold Theorem that applies to the class of singular nonlinear systems described in (8.5).

Lemma 8.9. *Assume that the exosystem (8.2) satisfies Assumption 3.1′ and the closed-loop system (8.5) has Property 8.1. Then,*

(i) *there exists a sufficiently smooth function* $\mathbf{x}_c(v)$ *defined for* $v \in V$ *satisfying* $\mathbf{x}_c(0) = 0$ *and*

$$S_c \frac{\partial \mathbf{x}_c(v)}{\partial v} a(v) = f_c(\mathbf{x}_c(v), v); \tag{8.18}$$

(ii) *for any sufficiently small x_{c0} and v_0, the solution of (8.5) denoted by* $\mathrm{col}(x_c(t), v(t))$ *exists and is bounded for all $t > 0$ and satisfies*

$$\lim_{t\to\infty} [x_c(t) - \mathbf{x}_c(v(t))] = 0. \tag{8.19}$$

(iii) *The closed-loop system (8.5) satisfies Property 8.2 if and only if there exists a sufficiently smooth function $\mathbf{x}_c(v)$ locally defined in $v \in V$ with $\mathbf{x}_c(0) = 0$ such that*

$$S_c \frac{\partial \mathbf{x}_c(v)}{\partial v} a(v) = f_c(\mathbf{x}_c(v), v), \tag{8.20}$$

$$0 = h_c(\mathbf{x}_c(v), v). \tag{8.21}$$

Proof. Part (i). Rewrite the first two equations of system (8.5) into the following form:

$$S_c \dot{x}_c = A_c x_c + B_c v + \phi(x_c, v), \tag{8.22}$$

$$\dot{v} = A_1 v + \psi(v), \tag{8.23}$$

where $\phi(x_c, v)$ and $\psi(v)$ vanish at their origins together with their first-order derivatives.

Assume rank $S_c = r$. Then there exist two nonsingular matrices T_1 and T_2 such that

$$T_1 S_c T_2 = \begin{bmatrix} I_r & 0 \\ 0 & 0 \end{bmatrix}.$$

Let

$$\begin{bmatrix} \bar{A}_{c11} & \bar{A}_{c12} \\ \bar{A}_{c21} & \bar{A}_{c22} \end{bmatrix} = T_1 A_c T_2, \quad T_1 B_c = \begin{bmatrix} \bar{B}_{c1} \\ \bar{B}_{c2} \end{bmatrix}, \quad \bar{x}_c = T_2^{-1} x_c = \begin{bmatrix} \bar{x}_{c1} \\ \bar{x}_{c2} \end{bmatrix},$$

$$T_1 \phi(x_c, v) = \begin{bmatrix} \bar{\phi}_1(\bar{x}_{c1}, \bar{x}_{c2}, v) \\ \bar{\phi}_2(\bar{x}_{c1}, \bar{x}_{c2}, v) \end{bmatrix},$$

where $\bar{A}_{c11} \in \mathcal{R}^{r\times r}$, $\bar{B}_{c1} \in \mathcal{R}^{r\times q}$, $\bar{x}_{c1} \in \mathcal{R}^r$, and $\bar{\phi}_1 \in \mathcal{R}^r$. Then, premultiplying T_1 on both sides of (8.22) gives

$$\dot{\bar{x}}_{c1} = \bar{A}_{c11}\bar{x}_{c1} + \bar{A}_{c12}\bar{x}_{c2} + \bar{B}_{c1} v + \bar{\phi}_1(\bar{x}_{c1}, \bar{x}_{c2}, v), \tag{8.24}$$

$$0 = \bar{A}_{c21}\bar{x}_{c1} + \bar{A}_{c22}\bar{x}_{c2} + \bar{B}_{c2} v + \bar{\phi}_2(\bar{x}_{c1}, \bar{x}_{c2}, v), \tag{8.25}$$

where $\bar{\phi}_1(\bar{x}_{c1}, \bar{x}_{c2}, v)$ and $\bar{\phi}_2(\bar{x}_{c1}, \bar{x}_{c2}, v)$ vanish at $\mathrm{col}(\bar{x}_{c1}, \bar{x}_{c2}, v) = 0$ together with their first-order derivatives.

It follows from the strong stability of (S_c, A_c) that $\bar{A}_{c22}$ is nonsingular. By the Implicit Function Theorem, there exists a unique sufficiently smooth function $\alpha(\bar{x}_{c1}, v)$ defined in an open neighborhood of $(\bar{x}_{c1}, v) = (0, 0)$ that satisfies $\alpha(0, 0) = 0$ and

$$0 = \bar{A}_{c21}\bar{x}_{c1} + \bar{A}_{c22}\alpha(\bar{x}_{c1}, v) + \bar{B}_{c2} v + \bar{\phi}_2(\bar{x}_{c1}, \alpha(\bar{x}_{c1}, v), v). \tag{8.26}$$

Furthermore,

$$\left.\frac{\partial \alpha(\bar{x}_{c1}, v)}{\partial \bar{x}_{c1}}\right|_{\substack{\bar{x}_{c1}=0 \\ v=0}} = -\bar{A}_{c22}^{-1}\bar{A}_{c21}.$$

Substituting $\bar{x}_{c2} = \alpha(\bar{x}_{c1}, v)$ into (8.24) gives

$$\dot{\bar{x}}_{c1} = (\bar{A}_{c11} - \bar{A}_{c12}\bar{A}_{c22}^{-1}\bar{A}_{c21})\bar{x}_{c1} + \bar{B}_{c1}v + \bar{\phi}_3(\bar{x}_{c1}, v),$$

where $\bar{\phi}_3(\bar{x}_{c1}, v)$ vanishes at $(\bar{x}_{c1}, v) = 0$ with its first-order derivative.

By part (i) of Lemma 8.4, $\bar{A}_{c11} - \bar{A}_{c12}\bar{A}_{c22}^{-1}\bar{A}_{c21}$ is Hurwitz since (S_c, A_c) is strongly stable by assumption.

Now consider the following *normal* system:

$$\begin{aligned}
\dot{\bar{x}}_{c1} &= \bar{f}_1(\bar{x}_{c1}, v) = (\bar{A}_{c11} - \bar{A}_{c12}\bar{A}_{c22}^{-1}\bar{A}_{c21})\bar{x}_{c1} + \bar{B}_{c1}v + \bar{\phi}_3(\bar{x}_{c1}, v),\\
\dot{v} &= a(v),\\
e &= \bar{h}(\bar{x}_{c1}, v),
\end{aligned} \tag{8.27}$$

where

$$\bar{h}(\bar{x}_{c1}, v) = h_c(T_{21}\bar{x}_{c1}, T_{22}\alpha(\bar{x}_{c1}, v), v) \tag{8.28}$$

and $T_2 = [T_{21} \;\; T_{22}]$ with $T_{21} \in \mathcal{R}^{n_c \times r}$. Since all the eigenvalues of $(\bar{A}_{c11} - \bar{A}_{c12}\bar{A}_{c22}^{-1}\bar{A}_{c21})$ have negative real parts and all the eigenvalues of A_1 have zero real parts by Assumption 3.1$'$, by Theorem 2.25, system (8.27) has a stable center manifold defined in an open neighborhood of the origin of $\mathcal{R}^q$, or, equivalently, there exists a sufficiently smooth function $\bar{\mathbf{x}}_{c1}(v)$ defined for $v \in V$ that satisfies $\bar{\mathbf{x}}_{c1}(0) = 0$ and is such that

$$\frac{\partial \bar{\mathbf{x}}_{c1}(v)}{\partial v}a(v) = (\bar{A}_{c11} - \bar{A}_{c12}\bar{A}_{c22}^{-1}\bar{A}_{c21})\bar{\mathbf{x}}_{c1}(v) + \bar{B}_{c1}v + \bar{\phi}_3(\bar{\mathbf{x}}_{c1}(v), v). \tag{8.29}$$

Moreover, there exist positive constants δ and λ such that, for all sufficiently small $\bar{x}_{c1}(0)$ and $v(0)$, the solution of (8.27) satisfies

$$||\bar{x}_{c1}(t) - \bar{\mathbf{x}}_{c1}(v(t))|| \leq \delta e^{-\lambda t}||\bar{x}_{c1}(0) - \bar{\mathbf{x}}_{c1}(v(0))||, \quad t \geq 0. \tag{8.30}$$

Let

$$\mathbf{x}_c(v) = T_2 \begin{bmatrix} \bar{\mathbf{x}}_{c1}(v) \\ \alpha(\bar{\mathbf{x}}_{c1}(v), v) \end{bmatrix}. \tag{8.31}$$

Then it can be readily verified, using (8.26) and (8.29), that (8.31) satisfies (8.18).

Part (ii). In terms of the solution $\bar{x}_{c1}$ of (8.27), we can define

$$x_c(t) = \begin{cases} T_2 \begin{bmatrix} \bar{x}_{c1}(t) \\ \alpha(\bar{x}_{c1}(t), v(t)) \end{bmatrix}, & t > 0, \\ x_{c0}, & t = 0. \end{cases} \tag{8.32}$$

Clearly, for $t > 0$, $x_c(t)$ is bounded and $\mathrm{col}(x_c(t), v(t))$ satisfies (8.5). Moreover, by (8.30) and the sufficient smoothness of $\alpha(\cdot, \cdot)$, we have

$$\begin{aligned}
\lim_{t\to\infty}[x_c(t) - \mathbf{x}_c(v(t))] &= T_2 \begin{bmatrix} \lim_{t\to\infty}(\bar{x}_{c1}(t) - \bar{\mathbf{x}}_{c1}(v(t))) \\ \lim_{t\to\infty}(\alpha(\bar{x}_{c1}(t), v(t)) - \alpha(\bar{\mathbf{x}}_{c1}(v(t)), v(t))) \end{bmatrix}\\
&= 0.
\end{aligned} \tag{8.33}$$

Part (iii). *Sufficiency.* Assume (8.20) and (8.21) hold for some $\mathbf{x}_c(v)$. Then, by part (ii) of this lemma, for all sufficiently small x_{c0} and v_0, the solution of (8.5) satisfies (8.19). It follows from the sufficient smoothness of $h(\cdot, \cdot)$ as well as (8.21) and (8.19) that

$$\lim_{t\to\infty} e(t) = \lim_{t\to\infty} [h_c(x_c(t), v(t)) - h_c(\mathbf{x}_c(v(t)), v(t))] = 0. \tag{8.34}$$

Necessity. Since the closed-loop system (8.5) satisfies Property 8.1, by part (i) of this lemma, there exists some sufficiently smooth function $\mathbf{x}_c(v)$ for $v \in V$ with $\mathbf{x}_c(0) = 0$ satisfying (8.20). We will further show that the function $\mathbf{x}_c(v)$ also satisfies (8.21) if the closed-loop system (8.5) satisfies Property 8.2. For this purpose, we first show that the function $\bar{\mathbf{x}}_{c1}(v)$ defined by (8.29) also satisfies

$$0 = \bar{h}(\bar{\mathbf{x}}_{c1}(v), v). \tag{8.35}$$

In fact, by (8.32), (8.28), and the assumption that the closed-loop system (8.5) satisfies Property 8.2, we have

$$\lim_{t\to\infty} h_c(x_c(t), v(t)) = \lim_{t\to\infty} h_c(T_2\bar{x}_c(t), v(t)) = \lim_{t\to\infty} \bar{h}(\bar{x}_{c1}(t), v(t)) = 0. \tag{8.36}$$

Thus the reduced normal system (8.27) satisfies Properties 8.1 and 8.2. We now recall from the output regulation theory for the *normal system* as stated in Lemma 3.6 that if, in addition to Property 8.1, (8.27) also satisfies Property 8.2, then $\bar{\mathbf{x}}_{c1}(v)$ necessarily satisfies (8.35). Now noting that $\mathbf{x}_c(v)$ and $\bar{\mathbf{x}}_{c1}(v)$ are related by (8.31) gives

$$h_c(\mathbf{x}_c(v), v) = \bar{h}(\bar{\mathbf{x}}_{c1}(v), v) = 0. \tag{8.37}$$

That is, $\mathbf{x}_c(v)$ also satisfies (8.21). □

Remark 8.10. A distinct feature of singular linear systems from normal linear systems is that the zero input response of the system may contain an impulsive function. However, when the system is strongly stable, the zero input response of the system is impulse free. This nice property is also retained for the singular nonlinear system described by (8.5) if the linearization of $S_c\dot{x}_c = f_c(x_c, 0)$ is strongly stable. This is evident from the explicit expression given by (8.32). However, as opposed to the *normal* system, the response $x_c(t)$ may be discontinuous at $t = 0$. The magnitude of the discontinuity of $x_c(t)$ as given by (8.32) can be calculated as follows. Let

$$x_{c0}(0+) = \lim_{t\to 0+} x_c(t)$$

and

$$T_2^{-1} = \begin{bmatrix} \Gamma_1 \\ \Gamma_2 \end{bmatrix},$$

where $\Gamma_1 \in \mathcal{R}^{r\times n_c}$. Then the magnitude of the discontinuity of $x_c(t)$ at $t = 0$ is

$$x_c(0+) - x_{c0} = T_2 \begin{bmatrix} 0 \\ \alpha(\Gamma_1 x_{c0}, v_0) - \Gamma_2 x_{c0} \end{bmatrix}.$$

Clearly, this magnitude can be made arbitrarily small by having x_{c0} and v_0 sufficiently small. ■

Remark 8.11. A geometric interpretation of Lemma 8.9 can be given as follows. Let $x_a = \mathrm{col}(x_c,\ v)$ and rewrite the system (8.5) as follows:

$$S_a \dot{x}_a = \begin{bmatrix} f_c(x_c, v) \\ a(v) \end{bmatrix} = f_a(x_a), \tag{8.38}$$

where $S_a = $ block diag (S_c, I_q). Then equations (8.18) and (8.2) can be put into the following:

$$S_a \frac{\partial \mathbf{x}_a(v)}{\partial v} a(v) = f_a(\mathbf{x}_a(v)), \tag{8.39}$$

where $\mathbf{x}_a(v) = \mathrm{col}(\mathbf{x}_c(v),\ v)$. Thus the manifold defined by $x_a = \mathrm{col}(\mathbf{x}_c(v),\ v)$ for $v \in V$ is a locally invariant manifold for the singular system (8.38). What is more, we can show that $\mathbf{x}_a(v)$ is actually a center manifold for the system (8.38) in a meaningful sense. In fact, denote the Jacobian matrices of $f_a(x_a)$ and $\mathbf{x}_a(v)$ at their origins by A_a and X_a, respectively. It is not difficult to verify, by linearizing (8.39), that

$$S_a X_a A_1 = A_a X_a. \tag{8.40}$$

Since $\sigma(S_a, A_a) = \sigma(S_c, A_c) \cup \sigma(I_q, A_1)$ and the matrix A_1 has only zero-real-part eigenvalues, the eigenspace of (S_a, A_a) associated with the eigenvalues of (I_q, A_1) is the tangent space to the manifold $x_a = \mathbf{x}_a(v)$ at $x_a = 0$. Thus, the manifold $x_a = \mathbf{x}_a(v)$ can be reasonably called the local center manifold of the system (8.38) passing through $x_a = 0$. ∎

Having established Lemma 8.9, it is possible to obtain the solvability conditions of the output regulation problem for singular systems via both the state feedback controller and the output feedback controller as given in the following two theorems.

Theorem 8.12. *Assume that the exosystem (8.2) satisfies Assumption 3.1′ and the singular plant (8.1) satisfies Assumption 8.1. Then the output regulation problem for the singular system (8.1) and (8.2) is solvable by a state feedback controller if and only if there exist sufficiently smooth functions* $\mathbf{x}(v)$ *with* $\mathbf{x}(0) = 0$ *and* $\mathbf{u}(v)$ *with* $\mathbf{u}(0) = 0$*, both defined in an open neighborhood* V *of the origin of* $\mathcal{R}^q$*, satisfying the following:*

$$S \frac{\partial \mathbf{x}(v)}{\partial v} a(v) = f(\mathbf{x}(v), \mathbf{u}(v), v), \tag{8.41}$$

$$0 = h(\mathbf{x}(v), v). \tag{8.42}$$

Proof. *Necessity.* Assume the state feedback control $u = k(x, v)$ solves the state feedback output regulation problem. Then, by Lemma 8.9, there exists some sufficiently smooth function $\mathbf{x}_c(v)$ for $v \in V$ with $\mathbf{x}_c(0) = 0$ satisfying (8.20) and (8.21). Define $\mathbf{x}(v) = \mathbf{x}_c(v)$ and $\mathbf{u}(v) = k(\mathbf{x}(v), v)$. Then it is straightforward to verify that $\mathbf{x}(v)$ and $\mathbf{u}(v)$ satisfy (8.41) and (8.42).

Sufficiency. Observe that, by Assumption 8.1, there exists a matrix K_x such that $(S, A + BK_x)$ is strongly stable. Suppose equations (8.41) and (8.42) are satisfied for some $\mathbf{x}(v)$ and $\mathbf{u}(v)$. Let

$$k(x, v) = \mathbf{u}(v) + K_x(x - \mathbf{x}(v)). \tag{8.43}$$

This controller yields a closed-loop system with $x_c = x$, $S_c = S$, $f_c(x_c, v) = f(x, k(x, v), v)$, and $h_c(x_c, v) = h(x, v)$. Then, Property 8.1 is satisfied since the Jacobian matrix of $f_c(x_c, 0) = f(x, k(x, 0), 0)$ at the origin is equal to $A + BK_x$. Next, let $\mathbf{x}_c(v) = \mathbf{x}(v)$. Clearly, $k(\mathbf{x}_c(v), v) = \mathbf{u}(v)$. Thus (8.41) and (8.42) lead to (8.20) and (8.21). It follows from Lemma 8.9 that Property 8.2 is also fulfilled. □

Theorem 8.13. *Assume that the exosystem (8.2) satisfies Assumption 3.1′ and the singular plant (8.1) satisfies Assumptions 8.1 and 8.2. Then the output regulation problem for the singular system (8.1) and (8.2) is solvable by an output feedback controller if and only if there exist sufficiently smooth functions* $\mathbf{x}(v)$ *with* $\mathbf{x}(0) = 0$ *and* $\mathbf{u}(v)$ *with* $\mathbf{u}(0) = 0$, *both defined for* $v \in V$, *satisfying equations (8.41) and (8.42).*

Proof. *Necessity.* Assume that the output feedback control $u = k(z, e)$, $S_z\dot{z} = g(z, e)$ solves the output regulation problem. Then, by Lemma 8.9, there exists some sufficiently smooth function $\mathbf{x}_c(v)$ for $v \in V$ with $\mathbf{x}_c(0) = 0$ satisfying (8.20) and (8.21). Perform the partition $\mathbf{x}_c(v) = \mathrm{col}(\mathbf{x}_1(v), \mathbf{x}_2(v))$ such that $\mathbf{x}_1(v) \in \mathcal{R}^n$. Let $\mathbf{x}(v) = \mathbf{x}_1(v)$ and $\mathbf{u}(v) = k(\mathbf{x}_2(v), 0)$. Then it is possible to verify that $\mathbf{x}(v)$ and $\mathbf{u}(v)$ satisfy (8.41) and (8.42).

Sufficiency. By Assumptions 8.1 and 8.2, there exist matrices K_x, L_1, and L_2 such that

$$(S, A + BK_x) \text{ and } \left(\begin{bmatrix} S & 0 \\ 0 & I_q \end{bmatrix}, \begin{bmatrix} A - L_1C & E - L_1F \\ -L_2C & A_1 - L_2F \end{bmatrix} \right)$$

are strongly stable.

Suppose equations (8.41) and (8.42) are satisfied by some sufficiently smooth functions $\mathbf{x}(v)$ and $\mathbf{u}(v)$ satisfying $\mathbf{x}(0) = 0$ and $\mathbf{u}(0) = 0$. Now let $z = \mathrm{col}(z_1, z_2)$ with $z_1 \in \mathcal{R}^n$ and $z_2 \in \mathcal{R}^q$, and

$$\begin{aligned} u &= k(z, e) = \mathbf{u}(z_2) + K_x(z_1 - \mathbf{x}(z_2)), \\ \begin{bmatrix} S & 0 \\ 0 & I_q \end{bmatrix} \begin{bmatrix} \dot{z}_1 \\ \dot{z}_2 \end{bmatrix} &= \begin{bmatrix} f(z_1, \mathbf{u}(z_2) + K_x(z_1 - \mathbf{x}(z_2)), z_2) + L_1(e - h(z_1, z_2)) \\ a(z_2) + L_2(e - h(z_1, z_2)) \end{bmatrix}. \end{aligned} \tag{8.44}$$

This controller yields a closed-loop system with $x_c = \mathrm{col}(x, z_1, z_2)$:

$$f_c(x_c, v) = \begin{bmatrix} f(x, \mathbf{u}(z_2) + K_x(z_1 - \mathbf{x}(z_2)), v) \\ f(z_1, \mathbf{u}(z_2) + K_x(z_1 - \mathbf{x}(z_2)), z_2) + L_1(h(x, v) - h(z_1, z_2)) \\ a(z_2) + L_2(h(x, v) - h(z_1, z_2)) \end{bmatrix} \tag{8.45}$$

and

$$S_c = \begin{bmatrix} S & 0 & 0 \\ 0 & S & 0 \\ 0 & 0 & I_q \end{bmatrix}.$$

The Jacobian matrix of $f_c(x_c, 0)$ at the origin is given as follows:

$$A_c = \begin{bmatrix} A & BK_x & BK_v \\ 0 & A + BK_x & E + BK_v \\ 0 & 0 & A_1 \end{bmatrix} + \begin{bmatrix} 0 \\ L_1 \\ L_2 \end{bmatrix} \begin{bmatrix} C & -C & -F \end{bmatrix}, \qquad (8.46)$$

where $K_v = \frac{\partial k}{\partial z_2}(0, 0)$. Some elementary transformation shows that

$$\begin{aligned} \det(\lambda S_c - A_c) &= \det(\lambda S - (A + BK_x)) \\ &\quad \times \det\left(\begin{bmatrix} S & 0 \\ 0 & I_q \end{bmatrix} \lambda - \begin{bmatrix} A - L_1 C & E - L_1 F \\ -L_2 C & A_1 - L_2 F \end{bmatrix} \right). \end{aligned}$$

Thus (S_c, A_c) is also strongly stable. That is, Property 8.1 is satisfied.

To verify Property 8.2, let $\mathbf{x}_c(v) = \text{col}(\mathbf{x}(v),\ \mathbf{x}(v),\ v)$. Then it is clear that

$$h_c(\mathbf{x}_c(v), v) = h(\mathbf{x}(v), v) = 0, \qquad (8.47)$$

$$\mathbf{u}(v) = k(\bar{\mathbf{x}}(v), v, 0). \qquad (8.48)$$

Using (8.47) and (8.48) and then (8.41) successively in (8.45) gives

$$f_c(\mathbf{x}_c(v), v) = \begin{bmatrix} f(\mathbf{x}(v), \mathbf{u}(v), v) \\ f(\mathbf{x}(v), \mathbf{u}(v), v) \\ a(v) \end{bmatrix} = S_c \frac{\partial \mathbf{x}_c(v)}{\partial v} a(v).$$

That is, (8.20) and (8.21) are satisfied. □

Remark 8.14. It is seen that the solvability of the output regulation problem by both state feedback and output feedback control relies on the same set of equations given by (8.41) and (8.42). Clearly, this set of equations can be viewed as the singular analog of the regulator equations introduced in Chapter 3. For convenience, we will refer to (8.41) and (8.42) as *singular regulator equations* in what follows. ■

8.4 Output Regulation via Normal Output Feedback Control

The output feedback controller constructed in Theorem 8.13 is also singular due to the singularity assumption on S. It is known that singular controllers are of high order, and it is less easy to implement singular controllers physically. Thus, in this section we will consider how to synthesize *normal* controllers to solve the output regulation problem for singular systems. Our approach to studying this problem consists of three steps. In the first step, we apply the standard coordinate transformation to the singular plant (8.1) to yield a reduced-order normal system. In the second step, we give the solvability conditions of the output regulation problem for the reduced-order normal system by a normal output feedback controller. Finally, we show that this normal output feedback controller also solves the output regulation problem for the original system.

Before introducing Lemma 8.15, let us note that there exist two nonsingular matrices $T_1, T_2 \in \mathcal{R}^{n\times n}$ such that

$$T_1 S T_2 = \bar{S} = \begin{bmatrix} I_{n_s} & 0 \\ 0 & 0 \end{bmatrix}.$$

Let

$$T_1 A T_2 = \bar{A} = \begin{bmatrix} \bar{A}_{11} & \bar{A}_{12} \\ \bar{A}_{21} & \bar{A}_{22} \end{bmatrix}, \quad T_1 B = \bar{B} = \begin{bmatrix} \bar{B}_1 \\ \bar{B}_2 \end{bmatrix}, \quad T_1 E = \bar{E} = \begin{bmatrix} \bar{E}_1 \\ \bar{E}_2 \end{bmatrix},$$

$$C T_2 = \bar{C} = \begin{bmatrix} \bar{C}_1 & \bar{C}_2 \end{bmatrix}, \qquad T_2^{-1} x = \bar{x} = \begin{bmatrix} \bar{x}_1 \\ \bar{x}_2 \end{bmatrix},$$

where $\bar{A}_{11} \in \mathcal{R}^{n_s\times n_s}$, $\bar{B}_1 \in \mathcal{R}^{n_s\times m}$, $\bar{E}_1 \in \mathcal{R}^{n_s\times q}$, $\bar{C}_1 \in \mathcal{R}^{p\times n_s}$, $\bar{x}_1 \in \mathcal{R}^{n_s}$, and all other matrices have appropriate dimensions. This coordinate transformation leads to the following singular system:

$$\dot{\bar{x}}_1 = \bar{f}_1(\bar{x}, u, v) = \bar{A}_{11}\bar{x}_1 + \bar{A}_{12}\bar{x}_2 + \bar{B}_1 u + \bar{E}_1 v + o(\bar{x}, u, v), \tag{8.49}$$

$$0 = \bar{f}_2(\bar{x}, u, v) = \bar{A}_{21}\bar{x}_1 + \bar{A}_{22}\bar{x}_2 + \bar{B}_2 u + \bar{E}_2 v + o(\bar{x}, u, v), \tag{8.50}$$

$$e = \bar{h}(\bar{x}, v) = \bar{C}\bar{x} + Fv + o(\bar{x}, v), \tag{8.51}$$

where the notation $o(x)$ denotes higher-order terms in x, and

$$\begin{bmatrix} \bar{f}_1(\bar{x}, u, v) \\ \bar{f}_2(\bar{x}, u, v) \end{bmatrix} = T_1 f(T_2\bar{x}, u, v),$$

$$\bar{h}(\bar{x}, v) = h(T_2\bar{x}, v).$$

Lemma 8.15. *Assume that the exosystem (8.2) satisfies Assumption 3.1′ and the singular plant (8.1) satisfies Assumptions 8.1 and 8.2. Suppose (S, A) is standard. Then, the output regulation problem of system (8.1) and (8.2) via a* normal output feedback controller *is solvable if and only if there exist sufficiently smooth functions $\mathbf{x}(v)$ with $\mathbf{x}(0) = 0$ and $\mathbf{u}(v)$ with $\mathbf{u}(0) = 0$, both defined in an open neighborhood V of the origin of $\mathcal{R}^q$, satisfying the singular regulator equations (8.41) and (8.42).*

Proof. The necessity follows trivially from Theorem 8.13.

The proof of sufficiency can be divided into three steps. In the first step, we apply the standard coordinate transformation to the singular plant (8.1) to yield a reduced-order normal system.

Step 1. Let us begin with the system (8.49) to (8.51). By Lemma 8.4, the system described by (8.49) to (8.51) possesses two properties, namely, that $(\bar{S}, \bar{A}, \bar{B})$ is strongly stabilizable and that

$$\left(\begin{bmatrix} \bar{S} & 0 \\ 0 & I_q \end{bmatrix}, \begin{bmatrix} \bar{C} & F \end{bmatrix}, \begin{bmatrix} \bar{A} & \bar{E} \\ 0 & A_1 \end{bmatrix} \right)$$

is strongly detectable.

Moreover, $\bar{A}_{22}$ is nonsingular since (S, A) is standard.

By the Implicit Function Theorem, there exists a unique, sufficiently smooth function $\alpha(\bar{x}_1, u, v)$ defined in an open neighborhood of $(\bar{x}_1, u, v) = (0, 0, 0)$ that satisfies $\alpha(0, 0, 0) = 0$ and

$$0 = \bar{f}_2(\bar{x}_1, \alpha(\bar{x}_1, u, v), u, v). \tag{8.52}$$

It is easy to show that

$$\alpha(\bar{x}_1, u, v) = -\bar{A}_{22}^{-1}\left(\bar{A}_{21}\bar{x}_1 + \bar{B}_2 u + \bar{E}_2 v\right) + o(\bar{x}_1, u, v).$$

Substituting $\bar{x}_2 = \alpha(\bar{x}_1, u, v)$ into (8.49) and (8.51) gives a reduced-order normal system as follows:

$$\begin{aligned} \dot{\bar{x}}_1 &= f_r(\bar{x}_1, u, v) = \bar{f}_1(\bar{x}_1, \alpha(\bar{x}_1, u, v), u, v) \\ &= A_r\bar{x}_1 + B_r u + E_r v + o(\bar{x}_1, u, v), \\ e &= h_r(\bar{x}_1, u, v) = \bar{h}(\bar{x}_1, \alpha(\bar{x}_1, u, v), v) \\ &= C_r\bar{x}_1 + D_r u + F_r v + o(\bar{x}_1, u, v), \end{aligned} \tag{8.53}$$

where $A_r, B_r, C_r, E_r, F_r, D_r$ are as defined in (8.13).

We are now ready to carry out the second step, which will show that the output regulation problem for the normal system obtained in Step 1 is solvable.

Step 2. System (8.53) is a normal system. We will show in this step that the output regulation problem for this system is solvable. By Lemma 8.5, (A_r, B_r) is stabilizable and

$$\left(\begin{bmatrix} C_r & F_r \end{bmatrix}, \begin{bmatrix} A_r & E_r \\ 0 & A_1 \end{bmatrix}\right)$$

is detectable. By Theorem 3.16, it suffices to verify that the regulator equations associated with (8.53) are solvable.

In fact, let $\bar{\mathbf{x}}(v) = T_2^{-1}\mathbf{x}(v)$ and denote $\bar{\mathbf{x}}(v) = \begin{bmatrix} \bar{\mathbf{x}}_1(v) \\ \bar{\mathbf{x}}_2(v) \end{bmatrix}$ with $\bar{\mathbf{x}}_1(v) \in \mathcal{R}^{n_s}$. Then $\mathbf{u}(v)$ and $\bar{\mathbf{x}}(v)$ satisfy

$$\frac{\partial \bar{\mathbf{x}}_1(v)}{\partial v} a(v) = \bar{f}_1(\bar{\mathbf{x}}(v), \mathbf{u}(v), v), \tag{8.54}$$

$$0 = \bar{f}_2(\bar{\mathbf{x}}(v), \mathbf{u}(v), v), \tag{8.55}$$

$$0 = \bar{h}(\bar{\mathbf{x}}(v), v). \tag{8.56}$$

Also, it is clear from (8.52) that

$$\bar{\mathbf{x}}_2(v) = \alpha(\bar{\mathbf{x}}_1(v), \mathbf{u}(v), v). \tag{8.57}$$

Substituting (8.57) into (8.54) and (8.56) gives

$$\begin{aligned} \frac{\partial \bar{\mathbf{x}}_1(v)}{\partial v} a(v) &= f_r(\bar{\mathbf{x}}_1(v), \mathbf{u}(v), v), \\ 0 &= h_r(\bar{\mathbf{x}}_1(v), \mathbf{u}(v), v). \end{aligned} \tag{8.58}$$

Thus, the two functions $\bar{\mathbf{x}}_1(v)$ and $\mathbf{u}(v)$ are the solution of the regulator equations associated with system (8.53).

By Theorem 3.16, the output regulation problem for system (8.53) is solvable by a normal output feedback controller of the following form:

$$u = k(z), \qquad \dot{z} = g(z, e), \tag{8.59}$$

where $z \in \mathcal{R}^{n_z}$ for some integer n_z.

We are now ready to carry out the third step to show that this normal output feedback controller also solves the output regulation problem for the original system.

Step 3. To show that the controller (8.59) also solves the output regulation problem for the original system (8.1), we only need to show that the closed-loop system composed of (8.1) and (8.59) satisfies Properties 8.1 and 8.2. To this end, let the linear approximation of the controller (8.59) be given by $u = K_z z$, $\dot{z} = G_1 z + G_2 e$, let A_{cr} be the Jacobian matrix of the closed-loop system composed of (8.53) and (8.59), and let (S_c, A_c) be the linearization of the closed-loop system composed of (8.1) and (8.59). Then, similar to the proof of Lemma 8.6, we have

$$A_{cr} = \begin{bmatrix} A_r & B_r K_z \\ G_2 C_r & G_1 + G_2 D_r K_z \end{bmatrix}$$

and

$$S_c = \begin{bmatrix} S & 0 \\ 0 & I_{n_z} \end{bmatrix}, \quad A_c = \begin{bmatrix} A & BK_z \\ G_2 C & G_1 \end{bmatrix}.$$

It follows from Lemma 8.6 that (S_c, A_c) is strongly stable.

Finally, to verify the satisfaction of Property 8.2, one only needs to note that, for sufficiently small x_0 and v_0,

$$\begin{aligned} 0 &= \lim_{t\to\infty} h_r(\bar{x}_1(t), u(t), v(t)) \\ &= \lim_{t\to\infty} \bar{h}(\bar{x}(t), v(t)) \\ &= \lim_{t\to\infty} h(x(t), v(t)). \quad \square \end{aligned}$$

The assumption that (S, A) is standard is the key to the validity of Lemma 8.15. This assumption is of course undesirable and can actually be removed through a linear output feedback precompensator, as shown in Lemma 8.7. Thus, combining Lemmas 8.7 and 8.15 leads to the main result of this section, as follows.

Theorem 8.16. *Assume that the exosystem (8.2) satisfies Assumption 3.1′ and the singular plant (8.1) satisfies Assumptions 8.1 and 8.2. Then the output regulation problem of system (8.1) and (8.2) via a* normal output feedback controller *is solvable if and only if there exist sufficiently smooth functions* $\mathbf{x}(v)$ *with* $\mathbf{x}(0) = 0$ *and* $\mathbf{u}(v)$ *with* $\mathbf{u}(0) = 0$, *both defined in an open neighborhood* V *of the origin of* $\mathcal{R}^q$, *satisfying the singular regulator equations (8.41) and (8.42).*

Proof. The necessity follows trivially from Lemma 8.15. To establish the sufficient condition, applying a linear output feedback control

$$u = K_e e + \bar{u} \tag{8.60}$$

to (8.1) gives a new system, with $\bar{u}$ as an input:

$$\begin{aligned}
S\dot{x} &= \tilde{f}(x, \bar{u}, v) = f(x, K_e h(x, v) + \bar{u}, v),\\
\dot{v} &= a(v),\\
e &= \tilde{h}(x, v) = h(x, v).
\end{aligned} \tag{8.61}$$

By Lemma 8.7, under Assumptions 8.1 and 8.2, there exists a gain matrix K_e such that (8.61) satisfies the following:

(i) $(S, \tilde{A})$ is standard, where $\tilde{A}$ is the Jacobian matrix of $\tilde{f}(x, 0, 0)$ at $x = 0$;
(ii) $(S, \tilde{A}, B)$ is strongly stabilizable, and $\left(\begin{bmatrix} S & 0 \\ 0 & I_q \end{bmatrix}, [C\ F], \begin{bmatrix} \tilde{A} & E \\ 0 & A_1 \end{bmatrix}\right)$ is strongly detectable.

Now, suppose that $\mathbf{x}(v)$ and $\mathbf{u}(v)$ are the solution of the regulator equations associated with (8.1) and (8.2). Then

$$\begin{aligned}
\tilde{f}(\mathbf{x}(v), \mathbf{u}(v), v) &= f(\mathbf{x}(v), K_e e + \mathbf{u}(v), v)\\
&= f(\mathbf{x}(v), K_e h(\mathbf{x}(v), v) + \mathbf{u}(v), v)\\
&= f(\mathbf{x}(v), \mathbf{u}(v), v) = S\frac{\partial \mathbf{x}(v)}{\partial v} a(v),\\
\tilde{h}(\mathbf{x}(v), v) &= h(\mathbf{x}(v), v) = 0;
\end{aligned}$$

that is, $\mathbf{x}(v)$ and $\mathbf{u}(v)$ are also the solution of the regulator equations associated with (8.61). Thus, system (8.61) satisfies all assumptions of Lemma 8.15. As a result, there exists a normal output feedback controller of the form $\bar{u} = k(z)$, $\dot{z} = g(z, e)$ that solves the output regulation problem for the system (8.61). Therefore, the following normal output feedback controller:

$$u = k(z) + K_e e, \qquad \dot{z} = g(z, e) \tag{8.62}$$

solves the output regulation problem for the original system (8.1) and (8.2). □

Example 8.17. Consider the following singular nonlinear system:

$$\begin{aligned}
\dot{x}_1 &= 2x_2 + x_3,\\
\dot{x}_2 &= -4x_1 + x_2 + x_4 - v_1 - 2v_2,\\
0 &= x_2 + x_3,\\
0 &= -x_1 - \sin x_2 + u,\\
e &= x_4 - v_1,
\end{aligned} \tag{8.63}$$

with the exosystem

$$\dot{v}_1 = 2v_2, \quad \dot{v}_2 = -2v_1.$$

This system is already in the standard form (8.49) to (8.51). Linearizing (8.63) at the origin gives

$$S = \begin{bmatrix} 1 & 0 & 0 & 0 \\ 0 & 1 & 0 & 0 \\ 0 & 0 & 0 & 0 \\ 0 & 0 & 0 & 0 \end{bmatrix}, \quad A = \begin{bmatrix} 0 & 2 & 1 & 0 \\ -4 & 1 & 0 & 1 \\ 0 & 1 & 1 & 0 \\ -1 & -1 & 0 & 0 \end{bmatrix}, \quad B = \begin{bmatrix} 0 \\ 0 \\ 0 \\ 1 \end{bmatrix},$$

$$C = \begin{bmatrix} 0 & 0 & 0 & 1 \end{bmatrix}, \quad E = \begin{bmatrix} 0 & 0 \\ -1 & -2 \\ 0 & 0 \\ 0 & 0 \end{bmatrix}, \quad F = \begin{bmatrix} -1 & 0 \end{bmatrix}.$$

It is easy to verify that the plant and the exosystem satisfy Assumptions 8.1 and 8.2. Moreover, the regulator equations of (8.63) admit the following unique solution:

$$\begin{aligned} \mathbf{x}_1(v) &= v_1, \\ \mathbf{x}_2(v) &= 2v_2, \\ \mathbf{x}_3(v) &= -2v_2, \\ \mathbf{x}_4(v) &= v_1, \\ \mathbf{u}(v) &= v_1 + \sin(2v_2). \end{aligned}$$

By Theorem 8.16, the output regulation problem for the given plant is solvable by a normal output feedback controller.

To actually construct a normal output feedback controller, first note that (S, A) is not standard. Applying the output feedback compensator $u = e + \bar{u}$ to plant (8.63) gives

$$\begin{aligned} \dot{x}_1 &= 2x_2 + x_3, \\ \dot{x}_2 &= -4x_1 + x_2 + x_4 - v_1 - 2v_2, \\ 0 &= x_2 + x_3, \\ 0 &= -x_1 - \sin x_2 + x_4 - v_1 + \bar{u}, \\ e &= x_4 - v_1, \end{aligned} \tag{8.64}$$

which gives

$$\tilde{A} = \begin{bmatrix} 0 & 2 & 1 & 0 \\ -4 & 1 & 0 & 1 \\ 0 & 1 & 1 & 0 \\ -1 & -1 & 0 & 1 \end{bmatrix},$$

which clearly renders $(S, \tilde{A})$ standard. Eliminating x_3 and x_4 from equation (8.64) gives the following reduced-order normal system:

$$\begin{aligned} \dot{x}_1 &= x_2, \\ \dot{x}_2 &= -3x_1 + x_2 - 2v_2 + \sin x_2 - \bar{u}, \\ e &= x_1 + \sin x_2 - \bar{u}. \end{aligned} \tag{8.65}$$

This system is in the normal form (8.53) with $\bar{x}_1 = \text{col}(x_1, x_2)$ and

$$f_r(\bar{x}_1, \bar{u}, v) = \begin{bmatrix} x_2 \\ -3x_1 + x_2 - 2v_2 + \sin x_2 - \bar{u} \end{bmatrix},$$
$$h_r(\bar{x}_1, \bar{u}, v) = x_1 + \sin x_2 - \bar{u}.$$

By Theorem 3.16, the robust output regulation problem for this system is solvable by an output feedback control of the form (3.54) with $y_m = e$. To be more specific, linearizing (8.65) gives

$$A_r = \begin{bmatrix} 0 & 1 \\ -3 & 2 \end{bmatrix}, \quad B_r = \begin{bmatrix} 0 \\ -1 \end{bmatrix}, \quad E_r = \begin{bmatrix} 0 & 0 \\ 0 & -2 \end{bmatrix},$$
$$C_r = [1 \;\; 1], \qquad D_r = -1, \qquad F_r = [0 \;\; 0].$$

We are now ready to design a controller to solve the output regulation problem of the normal system according to the method described in Chapter 3 as follows. Letting K_x be such that the eigenvalues of $A_r + B_r K_x$ are

$$-0.7071 \pm 0.7071j$$

gives

$$K_x = \begin{bmatrix} -2.0000 & 3.4142 \end{bmatrix},$$

and letting L be such that the eigenvalues of

$$\begin{bmatrix} A_r & E \\ 0 & A_1 \end{bmatrix} - L \begin{bmatrix} C_r & F \end{bmatrix}$$

are

$$-1.2720 \pm 3.7890j, \quad -1.8780 \pm 1.2423j$$

gives

$$L = \begin{bmatrix} -0.1132 \\ 8.4132 \\ 8.2695 \\ -7.3105 \end{bmatrix}.$$

Then, by (3.54), the following controller:

$$\bar{u} = k(z_1, z_2) = \mathbf{u}(z_2) + K_x(z_1 - \bar{\mathbf{x}}_1(z_2)),$$
$$\dot{z} = \begin{bmatrix} f_r(z_1, k(z_1, z_2), z_2) \\ a(z_2) \end{bmatrix} + L \begin{bmatrix} e - h_r(z_1, k(z_1, z_2), z_2) \\ e - h_r(z_1, k(z_1, z_2), z_2) \end{bmatrix},$$

where $z_1 \in \mathcal{R}^2$, $z_2 \in \mathcal{R}^2$, $\bar{\mathbf{x}}_1(z_2) = \text{col}\,(\mathbf{x}_1(z_2), \mathbf{x}_2(z_2))$, solves the output regulation problem for the normal system. Composition of this controller with the precompensator $u = e + \bar{u}$ gives the normal output feedback controller, which solves the output regulation problem of the original system. ∎

8.5 Approximate Solution of the Output Regulation Problem for Singular Systems

Like *normal* systems, the key to the existence of either state feedback or output feedback controller is the solvability of the singular regulator equations (8.41) and (8.42). Due to the nonlinearity of the plant and the exosystem, it is difficult to obtain the exact solution $\mathbf{x}(v)$ and $\mathbf{u}(v)$ for the *singular* or *normal* regulator equations. Thus, it is interesting to study the approximate solution of the singular regulator equations by Taylor series. In fact, by employing the technique similar to the one detailed in Chapter 4, we can also seek series of the form

$$\mathbf{x}(v) = \sum_{l\geq 1} X_l v^{[l]}, \quad \mathbf{u}(v) = \sum_{l\geq 1} U_l v^{[l]} \tag{8.66}$$

such that the singular regulator equations are satisfied formally.

For this purpose, expand the functions $f(x, u, v)$, $h(x, v)$, and $a(v)$ as follows:

$$\begin{aligned} f(x,u,v) &= \sum_{l\geq 1} \sum_{\substack{i+j+k=l \\ i,j,k\geq 0}} F_{ijk} x^{(i)} \otimes u^{(j)} \otimes v^{(k)}, \\ h(x,v) &= \sum_{l\geq 1} \sum_{\substack{i+k=l \\ i,k\geq 0}} H_{ik} x^{(i)} \otimes v^{(k)}, \\ a(v) &= \sum_{i\geq 1} A_i v^{(i)}. \end{aligned} \tag{8.67}$$

Substituting (8.67) and (8.66) into (8.41) and (8.42) and identifying the coefficients of $v^{[l]}$, $l = 1, 2, \ldots$, yields the following result.

Lemma 8.18. *The power series (8.66) formally satisfy the singular regulator equations (8.41) and (8.42) if and only if the following equations are satisfied for* $l = 1, 2, \ldots$:

$$\begin{aligned} SX_l M_l \left[\sum_{i=1}^{l} I_q^{(i-1)} \otimes A_1 \otimes I_q^{(l-i)} \right] N_l &= AX_l + BU_l + E_l, \\ 0 &= CX_l + F_l, \end{aligned} \tag{8.68}$$

where

$$\begin{aligned} &A = F_{100}, \quad B = F_{010}, \quad E_1 = E = F_{001}, \\ &C = H_{10}, \quad F_1 = F = H_{01}, \end{aligned}$$

and, for $l = 2, 3, \ldots,$

$$E_l = \sum_{n=2}^{l} \sum_{\substack{i+j+k=n \\ i,j,k \geq 0}} F_{ijk} G_{l-n}^{ij} N_l$$

$$- S \sum_{k=1}^{l-1} X_k M_k \left[\sum_{i=1}^{k} I_q^{(i-1)} \otimes A_{l-k+1} \otimes I_q^{(k-i)} \right] N_l, \tag{8.69}$$

$$F_l = \left[\sum_{n=2}^{l} \sum_{\substack{i+k=n \\ i,k \geq 0}} H_{ik} G_{l-n}^{ij} \right] N_l, \tag{8.70}$$

$$G_m^{ij} = \begin{cases} 0, & i = j = 0, m > 0, \\ 1, & i = j = 0, m > 0, \\ \delta_{i,j+m}, & j = 0, i = 1, 2, \ldots, \\ \lambda_{i,j+m}, & i = 0, j = 1, 2, \ldots, \\ \sum_{k=0}^{m} \delta_{i,i+k} \otimes \lambda_{j,j+m-k}, & i, j = 1, 2, \ldots, \end{cases}$$

$$\delta_{i,j} = \sum_{\substack{j_1+j_2+\cdots+j_i=j \\ j_1,j_2,\ldots,j_i \geq 1}} X_{j_1} M_{j_1} \otimes X_{j_2} M_{j_2} \otimes \cdots \otimes X_{j_i} M_{j_i}, \tag{8.71}$$

$$\lambda_{i,j} = \sum_{\substack{j_1+j_2+\cdots+j_i=j \\ j_1,j_2,\ldots,j_i \geq 1}} U_{j_1} M_{j_1} \otimes U_{j_2} M_{j_2} \otimes \cdots \otimes U_{j_i} M_{j_i}. \tag{8.72}$$

Proof. The proof is quite similar to that given in Lemma 4.7 of Chapter 4 and is therefore omitted. □

Equation (8.68) is an iterative sequence of the singular Sylvester equations. The following result establishes the solvability condition for these equations.

Theorem 8.19. *There exists a solution (unique if p=m) of (8.68) for any* E_l *and* F_l, $l = 1, 2, \ldots,$ *if and only if*

$$\text{rank} \begin{bmatrix} A - \lambda S & B \\ C & 0 \end{bmatrix} = n + p \tag{8.73}$$

for all $\lambda \in \Lambda_l$, *where*

$$\Lambda_l = \{ \lambda \mid \lambda = l_1 \lambda_1 + \cdots + l_q \lambda_l, l_1 + \cdots + l_q = l, \; l_1, \ldots, l_q = 0, 1, \ldots, l \},$$

with $\lambda_1, \ldots, \lambda_q$ *being the eigenvalues of* A_1.

Remark 8.20. Assume that the transmission zeros condition described in equation (8.73) holds up to some positive integer k. Let

$$\mathbf{x}^{(k)}(v) = \sum_{l=1}^{k} X_l v^{[l]}, \quad \mathbf{u}^{(k)}(v) = \sum_{l=1}^{k} U_l v^{[l]}. \tag{8.74}$$

Then, it is not difficult to see from the proof of Lemma 4.7 that $\mathbf{x}^{(k)}(v)$ and $\mathbf{u}^{(k)}(v)$ are such that

$$\begin{aligned} S\frac{\partial \mathbf{x}^{(k)}(v)}{\partial v}a(v) &= f(\mathbf{x}^{(k)}(v), \mathbf{u}^{(k)}(v), v) + o^k(v), \\ o^k(v) &= h(\mathbf{x}^{(k)}(v), \mathbf{u}^{(k)}(v), v). \end{aligned} \tag{8.75}$$

Moreover, if we replace $\mathbf{x}(v)$ and $\mathbf{u}(v)$ in the state feedback controller (8.43), the singular output feedback controller (8.44), and the normal output feedback controller (8.62) by $\mathbf{x}^{(k)}(v)$ and $\mathbf{u}^{(k)}(v)$, then it is not difficult to show that each of these controllers will result in a closed-loop system that satisfies Property 8.1 and admits a sufficiently smooth function $\mathbf{x}_c^{(k)}(v)$ with $\mathbf{x}_c^{(k)}(0) = 0$ such that

$$S_c\frac{\partial \mathbf{x}_c^{(k)}(v)}{\partial v}a(v) = f_c(\mathbf{x}_c^{(k)}(v), v), \tag{8.76}$$

$$o^{(k)}(v) = h_c(\mathbf{x}_c^{(k)}(v), v). \tag{8.77}$$

It can be readily shown, using the argument similar to what was used in Lemma 4.7, that the closed-loop system resulting from these controllers has the property that, for all sufficiently small x_{c0} and v_0, the trajectories $\mathrm{col}(x_c(t), v(t))$ of the closed-loop system satisfy

$$\lim_{t\to\infty}(e(t) - o^k(v(t))) = \lim_{t\to\infty}(h_c(x_c(t), v(t)) - o^k(v(t))) = 0.$$

Therefore, we say that these controllers solve the kth-order output regulation problem for the singular systems (8.1) and (8.2). ∎

8.6 Robust Output Regulation of Uncertain Singular Nonlinear Systems

In this section, we turn to the problem of the robust output regulation problem for uncertain singular nonlinear systems described by

$$\begin{aligned} S\dot{x}(t) &= f(x(t), u(t), v(t), w), x(0) = x_0, \\ e(t) &= h(x(t), v(t), w), \quad t \geq 0, \end{aligned} \tag{8.78}$$

where $x(t) \in \mathcal{R}^n$ is the plant state, $u(t) \in \mathcal{R}^m$ the plant input, $e(t) \in \mathcal{R}^m$ the plant output representing the tracking error, and $v(t) \in \mathcal{R}^q$ the exogenous signal representing the disturbance and/or the reference input generated by the following exosystem:

$$\dot{v} = A_1 v. \tag{8.79}$$

In (8.78), $w \in \mathcal{R}^{n_w}$ is the plant unknown parameter and $S \in \mathcal{R}^{n \times n}$ a singular constant matrix, and rank $S = n_s < n$. Also it is assumed that 0 is the nominal value of the uncertain parameter w.

As in Section 8.4, we will seek a normal dynamic output feedback controller as follows:

$$\begin{aligned} u(t) &= k(z(t), e(t)), \\ \dot{z}(t) &= g(z(t), e(t)), \end{aligned} \tag{8.80}$$

where $z(t)$ is the compensator state vector of dimension n_z.

The closed-loop composite system composed of the singular plant (8.78), the exosystem (8.79), and the control law (8.80) can be put into the following form:

$$\begin{aligned} S_c\dot{x}_c(t) &= f_c(x_c(t), v(t), w), \\ \dot{v}(t) &= A_1 v(t), \\ e(t) &= h_c(x_c(t), v(t), w), \end{aligned} \tag{8.81}$$

where

$$x_c = \begin{bmatrix} x \\ z \end{bmatrix}, \; S_c = \begin{bmatrix} S & 0 \\ 0 & I_{n_z} \end{bmatrix},$$

$$f_c(x_c, v, w) = \begin{bmatrix} f(x, k(z, h(x, v, w)), v, w) \\ g(z, h(x, v, w)) \end{bmatrix},$$

$$h_c(x_c, v, w) = h(x, v, w).$$

Again, it is assumed that all the functions involved in this setup are sufficiently smooth and defined globally on the appropriate Euclidean spaces, and $f(0, 0, 0, w) = 0$ and $h(0, 0, w) = 0$ for any $w \in W$, with W an open neighborhood of the origin of $\mathcal{R}^{n_w}$. Our results will be stated locally in terms of V and W, with V an open neighborhood of the origin in $\mathcal{R}^q$. In what follows, V and W are implicitly permitted to be made smaller to accommodate subsequent local arguments.

The linearization of the system (8.78) at $(x, u, v) = (0, 0, 0)$ will be frequently used, which entails the following notations:

$$A(w) = \frac{\partial f}{\partial x}(0, 0, 0, w), \quad B(w) = \frac{\partial f}{\partial u}(0, 0, 0, w), \quad E(w) = \frac{\partial f}{\partial v}(0, 0, 0, w),$$

$$C(w) = \frac{\partial h}{\partial x}(0, 0, w), \qquad F(w) = \frac{\partial h}{\partial v}(0, 0, w).$$

As a result, the system composed of (8.78) and (8.79) can also be written as

$$\begin{aligned} S\dot{x} &= A(w)x + B(w)u + E(w)v + o(x, u, v, w), \\ \dot{v} &= A_1 v, \\ e &= C(w)x + F(w)v + o(x, v, w), \end{aligned}$$

where $o(x, u, v, w)$ $(o(x, v, w))$ is a sufficiently smooth function vanishing at $(x, u, v) = (0, 0, 0)$ $((x, v) = (0, 0))$ together with its first-order derivatives with respect to (x, u, v) $((x, v))$ for any $w \in W$. For convenience, let $A, B, \ldots,$ denote $A(0), B(0), \ldots,$ respectively.

As in Chapter 5, we can list two desirable properties of the closed-loop system as follows.

Property 8.3. The linearization of $S_c\dot{x}_c = f_c(x_c, 0, 0)$ at $x_c = 0$ is strongly stable.

Property 8.4. The trajectory starting from any sufficiently small initial state (x_{c0}, v_0) satisfies

$$\lim_{t\to\infty} e(t) = \lim_{t\to\infty} h_c(x_c(t), v(t), w) = 0. \tag{8.82}$$

The Robust Output Regulation Problem. Find a controller of the form (8.80) such that the closed-loop composite system (8.81) satisfies Properties 8.3 and 8.4.

The above problem is clearly the extension of the robust output regulation problem for the normal systems studied in Chapter 5 to the singular systems. It can also be viewed as an extension of the output regulation problem of singular systems studied in Sections 8.1 to 8.4 by taking into account the uncertainty. Viewing w as being generated by an exosystem of the form $\dot{w} = 0$, a solvability condition can be obtained by slightly modifying Lemma 8.9, as follows.

Lemma 8.21. *Assume that the exosystem (8.79) satisfies Assumption 8.3 below and that the closed-loop system (8.81) has Property 8.3. Then the closed-loop system (8.81) also has Property 8.4 if and only if it has the following property.*

Property 8.5. There exists a sufficiently smooth function $\mathbf{x}_c(v, w)$ with $\mathbf{x}_c(0, 0) = 0$ that satisfies, for $v \in V$ and $w \in W$, the following partial differential equations:

$$S_c \frac{\partial \mathbf{x}_c(v, w)}{\partial v} A_1 v = f_c(\mathbf{x}_c(v, w), v, w), \tag{8.83}$$

$$0 = h_c(\mathbf{x}_c(v, w), v, w). \tag{8.84}$$

Various assumptions needed for the solvability of the above problem are listed as follows.

Assumption 8.3. All the eigenvalues of the matrix A_1 are simple and have zero real parts.

Assumption 8.4. The triple (S, A, B) is strongly stabilizable.

Assumption 8.5. The triple (S, C, A) is strongly detectable.

Assumption 8.6. There exist two sufficiently smooth functions $\mathbf{x}(v, w)$ and $\mathbf{u}(v, w)$ satisfying $\mathbf{x}(0, 0) = 0$ and $\mathbf{u}(0, 0) = 0$ such that, for $v \in V$, $w \in W$,

$$\begin{aligned} S\frac{\partial \mathbf{x}(v, w)}{\partial v} A_1 v &= f(\mathbf{x}(v, w), \mathbf{u}(v, w), v, w), \\ 0 &= h(\mathbf{x}(v, w), v, w). \end{aligned} \tag{8.85}$$

Remark 8.22. Assumptions 8.4 and 8.5 guarantee the existence of a linear *normal output feedback control* to achieve Property 8.3, and Assumption 8.3, together with Property 8.3, guarantees the boundedness of the solution of the closed-loop system for sufficiently small initial state $x_c(0)$ and $v(0)$. ■

We will study the above robust output regulation problem by an approach similar to what has been used to solve the output regulation problem by a normal output feedback control. For this purpose, let us tentatively assume that (S, A) is standard. Then we can perform the same coordinate transformation on (8.78) as was done on (8.1) in Section 8.4, which yields a system of the form

$$\begin{aligned} \dot{\bar{x}}_1 &= \bar{f}_1(\bar{x}_1, \bar{x}_2, u, v, w), \\ 0 &= \bar{f}_2(\bar{x}_1, \bar{x}_2, u, v, w), \\ e &= \bar{h}(\bar{x}_1, \bar{x}_2, v, w), \end{aligned} \tag{8.86}$$

where $\bar{x} = T_2^{-1}x = \text{col}(\bar{x}_1, \bar{x}_2)$,

$$\begin{bmatrix} \bar{f}_1(\bar{x}_1, \bar{x}_2, u, v, w) \\ \bar{f}_2(\bar{x}_1, \bar{x}_2, u, v, w) \end{bmatrix} = T_1 f(T_2\bar{x}, u, v, w),$$
$$\bar{h}(\bar{x}, v, w) = h(T_2\bar{x}, v, w),$$

and $\frac{\partial \bar{f}_2}{\partial \bar{x}_2}(0, 0, 0, 0, 0)$ is nonsingular. By the Implicit Function Theorem, there exists a unique, sufficiently smooth function $\alpha(\bar{x}_1, u, v, w)$ defined in an open neighborhood of $(\bar{x}_1, u, v, w) = (0, 0, 0, 0)$ that satisfies $\alpha(0, 0, 0, 0) = 0$ and

$$0 = \bar{f}_2(\bar{x}_1, \alpha(\bar{x}_1, u, v, w), u, v, w). \tag{8.87}$$

Substituting $\bar{x}_2 = \alpha(\bar{x}_1, u, v, w)$ into the first and third equations of (8.86) gives a reduced-order normal system

$$\begin{aligned} \dot{\bar{x}}_1 &= f_r(\bar{x}_1, u, v, w) = \bar{f}_1(\bar{x}_1, \alpha(\bar{x}_1, u, v, w), u, v, w), \\ e &= h_r(\bar{x}_1, u, v, w) = \bar{h}(\bar{x}_1, \alpha(\bar{x}_1, u, v, w), v, w). \end{aligned} \tag{8.88}$$

It is now possible to see that the linear approximation of (8.88) at $(\bar{x}_1, u, v, w) = (0, 0, 0, w)$ takes the following form:

$$\begin{aligned} \dot{\bar{x}}_1 &= A_r(w)\bar{x}_1 + B_r(w)u + E_r(w)v, \\ e &= C_r(w)\bar{x}_1 + D_r(w)u + F_r(w)v, \end{aligned}$$

where all the matrices in the above two equations are defined in Section 8.2. We will first establish the following result.

Lemma 8.23. *Assume that the exosystem (8.79) satisfies Assumption 8.3 and the plant (8.78) is standard, that is, (S, A) is standard. Then, if a controller of the form (8.59) solves the robust output regulation problem for the normal system (8.88), it also solves the robust output regulation problem for the singular system (8.78).*

Proof. Assume a controller of the form (8.59) solves the robust output regulation problem for the normal system (8.88). We need to show that the closed-loop system composed of (8.78) and (8.59) also satisfies Properties 8.3 and 8.4. For this purpose, let the closed-loop system composed of (8.88) and (8.59) be denoted by

$$\begin{aligned} \dot{\bar{x}}_{cr} &= \bar{f}_{cr}(\bar{x}_{cr}, v, w), \\ e &= \bar{h}_{cr}(\bar{x}_{cr}, v, w), \end{aligned} \tag{8.89}$$

where $\bar{x}_{cr} = \text{col}(\bar{x}_1, z)$ and

$$\bar{f}_{cr}(\bar{x}_{cr}, v, w) = \begin{bmatrix} f_r(\bar{x}_1, k(z), v, w) \\ g(z, e) \end{bmatrix},$$
$$\bar{h}_{cr}(\bar{x}_{cr}, v, w) = h_r(\bar{x}_1, k(z), v, w). \tag{8.90}$$

Also, let the linearization of (8.89) with $w = 0$ be denoted by $\dot{\bar{x}}_{cr} = \bar{A}_{cr}\bar{x}_{cr} + \bar{B}_{cr}v$, and the linearization of the closed-loop system composed of (8.78) and (8.59) with $w = 0$ by $S_c\dot{x}_c = A_c x_c + B_c v$ with $x_c = \text{col}(x, z)$. We will first show that the stability of $\bar{A}_{cr}$ implies the strong stability of (S_c, A_c). To this end, let the linearization of the controller (8.59) be denoted by $u = K_z z, \ \ \dot{z} = G_1 z + G_2 e$. Then

$$\bar{A}_{cr} = \begin{bmatrix} A_r & B_r K_z \\ G_2 C_r & G_1 + G_2 D_r K_z \end{bmatrix}$$

and

$$S_c = \begin{bmatrix} S & 0 \\ 0 & I_{n_z} \end{bmatrix}, \quad A_c = \begin{bmatrix} A & BK_z \\ G_2 C & G_1 \end{bmatrix}.$$

It follows from Lemma 8.6 that (S_c, A_c) is strongly stable.

Next we will show that the closed-loop system composed of (8.78) and (8.59) satisfies Property 8.5. Let $\bar{\mathbf{x}}_{cr}(v, w)$ be a sufficiently smooth function with $\bar{\mathbf{x}}_{cr}(0, 0) = 0$ that satisfies

$$\frac{\partial \bar{\mathbf{x}}_{cr}(v, w)}{\partial v} A_1 v = \bar{f}_{cr}(\bar{\mathbf{x}}_{cr}(v, w), v, w),$$
$$0 = \bar{h}_{cr}(\bar{\mathbf{x}}_{cr}(v, w), v, w). \tag{8.91}$$

Perform a partition $\bar{\mathbf{x}}_{cr}(v, w) = \text{col}(\bar{\mathbf{x}}_1(v, w), \mathbf{z}(v, w))$ with $\bar{\mathbf{x}}_1(v, w) \in \mathcal{R}^{n_s}$. Then using (8.90) leads to an expansion of (8.91) into the following:

$$\frac{\partial \bar{\mathbf{x}}_1(v, w)}{\partial v} A_1 v = f_r(\bar{\mathbf{x}}_1(v, w), k(\mathbf{z}(v, w)), v, w),$$
$$\frac{\partial \mathbf{z}(v, w)}{\partial v} A_1 v = g(k(\mathbf{z}(v, w)), 0),$$
$$0 = h_r(\bar{\mathbf{x}}_1(v, w), k(\mathbf{z}(v, w)), v, w). \tag{8.92}$$

Now let $\bar{\mathbf{x}}_2(v, w) = \alpha(\bar{\mathbf{x}}_1(v, w), k(\mathbf{z}(v, w)), v, w)$, where the function α is defined in (8.87). Then equation (8.87) implies

$$0 = \bar{f}_2(\bar{\mathbf{x}}_1(v, w), \alpha(\bar{\mathbf{x}}_1(v, w), k(\mathbf{z}(v, w)), v, w), k(\mathbf{z}(v, w)), v, w) \tag{8.93}$$

and equation (8.88) implies

$$f_r(\bar{\mathbf{x}}_1(v, w), k(\mathbf{z}(v, w)), v, w) = \bar{f}_1(\bar{\mathbf{x}}_1(v, w), \bar{\mathbf{x}}_2(v, w), k(\mathbf{z}(v, w)), v, w),$$
$$h_r(\bar{\mathbf{x}}_1(v, w), k(\mathbf{z}(v, w)), v, w) = \bar{h}(\bar{\mathbf{x}}_1(v, w), \bar{\mathbf{x}}_2(v, w), v, w). \tag{8.94}$$

Thus combining (8.92), (8.93), and (8.94) shows

$$\begin{aligned}\frac{\partial \bar{\mathbf{x}}_1(v, w)}{\partial v} A_1 v &= \bar{f}_1(\bar{\mathbf{x}}_1(v, w), \bar{\mathbf{x}}_2(v, w), k(\mathbf{z}(v, w)), v, w),\\ 0 &= \bar{f}_2(\bar{\mathbf{x}}_1(v, w), \bar{\mathbf{x}}_2(v, w), k(\mathbf{z}(v, w)), v, w),\\ \frac{\partial \mathbf{z}(v, w)}{\partial v} A_1 v &= g(k(\mathbf{z}(v, w)), 0),\\ 0 &= \bar{h}(\bar{\mathbf{x}}_1(v, w), \bar{\mathbf{x}}_2(v, w), v, w). \qquad (8.95)\end{aligned}$$

Finally, let

$$\mathbf{x}_c(v, w) = \begin{bmatrix} T_2 & 0 \\ 0 & I_{n_z} \end{bmatrix} \begin{bmatrix} \bar{\mathbf{x}}_1(v, w) \\ \bar{\mathbf{x}}_2(v, w) \\ \mathbf{z}(v, w) \end{bmatrix}. \qquad (8.96)$$

Then it is possible to verify, using (8.95), that $\mathbf{x}_c(v, w)$ satisfies (8.83) and (8.84). □

The solvability of the robust output regulation for normal systems of the form (8.88) has been established in Theorem 6.23. Combining Theorem 6.23 and Lemma 8.23 establishes the main result of this section.

Theorem 8.24. *Assume that the exosystem (8.79) satisfies Assumption 8.3, the singular plant (8.78) satisfies Assumptions 8.4 to 8.6, and the function* $\mathbf{u}(v, w)$ *satisfies conditions (6.31) and (6.33) with* $g_o(x, u) = u$. *Further, assume the following assumption.*

Assumption 8.7. For all λ such that $P_i^j(\lambda) = 0$ for some $i = 1, \dots, m$ and some $j = 1, \dots, I_i$,

$$\operatorname{rank} \begin{bmatrix} A - \lambda S & B \\ C & 0 \end{bmatrix} = n + m. \qquad (8.97)$$

Then the robust output regulation problem of the singular system is solvable by a controller of the form (8.80).

Proof. Let us divide the proof into two steps. In the first step, we assume that (S, A) is standard, and in the second step, we remove this assumption.

Step 1. By Lemma 8.23, it suffices to show that the robust output regulation problem of the reduced-order normal system (8.88) is solvable. By Theorem 6.23, we need to show that the reduced-order normal system (8.88) satisfies Assumptions 5.1 to 5.3 and the function $\mathbf{u}(v, w)$ satisfies the conditions (6.31) and (6.33) with $g_o(x, u) = u$; moreover, for all λ such that $P_i^j(\lambda) = 0$ for some $i = 1, \dots, m$ and some $j = 1, \dots, I_i$,

$$\operatorname{rank} \begin{bmatrix} A_r - \lambda I_{n_s} & B_r \\ C_r & D_r \end{bmatrix} = n_s + m. \qquad (8.98)$$

By Lemma 8.5, satisfaction of Assumptions 8.4 and 8.5 by (8.78) implies stabilizability of (A_r, B_r) and detectability of (C_r, A_r). Next, we verify that (8.88) satisfies

Assumption 5.1; that is, the regulator equations associated with the reduced-order normal system (8.88) admit a solution. To this end, let $\mathbf{x}(v, w)$ and $\mathbf{u}(v, w)$ be the solution of the regulator equations of the singular plant (8.78). Let $\bar{\mathbf{x}}(v, w) = T_2^{-1}\mathbf{x}(v, w)$ and denote $\bar{\mathbf{x}}(v, w) = \begin{bmatrix} \bar{\mathbf{x}}_1(v,w) \\ \bar{\mathbf{x}}_2(v,w) \end{bmatrix}$ with $\bar{\mathbf{x}}_1(v, w) \in \mathcal{R}^{n_s}$. Then $\mathbf{u}(v, w)$ and $\bar{\mathbf{x}}(v, w)$ satisfy

$$\frac{\partial \bar{\mathbf{x}}_1(v, w)}{\partial v} A_1 v = \bar{f}_1(\bar{\mathbf{x}}(v, w), \mathbf{u}(v, w), v, w), \tag{8.99}$$

$$0 = \bar{f}_2(\bar{\mathbf{x}}(v, w), \mathbf{u}(v, w), v, w), \tag{8.100}$$

$$0 = \bar{h}(\bar{\mathbf{x}}(v, w), v, w). \tag{8.101}$$

Also, (8.87) implies that

$$\bar{\mathbf{x}}_2(v, w) = \alpha(\bar{\mathbf{x}}_1(v, w), \mathbf{u}(v, w), v, w). \tag{8.102}$$

Substituting (8.102) into (8.99) and (8.101) gives

$$\begin{aligned} \frac{\partial \bar{\mathbf{x}}_1(v, w)}{\partial v} A_1 v &= f_r(\bar{\mathbf{x}}_1(v, w), \mathbf{u}(v, w), v, w), \\ 0 &= h_r\left(\bar{\mathbf{x}}_1(v, w), \mathbf{u}(v, w), v, w\right). \end{aligned} \tag{8.103}$$

Thus, the two functions $\bar{\mathbf{x}}_1(v, w)$ and $\mathbf{u}(v, w)$ are the solution of the regulator equations associated with the normal system (8.88).

Clearly, $\mathbf{u}(v, w)$ still satisfies conditions (6.31) and (6.33) with $g_o(x, u) = u$. Finally, it follows from part (iii) of Lemma 8.5 and Assumption 8.7 that the reduced-order normal system (8.88) satisfies (8.98).

Step 2. In this step, we will remove the assumption that (S, A) be standard. To this end, applying a linear output feedback control

$$u = K_e e + \bar{u} \tag{8.104}$$

to the plant (8.78) gives

$$\begin{aligned} S\dot{x} &= \tilde{f}(x, \bar{u}, v, w) = f(x, K_e h(x, v, w) + \bar{u}, v, w), \\ e &= \tilde{h}(x, v, w) = h(x, v, w). \end{aligned} \tag{8.105}$$

Suppose $\mathbf{x}(v, w)$ and $\mathbf{u}(v, w)$ are the solution of the regulator equations associated with (8.78) and (8.79). Then

$$\begin{aligned} \tilde{f}(\mathbf{x}(v, w), \mathbf{u}(v, w), v, w) &= f(\mathbf{x}(v, w), K_e e + \mathbf{u}(v, w), v, w) \\ &= f(\mathbf{x}(v, w), K_e h(\mathbf{x}(v, w), v, w) + \mathbf{u}(v, w), v, w) \\ &= f(\mathbf{x}(v, w), \mathbf{u}(v, w), v, w) \\ &= S\frac{\partial \mathbf{x}(v, w)}{\partial v} A_1 v, \\ \tilde{h}(\mathbf{x}(v, w), v) &= h(\mathbf{x}(v, w), v, w) = 0; \end{aligned}$$

that is, $\mathbf{x}(v, w)$ and $\mathbf{u}(v, w)$ are also the solution of the regulator equations associated with (8.105) and (8.79). Thus, the system (8.105) satisfies Assumption 8.6 and $\mathbf{u}(v, w)$ satisfies the conditions (6.31) and (6.33) with $g_o(x, \bar{u}) = \bar{u}$.

The linear approximation of (8.105) at $(x, u, v, w) = (0, 0, 0, w)$ can be expressed as

$$S\dot{x} = \tilde{A}(w)x + B(w)\bar{u} + \tilde{E}(w)v,$$
$$e = C(w)x + F(w)v, \tag{8.106}$$

where $\tilde{A}(w) = A(w) + B(w)K_eC(w)$ and $\tilde{E}(w) = E(w) + B(w)K_eF(w)$.

By Lemma 8.7, under Assumptions 8.4 and 8.5, there exists a matrix K_e such that

(a) $(S, \tilde{A})$ is standard,

(b) $(S, \tilde{A}, B)$ is strongly stabilizable and $(S, C, \tilde{A})$ is strongly detectable.

That is, system (8.105) also satisfies Assumptions 8.4 and 8.5.

Finally, note that, for all $\lambda \in \mathcal{C}$,

$$\begin{bmatrix} A + BK_eC - \lambda S & B \\ C & 0 \end{bmatrix} = \begin{bmatrix} A - \lambda S & B \\ C & 0 \end{bmatrix} \begin{bmatrix} I_n & 0 \\ K_eC & I_m \end{bmatrix}.$$

Thus, system (8.105) also satisfies condition (8.97). Since $(S, \tilde{A})$ is standard, by the first step of the proof of this theorem, the robust output regulation problem for system (8.105) and exosystem (8.79) can be solved by a controller of the form (8.59). Therefore, the robust output regulation problem for the original plant (8.78) and exosystem (8.79) can be solved by the composition of (8.104) and (8.59), that is, by

$$u(t) = k(z(t)) + K_ee(t),$$
$$\dot{z}(t) = g(z(t), e(t)), \tag{8.107}$$

which is clearly in the form of (8.80). □

Example 8.25. Let us slightly modify Example 8.17 by introducing an unknown parameter w in the second equation of (8.63) to yield the following uncertain singular nonlinear system:

$$\dot{x}_1 = 2x_2 + x_3,$$
$$\dot{x}_2 = -4x_1 + x_2 + x_4 - v_1 - 2(1 + w)v_2,$$
$$0 = x_2 + x_3,$$
$$0 = -x_1 - \sin x_2 + u,$$
$$e = x_4 - v_1, \tag{8.108}$$

with the same exosystem:

$$\dot{v}_1 = 2v_2, \quad \dot{v}_2 = -2v_1.$$

Correspondingly, the solution of the regulator equations of (8.108) is modified into the following:

$$\mathbf{x}_1(v, w) = (1 + w)v_1,$$
$$\mathbf{x}_2(v, w) = 2(1 + w)v_2,$$
$$\mathbf{x}_3(v, w) = -2(1 + w)v_2,$$
$$\mathbf{x}_4(v, w) = v_1,$$
$$\mathbf{u}(v, w) = (1 + w)v_1 + \sin(2(1 + w)v_2).$$

The linearization of (8.108) at the origin with $w = 0$ is the same as that in Example 8.17. Therefore, the plant and the exosystem satisfy Assumptions 8.3 to 8.6. Moreover, let $\pi(v, w) = (1 + w)v_1$. A simple calculation gives $\ddot{\pi}(v, w) + 4\pi(v, w) = 0$. Thus $\pi(v, w)$ has a minimal zeroing polynomial $P(\lambda) = \lambda^2+4$. As a result, there exists a smooth function $\Gamma : \mathcal{R}^2 \to \mathcal{R}$ such that

$$\mathbf{u}(v, w) = \Gamma(\pi(v, w), \dot{\pi}(v, w)) = \pi(v, w) + \sin(\dot{\pi}(v, w)).$$

Also, $\Psi = [1,\ 1]$ and

$$\Phi = \begin{bmatrix} 0 & 1 \\ -4 & 0 \end{bmatrix}.$$

Thus, conditions (6.31) and (6.33) with $g_o(x, u) = u$ are also satisfied. It remains to verify Assumption 8.7. Note that

$$\det \begin{bmatrix} A - \lambda S & B \\ C & 0 \end{bmatrix} = -\lambda^2 + \lambda - 4,$$

which has two roots $\lambda_{1,2} = 0.5 \pm j\sqrt{3.75}$. Thus Assumption 8.7 is satisfied. By Theorem 8.24, the robust output regulation problem for the given plant is solvable.

The desirable normal controller can be constructed based on the following reduced-order normal system:

$$\begin{aligned} \dot{x}_1 &= x_2, \\ \dot{x}_2 &= -3x_1 + x_2 - 2(1 + w)v_2 + \sin x_2 - \bar{u}, \\ e &= x_1 + \sin x_2 - \bar{u}, \end{aligned} \tag{8.109}$$

which is modified from (8.65) by taking into account the uncertain parameter w.

By Theorem 6.23, the robust output regulation problem for this system is solvable by a normal output feedback control of the form (6.44). To actually construct a controller, linearizing (8.109) gives

$$A_r = \begin{bmatrix} 0 & 1 \\ -3 & 2 \end{bmatrix}, \quad B_r = \begin{bmatrix} 0 \\ -1 \end{bmatrix}, \quad C_r = [1 \ \ 1], \quad D_r = -1.$$

Let

$$M = \begin{bmatrix} 0 & 1 \\ -3 & -2 \end{bmatrix}, \quad N = \begin{bmatrix} 0 \\ 1 \end{bmatrix}.$$

Solving the Sylvester equation $T\Phi - MT = N\Psi$ gives

$$T = \begin{bmatrix} 0.4118 & -0.1765 \\ 0.7059 & 0.4118 \end{bmatrix}.$$

Letting the eigenvalues of the following matrices:

$$\begin{bmatrix} A_r & B_r\Psi T^{-1} \\ 0 & M + N\Psi T^{-1} \end{bmatrix} + \begin{bmatrix} B_r \\ N \end{bmatrix} K$$

and

$$\begin{bmatrix} A_r & B_r \Psi T^{-1} \\ 0 & M + N\Psi T^{-1} \end{bmatrix} + L \begin{bmatrix} C_r & D_r \Psi T^{-1} \end{bmatrix}$$

be given by

$$-0.4240 \pm 1.2630j, \quad -0.6260 \pm 0.4141j$$

and

$$-1.2720 \pm 3.7890j, \quad -1.8780 \pm 1.2423j,$$

respectively, gives the control gain

$$K = \begin{bmatrix} -2.2334 & 2.0667 & 1.4333 & -2.0333 \end{bmatrix}$$

and the observer gain

$$L = \begin{bmatrix} -8.3826 \\ 23.0341 \\ -5.9852 \\ 0.1831 \end{bmatrix}.$$

Finally, the controller is given by

$$\begin{aligned} \bar{u} &= \beta(\eta) + K\xi, \\ \dot{\eta} &= M\eta + N(K\xi + \Psi T^{-1}\eta), \\ \dot{\xi} &= Q\xi + Le, \end{aligned}$$

where $\eta \in \mathcal{R}^2$,

$$Q = \begin{bmatrix} 27.1040 & -7.9415 & -3.6325 & 0.2794 \\ -75.2442 & 24.5032 & 9.5482 & -0.7344 \\ 19.3524 & -6.3843 & -2.5936 & 1.1995 \\ -2.8255 & 2.2620 & -2.4873 & -2.0394 \end{bmatrix},$$

and $\beta(\eta) = [1\ 0]T^{-1}\eta + \sin([0\ 1]T^{-1}\eta)$.

Composition of this controller with the precompensator $u = \bar{u} + e$ gives the normal output feedback controller that solves the output regulation problem of the original system. ∎

Chapter 9

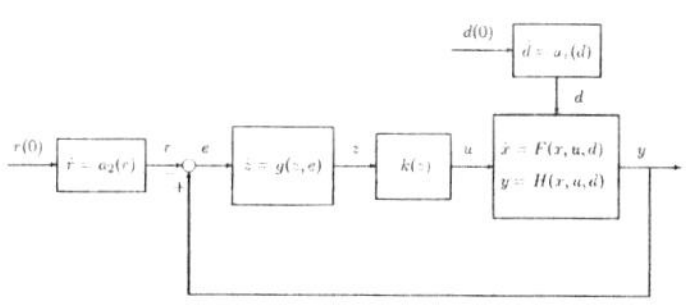

Output Regulation for Discrete-Time Nonlinear Systems

In this chapter, we will study the output regulation problem for discrete-time nonlinear systems. The contents of this chapter are basically the discrete-time counterparts of what are covered in Chapters 3 to 5 for continuous-time systems. Whereas in linear systems, the technicalities for dealing with discrete-time and continuous-time systems are quite similar, for nonlinear systems, there are some subtle differences between the discrete-time output regulation problem and the continuous-time output regulation problem. Most notably, as we will see in the next section, the regulator equations associated with the discrete-time systems are a set of algebraic functional equations, in contrast with the regulator equations associated with the continuous-time systems.

Technically, the major tool used for handling the output regulation problem for continuous-time nonlinear systems is the center manifold theory for differential equations as summarized in Section 2.4, while the major tool used for handling the output regulation problem for discrete-time nonlinear systems is the center manifold theory for maps summarized in Section 2.5.

The chapter is organized as follows. In Section 9.1, we formulate and solve the output regulation problem for discrete-time systems without involving uncertain parameters. In Section 9.2, we present an approximation method for a discrete-time output regulation problem based on Taylor series expansion. In Section 9.3, we study the robust output regulation problem for a discrete-time systems with uncertain parameters. In Section 9.4, an example is given to illustrate the discrete-time robust output regulation problem.

9.1 Discrete-Time Output Regulation

We consider a class of discrete-time nonlinear systems of the form described by

$$\begin{aligned} x(t+1) &= f(x(t), u(t), v(t)), \quad x(0) = x_0, \\ e(t) &= h(x(t), u(t), v(t)), \quad t = 0, 1, \ldots, \end{aligned} \tag{9.1}$$

where $x(t)$ is the n-dimensional plant state, $u(t)$ the m-dimensional plant input, $e(t)$ the p-dimensional plant output representing tracking error, and $v(t)$ the q-dimensional disturbance signal, which can represent either disturbance signal or the reference input or both. $v(t)$ is generated by a q-dimensional autonomous difference equation of the following form:

$$v(t+1) = a(v(t)), \quad v(0) = v_0, \quad t = 0, 1, \ldots. \tag{9.2}$$

For simplicity, all the functions involved in this setup are assumed to be sufficiently smooth and defined globally on the appropriate Euclidean spaces, with the value zero at the respective origins. Our results will be stated locally in terms of an open neighborhood V of the origin in $\mathcal{R}^q$, and we implicitly permit V to be made smaller to accommodate subsequent local arguments.

We will also consider two classes of control laws, namely,

1. **Static State Feedback:**

$$u(t) = k(x(t), v(t)), \tag{9.3}$$

where the function $k(\cdot, \cdot)$ is required to be sufficiently smooth and satisfies $k(0, 0) = 0$.

2. **Dynamic Measurement Output Feedback:**

$$\begin{aligned} u(t) &= k(z(t)), \\ z(t+1) &= g(z(t), y_m(t)), \end{aligned} \tag{9.4}$$

where $z(t)$ is the compensator state of dimension n_z to be specified later; $y_m(t) = h_m(x(t), u(t), v(t))$, where $h_m : \mathcal{R}^{n+m+q} \to \mathcal{R}^{p_m}$ for some integer p_m, and is called the measurement output; and the functions $k(\cdot)$ and $g(\cdot, \cdot)$ are required to be sufficiently smooth and satisfy $k(0) = 0$ and $g(0, 0) = 0$.

To formulate the requirements on the closed-loop system, we denote the closed-loop system consisting of the plant (9.1), the exosystem (9.2), and the controller (9.3) or (9.4) as follows:

$$\begin{aligned} x_c(t+1) &= f_c(x_c(t), v(t)), \quad x_c(0) = x_{c0}, \\ v(t+1) &= a(v(t)), \\ e(t) &= h_c(x_c(t), v(t)), \quad t = 0, 1, \ldots, \end{aligned} \tag{9.5}$$

where, under the static state feedback control, $x_c = x$, and $h_c(\cdot, \cdot)$ and $f_c(\cdot, \cdot)$ are described as follows:

$$\begin{aligned} f_c(x_c, v) &= f(x, k(x, v), v), \\ h_c(x_c, v) &= h(x, k(x, v), v), \end{aligned} \tag{9.6}$$

and, under the dynamic measurement output feedback control, $x_c = \mathrm{col}(x, z)$ and $h_c(\cdot, \cdot)$ and $f_c(\cdot, \cdot)$ are described as follows:

$$\begin{aligned} h_c(x_c, v) &= h(x, k(z), v), \\ f_c(x_c, v) &= \begin{bmatrix} f(x, k(z), v) \\ g(z, h_m(x, k(z), v)) \end{bmatrix}. \end{aligned} \tag{9.7}$$

Discrete-Time Nonlinear Output Regulation Problem (DNORP): Design a control law (9.3) or (9.4) such that the closed loop composite system (9.5) has the following properties.

Property 9.1. The equilibrium point of the closed-loop system (9.5) at $(x_c, v) = (0, 0)$ is stable in the sense of Lyapunov, and

Property 9.2. For all sufficiently small $x_c(0)$ and $v(0)$, the trajectory $\mathrm{col}(x_c(t), v(t))$ of (9.5) satisfies

$$\lim_{t\to\infty} e(t) = \lim_{t\to\infty} h_c(x_c(t), v(t)) = 0. \tag{9.8}$$

Remark 9.1. As a result of Property 9.1, for all sufficiently small $x_c(0)$ and $v(0)$, the trajectories $(x_c(t), v(t))$ of the closed-loop system (9.5) exist and are bounded for all $t = 0, 1, \ldots$. By Theorem 2.33 and Assumption 9.1, to be introduced later, Property 9.1 is automatically satisfied if the closed-loop system has the following property:

Property 9.3. All the eigenvalues of the matrix

$$\frac{\partial f_c}{\partial x_c}(0, 0) \tag{9.9}$$

have modulus smaller than 1.

Like the continuous-time case, it is quite straightforward to achieve Property 9.3 by using a linear feedback control under Assumption 9.2 and/or 9.3 to be given below. We often impose Property 9.3 instead of Property 9.1 on the closed-loop system. In analogy to the continuous-time case, we will call the problem of synthesizing a feedback control law such that the closed-loop system satisfies Properties 9.2 and 9.3 as the discrete-time output regulation problem with exponential stability. ∎

If there exists a control law such that the closed-loop system satisfies Properties 9.1 and 9.2, we say that the nonlinear output regulation problem is (locally) solvable and the control law is called a nonlinear *servoregulator*. In particular, the control law given by equation (9.3) is called a *state feedback servoregulator*, and the control law given by equation (9.4) is called a *measurement output feedback servoregulator*. Alternatively, we say that the control law achieves asymptotic tracking and disturbance rejection in the plant.

Various assumptions needed for the solvability of the problem are listed below.

Assumption 9.1. The equilibrium point of exosystem (9.2) at $v = 0$ is Lyapunov stable, and all the eigenvalues of $\frac{\partial a}{\partial v}(0)$ are on the unit circle.

Assumption 9.1′. The equilibrium point of the exosystem (9.2) at $v = 0$ is Lyapunov stable and there is an open neighborhood of $v = 0$ in which every point is Poisson stable in the sense to be described in Remark 9.2.

Assumption 9.2. The pair

$$\left(\frac{\partial f}{\partial x}(0, 0, 0), \quad \frac{\partial f}{\partial u}(0, 0, 0)\right)$$

is stabilizable.

Assumption 9.3. The pair

$$\left(\begin{bmatrix} \frac{\partial h_m}{\partial x}(0,0,0) & \frac{\partial h_m}{\partial v}(0,0,0) \end{bmatrix}, \begin{bmatrix} \frac{\partial f}{\partial x}(0,0,0) & \frac{\partial f}{\partial v}(0,0,0) \\ 0 & \frac{\partial a}{\partial v}(0) \end{bmatrix} \right)$$

is detectable.

Remark 9.2. A point $v^0 \in \mathcal{R}^q$ is said to be Poisson stable if the solution $v(t, v^0)$ of the exosystem (9.2) exists for all $t = 0, 1, 2, \ldots$, and for each open neighborhood V^0 of v^0 and for any integer $N > 0$, there exists an integer $n_1 > N$ such that $v(n_1, v^0) \in V^0$, and an integer $n_2 < -N$ such that $v(n_2, v^0) \in V^0$. ∎

Remark 9.3. Assumptions 9.1 to 9.3 are clearly the discrete-time counterparts of Assumptions 3.1 to 3.3. They will play the same role in dealing with discrete-time systems as Assumptions 3.1 to 3.3 do to continuous-time systems. Also, Assumption 9.1′ is the discrete-time counterpart of Assumption 3.1′. This assumption is only needed when the necessary condition of the solvability of the discrete-time output regulation problem is concerned. ∎

We first establish a result parallel to Lemma 3.6.

Lemma 9.4. *Under Assumption 9.1′, suppose that the closed-loop system (9.5) resulting from the controller (9.3) or (9.4) has Property 9.3. Then, it also has Property 9.2 if and only if there exists a sufficiently smooth function $\mathbf{x}_c(v)$ with $\mathbf{x}_c(0) = 0$ that satisfies, for $v \in V$, where V is an open neighborhood of $0 \in \mathcal{R}^q$, the following algebraic equations:*

$$\mathbf{x}_c(a(v)) = f_c(\mathbf{x}_c(v), v), \tag{9.10}$$

$$0 = h_c(\mathbf{x}_c(v), v). \tag{9.11}$$

Proof. First note that Assumption 9.1′ implies Assumption 9.1, and thus the exosystem has a stable equilibrium at the origin and all the eigenvalues of its Jacobian matrix have modulus 1. Since the closed-loop system has Property 9.3, by Theorem 2.31, there exists a center manifold for the closed-loop system (9.5). That is, there exists a sufficiently smooth function $\mathbf{x}_c(v)$ with $\mathbf{x}_c(0) = 0$ that satisfies (9.10) for $v \in V$. Moreover, by Theorem 2.33, the equilibrium of the closed-loop system (9.5) at the origin is Lyapunov stable. Thus, the solution of the closed-loop system (9.5) starting from sufficiently small initial state exists for all $t = 0, 1, 2, \ldots$.

If part. Since the function $\mathbf{x}_c(v)$ with $\mathbf{x}_c(0) = 0$ that satisfies (9.10) for $v \in V$ defines a center manifold $x_c = \mathbf{x}_c(v)$ for the closed-loop system (9.5), by Theorem 2.34, there exist positive constants δ and $\lambda < 1$ such that, for all sufficiently small $x_c(0)$ and $v(0)$, the trajectories $x_c(t)$ of the closed-loop system (9.5) satisfy

$$||x_c(t) - \mathbf{x_c}(v(t))|| \leq \delta\lambda^t, \quad t = 0, 1, 2, \ldots. \tag{9.12}$$

Furthermore, there exists a compact set S_c in $\mathcal{R}^{n+n_z+q}$, where $n_z = 0$ for state feedback, such that, for $t = 0, 1, 2, \ldots$, $\text{col}(x_c(t),\ v(t)) \in S_c$, $\text{col}(\mathbf{x}_c(v(t)),\ v(t)) \in S_c$; therefore, there exists a finite constant L such that

$$\left\| \frac{\partial h_c}{\partial x_c}(x_c, v) \right\| < L \tag{9.13}$$

for $\mathrm{col}(x_c,\ v) \in S_c$. Thus, if the function $\mathbf{x}_c(v)$ also satisfies (9.11), then

$$\begin{aligned} \lim_{t\to\infty} \|e(t)\| &= \lim_{t\to\infty} \|h_c(x_c(t), v(t))\| \\ &= \lim_{t\to\infty} \|h_c(x_c(t), v(t)) - h_c(\mathbf{x_c}(v(t)), v(t))\| \\ &\le \lim_{t\to\infty} L\|x_c(t) - \mathbf{x_c}(v(t))\| = 0; \end{aligned} \tag{9.14}$$

that is, the closed-loop system also has Property 9.2.

Only if part. Assume that the closed-loop system has both Properties 9.2 and 9.3, yet (9.11) is not true. Then there exists a sufficiently small $v_0 \in V$ such that the solution of the closed-loop system (9.5) satisfying $\mathrm{col}(x_c(0), v(0)) = \mathrm{col}(\mathbf{x}_c(v_0), v_0)$, denoted by $\mathrm{col}(x_c(t, \mathbf{x}_c(v_0)), v(t, v_0))$, exists for all $t = 0, 1, 2, \ldots$ and satisfies

$$\lim_{t\to\infty} \|h_c(x_c(t, \mathbf{x}_c(v_0)), v(t, v_0))\| = 0, \tag{9.15}$$

yet

$$\|h_c(\mathbf{x}_c(v_0), v_0)\| > 0.$$

Thus there exists a neighborhood $V_0 \subset V$ of v_0 and some real number $R > 0$ such that

$$\|h_c(\mathbf{x}_c(v), v)\| > R$$

for all $v \in V_0$. Clearly, $x_c(t, \mathbf{x}_c(v_0)) = \mathbf{x}_c(v(t, v_0))$, since $x_c(0, \mathbf{x}_c(v_0)) = \mathbf{x}_c(v_0) = \mathbf{x}_c(v(0, v_0))$ and (9.10) implies

$$\mathbf{x}_c(v(t+1), v_0) = f_c(\mathbf{x}_c(v(t, v_0)), v(t, v_0)), \quad t = 0, 1, 2, \ldots.$$

But, since the exosystem satisfies Assumption 9.1′, we can assume that v_0 is small enough so that it is Poisson stable, and therefore, given any integer $N > 0$, there exists an integer $n_1 > N$ such that $v(n_1, v_0) \in V_0$. Thus,

$$\|h_c(x_c(n_1, \mathbf{x}_c(v_0)), v(n_1, v_0))\| = \|h_c(\mathbf{x}_c(v(n_1, v_0)), v(n_1, v_0))\| > R,$$

which contradicts (9.15). □

Next we will establish the solvability of the state feedback output regulation problem in terms of the given plant.

Theorem 9.5. *Under Assumptions 9.1′ and 9.2, the discrete-time nonlinear output regulation problem with exponential stability is solvable by a static state feedback control of the form (9.3) if and only if there exist two sufficiently smooth functions* $\mathbf{x}(v)$ *and* $\mathbf{u}(v)$ *satisfying* $\mathbf{x}(0) = 0$ *and* $\mathbf{u}(0) = 0$ *such that*

$$\begin{aligned} \mathbf{x}(a(v)) &= f(\mathbf{x}(v), \mathbf{u}(v), v), \\ 0 &= h(\mathbf{x}(v), \mathbf{u}(v), v). \end{aligned} \tag{9.16}$$

Proof. Assume that a controller of the form $u = k(x, v)$ solves the discrete-time nonlinear output regulation problem. Then, by Lemma 9.4 , there exists a sufficiently smooth function $\mathbf{x}_c(v)$ that satisfies (9.10) and (9.11) for $v \in V$. Let $\mathbf{x}(v) = \mathbf{x}_c(v)$ and $\mathbf{u}(v) = k(\mathbf{x}(v), v)$. Then, $\mathbf{x}(v)$ and $\mathbf{u}(v)$ satisfy (9.16). On the other hand, assume that $\mathbf{x}(v)$ and $\mathbf{u}(v)$ satisfy (9.16) for $v \in V$. Let $K_x \in \mathcal{R}^{m\times n}$ be any constant matrix such that the eigenvalues of the following matrix:

$$\frac{\partial f}{\partial x}(0, 0, 0) + \frac{\partial f}{\partial u}(0, 0, 0)K_x \tag{9.17}$$

have modulus smaller than 1. Due to Assumption 9.2, K_x always exists. Let

$$k(x, v) = \mathbf{u}(v) + K_x(x - \mathbf{x}(v)). \tag{9.18}$$

Then, under (9.18), the closed-loop system (9.5) satisfies Property 9.3. Moreover, letting $\mathbf{x}_c(v) = \mathbf{x}(v)$ leads to

$$\begin{aligned} f_c(\mathbf{x}_c(v), v) &= f(\mathbf{x}_c(v), k(\mathbf{x}_c(v), v), v) = f(\mathbf{x}(v), \mathbf{u}(v), v) = \mathbf{x}(a(v)) = \mathbf{x}_c(a(v)), \\ h_c(\mathbf{x}_c(v), v) &= h(\mathbf{x}_c(v), k(\mathbf{x}_c(v), v), v) = h(\mathbf{x}(v), \mathbf{u}(v), v) = 0 \end{aligned}$$

as $\mathbf{x}(v)$ and $\mathbf{u}(v)$ satisfy equations (9.16). By Lemma 9.4, the controller as defined by (9.18) solves the discrete-time nonlinear output regulation problem. □

Remark 9.6. Equations (9.16) play the same role for the discrete-time nonlinear output regulation problem as equations (3.30) do for the continuous-time nonlinear output regulation problem and are thus called the *discrete-time regulator equations.* In contrast to the linear case, in which both the continuous-time and discrete-time regulator equations take exactly the same form as follows:

$$\begin{aligned} XA_1 &= AX + BU + E, \\ 0 &= CX + DU + F, \end{aligned} \tag{9.19}$$

the discrete-time regulator equations are a set of nonlinear algebraic equations that are distinctly different from the continuous-time regulator equations, which are a set of nonlinear partial differential and algebraic equations. It is this difference that necessitates an independent treatment of the nonlinear discrete-time output regulation problem. ∎

By the same token as Remark 1.8, we will call the functions $\mathbf{u}(v)$ and $\mathbf{x}(v)$ *zero-error constrained input* and *zero-error constrained state* for the plant and the exosystem, respectively.

When the plant state and/or disturbance state are not available, one can consider using the measurement output feedback to solve the output regulation problem.

Theorem 9.7. *Under Assumptions 9.1′, 9.2, and 9.3, the discrete-time nonlinear output regulation problem is solvable by a dynamic measurement output feedback controller if and only if there exist two sufficiently smooth functions* $\mathbf{x}(v)$ *and* $\mathbf{u}(v)$ *with* $\mathbf{x}(0) = 0$ *and* $\mathbf{u}(0) = 0$ *that satisfy the discrete-time nonlinear regulator equations (9.16).*

Proof. *Necessity.* Assume that the output feedback control $u(t) = k(z(t))$, $z(t+1) = g(z(t), y_m(t))$ solves the output regulation problem. Then, by Lemma 9.4, there exists some sufficiently smooth function $\mathbf{x}_c(v)$ for $v \in V$ with $\mathbf{x}_c(0) = 0$ satisfying (9.10) and (9.11). Partition $\mathbf{x}_c(v)$ as

$$\mathbf{x}_c(v) = \begin{bmatrix} \mathbf{x}_{c1}(v) \\ \mathbf{x}_{c2}(v) \end{bmatrix},$$

where $\mathbf{x}_{c1}(\cdot) \in \mathcal{R}^n$ and $\mathbf{x}_{c2}(v) \in \mathcal{R}^{n_z}$. Substituting (9.7) into (9.10) and (9.11) gives

$$\begin{aligned} \mathbf{x}_{c1}(a(v)) &= f(\mathbf{x}_{c1}(v), k(\mathbf{x}_{c2}(v)), v), \\ \mathbf{x}_{c2}(a(v)) &= g(\mathbf{x}_{c2}(v), h_m(\mathbf{x}_{c1}(v), k(\mathbf{x}_{c2}(v)), v)), \\ 0 &= h(\mathbf{x}_{c1}(v), k(\mathbf{x}_{c2}(v)), v). \end{aligned} \tag{9.20}$$

Letting $\mathbf{x}(v) = \mathbf{x}_{c1}(v)$ and $\mathbf{u}(v) = k(\mathbf{x}_{c2}(v))$ shows that $\mathbf{x}(v)$ and $\mathbf{u}(v)$ satisfy (9.16).

Sufficiency. Note that, under Assumption 9.2, there exists a state feedback gain K_x such that all the eigenvalues of $\frac{\partial f}{\partial x}(0,0,0) + \frac{\partial f}{\partial u}(0,0,0)K_x$ have modulus smaller than 1. By Assumption 9.3, there exist constant matrices L_1 and L_2 such that all the eigenvalues of the matrix

$$\begin{bmatrix} \frac{\partial f}{\partial x}(0,0,0) & \frac{\partial f}{\partial v}(0,0,0) \\ 0 & \frac{\partial a}{\partial v}(0) \end{bmatrix} - \begin{bmatrix} L_1 \\ L_2 \end{bmatrix} \begin{bmatrix} \frac{\partial h_m}{\partial x}(0,0,0) & \frac{\partial h_m}{\partial v}(0,0,0) \end{bmatrix}$$

have modulus smaller than 1.

Suppose equations (9.16) are satisfied by some sufficiently smooth functions $\mathbf{x}(v)$ and $\mathbf{u}(v)$ satisfying $\mathbf{x}(0) = 0$ and $\mathbf{u}(0) = 0$. Let $z = \mathrm{col}(z_1, z_2)$ with $z_1 \in \mathcal{R}^n$ and $z_2 \in \mathcal{R}^q$, and

$$k(z) = k(z_1, z_2) = \mathbf{u}(z_2) + K_x(z_1 - \mathbf{x}(z_2)), \tag{9.21}$$

$$g(z, y_m) = \begin{bmatrix} f(z_1, k(z_1, z_2), z_2) + L_1(y_m - h_m(z_1, k(z_1, z_2), z_2)) \\ a(z_2) + L_2(y_m - h_m(z_1, k(z_1, z_2), z_2)) \end{bmatrix}. \tag{9.22}$$

This controller yields a closed-loop system with

$$x_c = \mathrm{col}(x, z_1, z_2), \quad h_c(x_c, v) = h(x, k(z_1, z_2), v),$$

and

$$f_c(x_c, v) = \begin{bmatrix} f(x, k(z_1, z_2), v) \\ f(z_1, k(z_1, z_2), z_2) + L_1(y_m - h_m(z_1, k(z_1, z_2), z_2)) \\ a(z_2) + L_2(y_m - h_m(z_1, k(z_1, z_2), z_2)) \end{bmatrix}. \tag{9.23}$$

The function defined in (9.23) takes a form similar to that given in (3.55). Therefore, the Jacobian matrix $f_c(x_c, 0)$ at the origin takes the same expression as the matrix A_c calculated in (3.56) and is thus exponentially stable, that is, is a Schur matrix.

To verify that the closed-loop system satisfies Property 9.2, let $\mathrm{col}(\mathbf{x}(v), \mathbf{u}(v))$ be the solution of the regulator equations (9.16), $\mathbf{z}_1(v) = \mathbf{x}(v)$ and $\mathbf{z}_2(v) = v$, and

$$\mathbf{z}(v) = \begin{bmatrix} \mathbf{z}_1(v) \\ \mathbf{z}_2(v) \end{bmatrix} = \begin{bmatrix} \mathbf{x}(v) \\ v \end{bmatrix}.$$

Then, from (9.21),

$$k(\mathbf{z}(v)) = k(\mathbf{z}_1(v), \mathbf{z}_2(v)) = k(\mathbf{x}(v), v) = \mathbf{u}(v) \tag{9.24}$$

and

$$\begin{aligned}\mathbf{z}(a(v)) &= \begin{bmatrix} \mathbf{x}(a(v)) \\ a(v) \end{bmatrix} \\ &= \begin{bmatrix} f(\mathbf{x}(v), \mathbf{u}(v), v) \\ a(v) \end{bmatrix} \\ &= \begin{bmatrix} f(\mathbf{x}(v), k(\mathbf{z}(v)), v) \\ a(v) \end{bmatrix}.\end{aligned} \tag{9.25}$$

Let $\mathbf{x}_c(v) = \text{col}(\mathbf{x}(v),\ \mathbf{x}(v),\ v)$. Then, by (9.24) and (9.25),

$$f_c(\mathbf{x}_c(v), v) = \begin{bmatrix} f(\mathbf{x}(v), \mathbf{u}(v), v) \\ f(\mathbf{x}(v), \mathbf{u}(v), v) \\ a(v) \end{bmatrix} = \begin{bmatrix} \mathbf{x}(a(v)) \\ \mathbf{x}(a(v)) \\ a(v) \end{bmatrix} = \mathbf{x}_c(a(v)),$$

$$h_c(\mathbf{x}_c(v), v) = h(\mathbf{x}(v), \mathbf{u}(v), v) = 0.$$

That is, (9.10) and (9.11) are satisfied. □

9.2 Approximation Method for the Discrete-Time Output Regulation

Similar to the continuous-time case, due to the nonlinear nature of the discrete-time regulator equations (9.16), it is usually impossible to obtain the exact solution of regulator equations (9.16). In this section, the kth-order output regulation problem formulated for the continuous-time systems will be extended to the discrete-time systems; then an approximation method for obtaining the solution of the discrete-time regulator equations by the Taylor series will be presented, which in turn leads to a method to synthesize both the state feedback and the output feedback control laws to approximately solve the discrete-time nonlinear output regulation problem in a similar way to what was done to continuous-time systems in Chapter 4.

Discrete-Time kth-Order Nonlinear Output Regulation Problem (DKNORP): Given some integer $k \geq 1$, design a control law of the form (9.3) or (9.4) such that the closed-loop system (9.5) has Property 9.3 and the following:

Property 9.4. For all sufficiently small x_{c0} and v_0, the trajectories $\text{col}(x_c(t), v(t))$ of the closed-loop system (9.5) satisfy

$$\lim_{t\to\infty} (e(t) - o^k(v(t))) = \lim_{t\to\infty} (h_c(x_c(t), v(t)) - o^k(v(t))) = 0, \tag{9.26}$$

where $o^k(v)$ is some sufficiently smooth function of v zero up to kth-order.

Let us first state some results that are discrete counterparts of Lemma 4.2, Theorem 4.3, and Theorem 4.5.

Lemma 9.8. *Under Assumption 9.1′, suppose the closed-loop system (9.5) has Property 9.3. Then the closed-loop system (9.5) also has Property 9.4 if and only if there exists a sufficiently smooth function* $\mathbf{x}_c^{(k)}(v)$ *with* $\mathbf{x}_c^{(k)}(0) = 0$ *that satisfies, for* $v \in V$*, the following equations:*

$$\mathbf{x}_c^{(k)}(a(v)) = f_c(\mathbf{x}_c^{(k)}(v), v), \tag{9.27}$$

$$o^{(k)}(v) = h_c(\mathbf{x}_c^{(k)}(v), v). \tag{9.28}$$

The proof of Lemma 9.8 is quite similar to that of Lemma 4.2 and is thus omitted.

Theorem 9.9.

(i) *Under Assumptions 9.1′ and 9.2, the discrete-time kth-order nonlinear output regulation problem is solvable by a static state feedback controller of the form (9.3) if and only if there exist two sufficiently smooth functions* $\mathbf{x}^{(k)}(v)$ *and* $\mathbf{u}^{(k)}(v)$ *satisfying* $\mathbf{x}^{(k)}(0) = 0$ *and* $\mathbf{u}^{(k)}(0) = 0$ *such that*

$$\begin{aligned} \mathbf{x}^{(k)}(a(v)) &= f(\mathbf{x}^{(k)}(v), \mathbf{u}^{(k)}(v), v) + o^k(v), \\ o^k(v) &= h(\mathbf{x}^{(k)}(v), \mathbf{u}^{(k)}(v), v). \end{aligned} \tag{9.29}$$

(ii) *Under Assumptions 9.1′, 9.2, and 9.3, the discrete-time kth-order nonlinear output regulation problem is solvable by a measurement output feedback controller of the form (9.4) if and only if there exist two sufficiently smooth functions* $\mathbf{x}^{(k)}(v)$ *and* $\mathbf{u}^{(k)}(v)$ *satisfying* $\mathbf{x}^{(k)}(0) = 0$*,* $\mathbf{u}^{(k)}(0) = 0$*, and (9.29).*

Proof. The proof of this theorem can be directly obtained from Lemma 9.8. Here we will only sketch the sufficient part of the proof. Consider the following state feedback controller:

$$u = \mathbf{u}^{(k)}(v) + K_x(x - \mathbf{x}^{(k)}(v)) \tag{9.30}$$

and the measurement output feedback controller of the form (9.3) with $z = \mathrm{col}(z_1, z_2)$:

$$\begin{aligned} k(z) = k(z_1, z_2) &= \mathbf{u}^{(k)}(z_2) + K_x(z_1 - \mathbf{x}^{(k)}(z_2)), \\ g(z, y_m) &= \begin{bmatrix} f(z_1, k(z_1, z_2), z_2) + L_1(y_m - h_m(z_1, k(z_1, z_2), z_2)) \\ a(z_2) + L_2(y_m - h_m(z_1, k(z_1, z_2), z_2)) \end{bmatrix}, \end{aligned} \tag{9.31}$$

which are obtained by replacing $\mathbf{x}(\cdot)$ and $\mathbf{u}(\cdot)$ in the state feedback controller (9.18) and the measurement output feedback controller (9.21) and (9.22) with $\mathbf{u}^{(k)}(\cdot)$ and $\mathbf{x}^{(k)}(\cdot)$. It is not difficult to verify that each of these controllers will result in a closed-loop system that satisfies Property 9.3 and induces a sufficiently smooth function $\mathbf{x}_c^{(k)}(v)$ with $\mathbf{x}_c^{(k)}(0) = 0$ such that (9.27) and (9.28) hold. Thus, it follows from Lemma 9.8 that (9.30) and (9.31) solve, respectively, the state feedback and the measurement output feedback kth-order output regulation problem for the discrete-time nonlinear systems (9.1) and (9.2). ☐

As indicated by Theorem 9.9, like the continuous-time kth-order output regulation problem, the key to solving the discrete-time kth-order output regulation problem is to obtain a kth-order solution of the discrete-time regulator equations. In what follows, we will present a method for approximately solving the discrete-time regulator equations by Taylor series. The approach is similar to what was developed in Chapter 4. Therefore, the same Kronecker product notation as used in Chapter 4 will be adopted. Let us first write the problem description in terms of the series expansions

$$f(x,u,v)=\sum_{l\ge 1}\sum_{\substack{i+j+k=l\\ i,j,k\ge 0}} F_{ijk}x^{(i)}\otimes u^{(j)}\otimes v^{(k)},$$

$$h(x,u,v)=\sum_{l\ge 1}\sum_{\substack{i+j+k=l\\ i,j,k\ge 0}} H_{ijk}x^{(i)}\otimes u^{(j)}\otimes v^{(k)},$$

$$a(v)=\sum_{i\ge 1} A_i v^{(i)}. \tag{9.32}$$

Also, for the $q\times 1$ vector $v=[v_1,\dots,v_q]^T$, let $v^{[l]}$ denote the vector

$$v^{[l]}=[v_1^l,\ v_1^{l-1}v_2,\dots,v_1^{l-1}v_q,\ v_1^{l-2}v_2^2,\ v_1^{l-2}v_2v_3,\dots,v_1^{l-2}v_2v_q,\dots,v_q^l]^T. \tag{9.33}$$

Then we seek series of the form

$$\mathbf{x}(v)=\sum_{l\ge 1} X_l v^{[l]},\quad \mathbf{u}(v)=\sum_{l\ge 1} U_l v^{[l]} \tag{9.34}$$

such that (9.16) is satisfied formally. Once again, note that there exist matrices M_l and N_l of appropriate dimensions such that

$$v^{[l]}=M_l v^{(l)},\quad v^{(l)}=N_l v^{[l]}. \tag{9.35}$$

Our approach involves substituting equations (9.32), (9.34), and (9.35) into the regulator equations (9.16) and identifying the coefficients of $v^{[l]}$, $l=1,2,\dots,$ which yields the following result.

Lemma 9.10. *The power series (9.34) formally satisfy the regulator equations (9.16) if and only if the following linear equations are satisfied for* $l=1,2,\dots$:

$$X_l M_l A_1^{(l)} N_l=\frac{\partial f}{\partial x}(0,0,0)X_l+\frac{\partial f}{\partial u}(0,0,0)U_l+E_l,$$

$$0=\frac{\partial h}{\partial x}(0,0,0)X_l+\frac{\partial h}{\partial u}(0,0,0)U_l+F_l, \tag{9.36}$$

where

$$E_1=F_{001}=\frac{\partial f}{\partial v}(0,0,0),\quad F_1=H_{001}=\frac{\partial h}{\partial v}(0,0,0)$$

and, for $l=2,3,\dots,$

$$E_l=\sum_{n=2}^{l}\sum_{\substack{i+j+k=n\\ i,j,k\ge 0}} F_{ijk}G_{l-n}^{ij}N_l-\sum_{k=1}^{l-1}X_kM_k\xi_{k,l}N_l, \tag{9.37}$$

$$F_l=\left[\sum_{n=2}^{l}\sum_{\substack{i+j+k=n\\ i,j,k\ge 0}} H_{ijk}G_{l-n}^{ij}\right]N_l, \tag{9.38}$$

where

$$\xi_{i,j} = \sum_{\substack{j_1+j_2+\cdots+j_i=j \\ j_1,j_2,\ldots,j_i\geq 1}} A_{j_1} \otimes A_{j_2} \otimes \cdots \otimes A_{j_i}, \quad j \geq i > 0, \tag{9.39}$$

$$G_m^{ij} = \begin{cases} 0, & i = j = 0, m > 0, \\ 1, & i = j = 0, m = 0, \\ \delta_{i,i+m}, & j = 0, i = 1, 2, \ldots, \\ \lambda_{j,j+m}, & i = 0, j = 1, 2, \ldots, \\ \sum_{k=0}^{m} \delta_{i,i+k} \otimes \lambda_{j,j+m-k}, & i, j = 1, 2, \ldots, \end{cases} \tag{9.40}$$

$$\delta_{i,j} = \sum_{\substack{j_1+j_2+\cdots+j_i=j \\ j_1,j_2,\ldots,j_i\geq 1}} X_{j_1} M_{j_1} \otimes X_{j_2} M_{j_2} \otimes \cdots \otimes X_{j_i} M_{j_i}, \quad j \geq i > 0, \tag{9.41}$$

$$\lambda_{i,j} = \sum_{\substack{j_1+j_2+\cdots+j_i=j \\ j_1,j_2,\ldots,j_i\geq 1}} U_{j_1} M_{j_1} \otimes U_{j_2} M_{j_2} \otimes \cdots \otimes U_{j_i} M_{j_i}, \quad j \geq i > 0. \tag{9.42}$$

Proof. Substituting equations (9.32) and (9.34) into equations (9.16) yields the following equations:

$$\sum_{i\geq 1} X_i \left(\sum_{j\geq 1} A_j v^{(j)} \right)^{[i]} = \sum_{l\geq 1} \sum_{\substack{i+j+k=l \\ i,j,k\geq 0}} F_{ijk} \mathbf{x}^{(i)}(v) \otimes \mathbf{u}^{(j)}(v) \otimes v^{(k)}, \tag{9.43}$$

$$0 = \sum_{l\geq 1} \sum_{\substack{i+j+k=l \\ i,j,k\geq 0}} H_{ijk} \mathbf{x}^{(i)}(v) \otimes \mathbf{u}^{(j)}(v) \otimes v^{(k)}. \tag{9.44}$$

The left-hand side of (9.43) can be written as

$$\begin{aligned} \sum_{i\geq 1} X_i \left(\sum_{j\geq 1} A_j v^{(j)} \right)^{[i]} &= \sum_{i\geq 1} X_i M_i \left(\sum_{j\geq 1} A_j v^{(j)} \right)^{(i)} \\ &= \sum_{i\geq 1} X_i M_i \left(\sum_{k\geq i} \sum_{\substack{j_1+\cdots+j_i=i \\ j_1,\ldots,j_i\geq 1}} (A_{j_1} \otimes \cdots \otimes A_{j_i}) v^{(k)} \right) \\ &= \sum_{l\geq 1} \sum_{k=1}^{l} X_k M_k \xi_{k,l} v^{(l)}, \end{aligned} \tag{9.45}$$

where $\xi_{k,l}$ is given by (9.39).

The right-hand sides of (9.43) and (9.44) are the same as those of (4.36) and (4.37) and hence are given by (4.43) and (4.44). For convenience, they are repeated below:

$$\sum_{l\geq 1}\left[F_{100}G_{l-1}^{10}+F_{010}G_{l-1}^{01}+F_{001}G_{l-1}^{00}+\sum_{n=2}^{l}\sum_{\substack{i+j+k=n\\ i,j,k\geq 0}}F_{ijk}G_{l-n}^{ij}\right]N_l v^{[l]} \tag{9.46}$$

and

$$\sum_{l\geq 1}\left[H_{100}G_{l-1}^{10}+H_{010}G_{l-1}^{01}+H_{001}G_{l-1}^{00}+\sum_{n=2}^{l}\sum_{\substack{i+j+k=n\\ i,j,k\geq 0}}H_{ijk}G_{l-n}^{ij}\right]N_l v^{[l]}. \tag{9.47}$$

Thus, we have, for $l \geq 1$,

$$\begin{aligned}\sum_{k=1}^{l}X_k M_k \xi_{k,l} N_l v^{[l]}\\ &= \left[F_{100}G_{l-1}^{10}+F_{010}G_{l-1}^{01}+F_{001}G_{l-1}^{00}+\sum_{n=2}^{l}\sum_{\substack{i+j+k=n\\ i,j,k\geq 0}}F_{ijk}G_{l-n}^{ij}\right]N_l v^{[l]},\\ 0 &= \left[H_{100}G_{l-1}^{10}+H_{010}G_{l-1}^{01}+H_{001}G_{l-1}^{00}+\sum_{n=2}^{l}\sum_{\substack{i+j+k=n\\ i,j,k\geq 0}}H_{ijk}G_{l-n}^{ij}\right]N_l v^{[l]}.\end{aligned}$$

Equating the coefficients of $v^{[l]}$ on both sides of the above two equations, and using $G_{l-1}^{10} = \delta_{1,l} = X_l M_l$, $G_{l-1}^{01} = \lambda_{1,l} = U_l M_l$, $G_0^{00} = 1$, $G_{l-1}^{00} = 0, l > 1$ along with the fact that $M_l N_l$ is an identity matrix completes the proof. $\square$

An examination of equations (9.36) to (9.42) shows that E_l and F_l depend only on $X_1, \ldots, X_{l-1}$ and $U_1, \ldots, U_{l-1}$. Therefore, equation (9.36) provides an iterative sequence of linear matrix equations.

Lemma 9.11. *There exists a solution (unique if $p = m$) of equations (9.36) for any E_l and F_l, $l = 1, 2, \ldots$, if and only if the plant satisfies the following assumption.*

Assumption 9.4.

$$\text{rank}\begin{bmatrix}\frac{\partial f}{\partial x}(0,0,0)-\lambda I & \frac{\partial f}{\partial u}(0,0,0)\\ \frac{\partial h}{\partial x}(0,0,0) & \frac{\partial h}{\partial u}(0,0,0)\end{bmatrix} = n + p \tag{9.48}$$

for all λ given by

$$\{\,\lambda \mid \lambda = \lambda_1^{l_1}\times\lambda_2^{l_2}\times\cdots\times\lambda_q^{l_q},\ \ l_1+\cdots+l_q = l,\ \ l_1,\ldots,l_q = 0,1,\ldots,l\,\}, \tag{9.49}$$

where $\lambda_1, \ldots, \lambda_q$ are eigenvalues of the matrix $\frac{\partial a}{\partial v}(0)$.

Proof. For a given l, equations (9.36) actually take the same form as the linear regulator equations (1.108). Thus, by Theorem 1.9, equations (9.36) have a solution for any E_l and F_l if and only if equality (9.48) holds for all λ in the spectrum of

$$A^{[l]} \stackrel{def}{=} M_l A_1^{(l)} N_l. \tag{9.50}$$

We now show that the eigenvalues of $A^{[l]}$ are precisely those described by (9.49). To this end, again define P^l as the vector space of all homogeneous polynomials in $v_1, \ldots, v_q$ of degree l; then the components of $v^{[l]}$ give a basis of P^l. Also define a linear mapping $L_{A_1 v} : P^l \to P^l$ such that, for each $\phi \in P^l$,

$$L_{A_1 v}(\phi) = \phi(A_1 v). \tag{9.51}$$

Note that

$$(A_1 v)^{[l]} = M_l (A_1 v)^{(l)} = M_l (A_1)^{(l)} v^{(l)} = M_l (A_1)^{(l)} N_l v^{[l]} = A^{[l]} v^{[l]}. \tag{9.52}$$

Thus $(A^{[l]})^T$ is the matrix of the linear mapping $L_{A_1 v} : P^l \to P^l$ under the ordered basis

$$\{v_1^l,\ v_1^{l-1} v_2, \ldots, v_1^{l-1} v_q,\ v_1^{l-2} v_2^2,\ v_1^{l-2} v_2 v_3, \ldots, v_1^{l-2} v_2 v_q, \ldots, v_q^l\}. \tag{9.53}$$

Thus, the spectrum of $A^{[l]}$ is the same as that of the linear mapping (9.51).

Now let the Jordan canonical form of A_1 be

$$A_1 = \begin{bmatrix} J_1 & 0 & \cdots & 0 \\ 0 & J_2 & \cdots & 0 \\ \cdots & \cdots & \cdots & \cdots \\ 0 & 0 & \cdots & J_k \end{bmatrix}_{q \times q}, \tag{9.54}$$

where

$$J_i = \begin{bmatrix} \lambda_i & 1 & 0 & \cdots & 0 \\ 0 & \lambda_i & 1 & \cdots & 0 \\ \cdots & \cdots & \cdots & \cdots & \cdots \\ 0 & 0 & 0 & \cdots & \lambda_i \end{bmatrix}_{n_i \times n_i}$$

is an $n_i \times n_i$ Jordan block with eigenvalue λ_i. Suppose the generalized row eigenvectors of A_1 are

$$\zeta_{11}, \zeta_{12}, \ldots, \zeta_{1n_1}, \zeta_{21}, \ldots, \zeta_{k1}, \ldots, \zeta_{kn_k}, \tag{9.55}$$

which satisfy

$$\zeta_{ij} A_1 = \begin{cases} \lambda_i \zeta_{ij}, & j = n_i, \\ \lambda_i \zeta_{ij} + \zeta_{i(j+1)}, & j < n_i. \end{cases} \tag{9.56}$$

Clearly,

$$\left\{ (\zeta_{11} v)^{\alpha_{11}} (\zeta_{12} v)^{\alpha_{12}} \cdots (\zeta_{k1} v)^{\alpha_{k1}} \cdots (\zeta_{kn_k} v)^{\alpha_{kn_k}} \;\middle|\; \sum_{i=1}^{k} \sum_{j=1}^{n_i} \alpha_{ij} = l \right\} \tag{9.57}$$

also constitutes a basis for P^l. Furthermore, for $j = n_i$,

$$L_{A_1 v}((\zeta_{ij} v)^s) = \lambda_i^s (\zeta_{ij} v)^s, \tag{9.58}$$

and for $j < n_i$,

$$L_{A_1 v}((\zeta_{ij} v)^s) = (\zeta_{ij} A_1 v)^s \tag{9.59}$$

$$= (\lambda_i \zeta_{ij} v + \zeta_{i(j+1)} v)^s$$

$$= \lambda_i^s (\zeta_{ij} v)^s + s(\lambda_i \zeta_{ij} v)^{s-1} \zeta_{i(j+1)} v + \cdots + (\zeta_{i(j+1)} v)^s. \tag{9.60}$$

Now define an order on (9.57) in the following "lexicographic" way:

$$(\zeta_{11} v)^{\alpha_{11}} \cdots (\zeta_{kn_k} v)^{\alpha_{kn_k}} > (\zeta_{11} v)^{\beta_{11}} \cdots (\zeta_{kn_k} v)^{\beta_{kn_k}}$$

if and only if there exist positive integers i_0 and j_0 ($\leq n_{i_0}$) such that

$$\alpha_{i_0 j_0} < \beta_{i_0 j_0}$$

and

$$\alpha_{ij} = \beta_{ij}$$

if $i < i_0$, $j \leq n_i$ or $i = i_0$, $j < j_0$. Then (9.57) constitutes an ordered basis of P^l. Using (9.58) and (9.60) gives

$$L_{A_1 v}((\zeta_{11} v)^{\alpha_{11}} \cdots (\zeta_{kn_k} v)^{\alpha_{kn_k}})$$

$$= (\zeta_{11} A_1 v)^{\alpha_{11}} \cdots (\zeta_{kn_k} A_1 v)^{\alpha_{kn_k}}$$

$$= \left(\lambda_1^{\sum_{j=1}^{n_1} \alpha_{1j}} \times \lambda_2^{\sum_{j=1}^{n_2} \alpha_{2j}} \times \cdots \times \lambda_k^{\sum_{j=1}^{n_k} \alpha_{kj}}\right) (\zeta_{11} v)^{\alpha_{11}} \cdots (\zeta_{kn_k} v)^{\alpha_{kn_k}}$$

$$+ \text{ terms greater than } (\zeta_{11} v)^{\alpha_{11}} \cdots (\zeta_{kn_k} v)^{\alpha_{kn_k}}.$$

Thus, the matrix of the linear mapping $L_{A_1 v}$ on P^l is upper triangular, with the diagonal elements being

$$\lambda = \lambda_1^{\sum_{j=1}^{n_1} \alpha_{1j}} \times \lambda_2^{\sum_{j=1}^{n_2} \alpha_{2j}} \times \cdots \times \lambda_k^{\sum_{j=1}^{n_k} \alpha_{kj}}.$$

Therefore, the eigenvalues of $L_{A_1 v}$ on P^l are exactly given by equation (9.49). □

Remark 9.12. If the solution of equations (9.36) is such that (9.34) has a positive convergent radius, (9.34) is an exact solution of equations (9.16) in power series form. In particular, if the solution of equations (9.16) is a polynomial in $v^{[l]}$, then Lemma 9.10 gives an approach to exactly solve equations (9.16). Note that equation (9.48) represents the constraints on the transmission zeros of the Jacobian linearization of the plant which can be viewed as the discrete-time counterpart of the transmission zeros condition for the continuous-time output regulation problem, as studied in Chapter 4. ∎

Assume that the transmission zeros condition in equation (9.48) holds up to some positive integer k. Let

$$\mathbf{x}^{(k)}(v) = \sum_{l=1}^{k} X_l v^{[l]}, \quad \mathbf{u}^{(k)}(v) = \sum_{l=1}^{k} U_l v^{[l]}. \tag{9.61}$$

Then, it is not difficult to see from the proof of Lemma 9.10 that there exist degree k polynomials $\mathbf{x}^{(k)}(v)$ and $\mathbf{u}^{(k)}(v)$ such that equations (9.29) are satisfied. By Theorem 9.9, we immediately obtain the following sufficient conditions for the solvability of the kth-order nonlinear output regulation problem.

Theorem 9.13.

(i) *Under Assumptions 9.1, 9.2, and 9.4, for any integer k, the kth-order nonlinear output regulation problem is solvable by the state feedback control law of the form (9.30).*

(ii) *Under additional Assumption 9.3, the kth-order nonlinear output regulation problem is solvable by the measurement output feedback control law (9.31).*

9.3 Robust Output Regulation for Discrete-Time Uncertain Nonlinear Systems

Consider a discrete-time nonlinear system described by

$$\begin{aligned} x(t+1) &= f(x(t), u(t), v(t), w), \qquad x(0) = x_0, \quad t = 0, 1, 2, \dots, \\ e(t) &= h(x(t), u(t), v(t), w), \end{aligned} \tag{9.62}$$

where $x(t) \in \mathcal{R}^n$ is the plant state, $u(t) \in \mathcal{R}^m$ the plant input, $e(t) \in \mathcal{R}^p$ the plant output representing the tracking error, $w \in \mathcal{R}^{n_w}$ the plant uncertain parameters, and $v(t) \in \mathcal{R}^q$ the exogenous signal representing the disturbance and/or the reference input. Again, it is assumed that $v(t)$ is generated by the autonomous system (9.2).

The class of control laws is described by

$$\begin{aligned} u(t) &= k(x(t), v(t), z(t)), \\ z(t+1) &= g(z(t), e(t)), \quad t = 0, 1, 2, \dots, \end{aligned} \tag{9.63}$$

where $z(t)$ is the compensator state vector of dimension n_z to be specified later. The above controller encompasses three cases.

3. **Dynamic State Feedback:** When $v(t)$ does not appear in (9.63), that is,

$$\begin{aligned} u(t) &= k(x(t), z(t)), \\ z(t+1) &= g(z(t), e(t)). \end{aligned} \tag{9.64}$$

4. **Dynamic Output Feedback:** When $x(t)$ and $v(t)$ do not appear in (9.63), that is,

$$\begin{aligned} u(t) &= k(z(t)), \\ z(t+1) &= g(z(t), e(t)). \end{aligned} \tag{9.65}$$

5. **Dynamic Output Feedback with Feedforward:** When $x(t)$ does not appear in (9.63), that is,

$$\begin{aligned} u(t) &= k(z(t), v(t)), \\ z(t+1) &= g(z(t), e(t)). \end{aligned} \tag{9.66}$$

Letting $x_c = \mathrm{col}(x, z)$, the resulting closed-loop system can be written as

$$\begin{aligned} x_c(t+1) &= f_c(x_c(t), v(t), w), \quad x_c(0) = x_{c0}, \; t = 0, 1, 2, \ldots, \\ v(t+1) &= a(t), \\ e(t) &= h_c(x_c(t), v(t), w), \end{aligned} \tag{9.67}$$

where

$$\begin{aligned} f_c(x_c, v, w) &= \begin{bmatrix} f(x, k(x, v, z), v, w) \\ g(z, h(x, k(x, v, z), v, w)) \end{bmatrix}, \\ h_c(x_c, v, w) &= h(x, k(x, v, z), v, w). \end{aligned} \tag{9.68}$$

For simplicity, all the functions involved in this setup are assumed to be sufficiently smooth and defined globally on the appropriate Euclidean spaces, with the value zero at the respective origins. Throughout this chapter, we use V and W to denote some open neighborhoods of the origins of $\mathcal{R}^q$ and $\mathcal{R}^{n_w}$, respectively. For convenience of presentation, we allow V and W to be made arbitrarily small.

The discrete-time kth-order robust output regulation problem and the discrete-time robust output regulation problem are formulated as follows.

Discrete-Time kth-Order Robust Nonlinear Output Regulation Problem (DKRNORP). Find a controller of the form (9.63) such that the closed-loop system (9.67) satisfies the following properties.

Property 9.5. The matrix $\frac{\partial f_c}{\partial x_c}(0, 0, 0)$ is Schur.

Property 9.6. For all sufficiently small x_{c0}, v_0, and w, the trajectory $x_c(t)$ of the closed-loop system (9.67) satisfies

$$\lim_{t \to \infty} (e(t) - o^k(v(t))) = 0, \tag{9.69}$$

where k is some given positive integer and $o^k(v)$ is some sufficiently smooth function of v zero up to kth order.

Discrete-Time Robust Nonlinear Output Regulation Problem (DRNORP). Find a controller of the form (9.63) such that the closed-loop system (9.67) satisfies Property 9.5 and the following:

Property 9.7. For all sufficiently small x_{c0}, v_0, and w, the trajectory $x_c(t)$ of the closed-loop system (9.67) satisfies

$$\lim_{t \to \infty} e(t) = 0. \tag{9.70}$$

The two problems defined above are discrete-time counterparts of the kth-order robust output regulation problem and the robust output regulation problem for continuous-time systems described in Chapter 5. They can also be viewed as extensions of the kth-order discrete-time output regulation problem and the discrete-time output regulation problem studied in the last two sections by further taking into account the model uncertainty. Viewing w as being generated by an exosystem of the form $w(t+1) = w(t)$, a solvability condition can be obtained by slightly modifying Lemma 9.4, as follows.

Lemma 9.14. *Assume the exosystem satisfies Assumption 9.1′, and the closed-loop system (9.67) has Property 9.5. Then*

(i) *The closed-loop system (9.67) has Property 9.6 if and only if*

> ***Property 9.8.*** There exists a sufficiently smooth function $\mathbf{x}_c^{(k)}(v, w)$ with $\mathbf{x}_c^{(k)}(0, 0) = 0$ that satisfies, for $v \in V$ and $w \in W$, the following algebraic equations:
>
> $$\mathbf{x}_c^{(k)}(a(v), w) = f_c(\mathbf{x}_c^{(k)}(v, w), v, w), \tag{9.71}$$
>
> $$o^k(v) = h_c(\mathbf{x}_c^{(k)}(v, w), v, w). \tag{9.72}$$

(ii) *The closed-loop system (9.67) has Property 9.7 if and only if*

> ***Property 9.9.*** There exists a sufficiently smooth function $\mathbf{x}_c(v, w)$ with $\mathbf{x}_c(0, 0) = 0$ that satisfies, for $v \in V$ and $w \in W$, the following algebraic equations:
>
> $$\mathbf{x}_c(a(v), w) = f_c(\mathbf{x}_c(v, w), v, w), \tag{9.73}$$
>
> $$0 = h_c(\mathbf{x}_c(v, w), v, w). \tag{9.74}$$

Various assumptions needed for the solvability of the above two problems are listed as follows.

Assumption 9.5. There exist sufficiently smooth functions $\mathbf{x}(v, w)$ and $\mathbf{u}(v, w)$ with $\mathbf{x}(0, 0) = 0$ and $\mathbf{u}(0, 0) = 0$ such that, for $v \in V$, $w \in W$,

$$\begin{aligned} \mathbf{x}(a(v), w) &= f(\mathbf{x}(v, w), \mathbf{u}(v, w), v, w), \\ 0 &= h(\mathbf{x}(v, w), \mathbf{u}(v, w), v, w), \end{aligned} \tag{9.75}$$

where $V \subset \mathcal{R}^q$, $W \subset \mathcal{R}^{n_w}$ are some open neighborhoods of the origin of $\mathcal{R}^q$ and $\mathcal{R}^{n_w}$, respectively.

Assumption 9.6. The pair $\left(\frac{\partial f}{\partial x}(0, 0, 0, 0), \frac{\partial f}{\partial u}(0, 0, 0, 0)\right)$ is stabilizable.

Assumption 9.7. The pair $\left(\frac{\partial h}{\partial x}(0, 0, 0, 0), \frac{\partial f}{\partial x}(0, 0, 0, 0)\right)$ is detectable.

Assumption 9.8. For $l = 1, 2, \ldots,$

$$\operatorname{rank} \begin{bmatrix} \frac{\partial f}{\partial x}(0, 0, 0, 0) - \lambda I & \frac{\partial f}{\partial u}(0, 0, 0, 0) \\ \frac{\partial h}{\partial x}(0, 0, 0, 0) & \frac{\partial h}{\partial u}(0, 0, 0, 0) \end{bmatrix} = n + p \tag{9.76}$$

for all λ given by

$$\begin{aligned} \{\lambda \mid \lambda = \lambda_1^{l_1} \times \lambda_2^{l_2} \times \cdots \times \lambda_q^{l_q},\ & l_1 + l_2 + \cdots + l_q = l, \\ & l_1, l_2, \ldots, l_q = 0, 1, 2, \ldots\}, \end{aligned} \tag{9.77}$$

where $\lambda_1, \lambda_2, \ldots, \lambda_q$ are eigenvalues of the matrix $\frac{\partial a}{\partial v}(0)$.

We will employ a discrete-time version of the internal model principle to solve the above two problems. Like continuous-time systems, we can also convert the discrete-time kth-order robust output regulation problem of the given nonlinear plant with the given exosystem into a discrete-time robust output regulation problem of a linearized plant with a k-fold exosystem. For this purpose, let

$$\begin{aligned}
f(x,u,v,w) &= A(w)x + B(w)u + E(w)v + f_2(x,u,v,w),\\
h(x,u,v,w) &= C(w)x + D(w)u + F(w)v + h_2(x,u,v,w),\\
a(v) &= A_1 v + a_2(v),\\
f(x,k(x,v,z),v,w) &= A_c(w)x + B_c(w)z + E_c(w)v + f_{c2}(x_c,v,w),\\
h_c(x_c,v,w) &= C_c(w)x + D_c(w)z + F_c(w)v + h_{c2}(x_c,v,w),
\end{aligned}$$

where $A(w)$, $B(w)$, $E(w)$, and so forth are given by

$$A(w) = \frac{\partial f}{\partial x}(0,0,0,w),\quad B(w) = \frac{\partial f}{\partial u}(0,0,0,w),\quad E(w) = \frac{\partial f}{\partial v}(0,0,0,w),\ldots.$$

For convenience, in what follows, we will use the shorthand notation A, B, E, and so forth to denote $A(0)$, $B(0)$, $E(0)$, and so forth.

Now, assume a control law of the form (9.63) with $g(z,e) = \mathcal{G}_1 z + \mathcal{G}_2 e$ that makes the closed-loop system (9.67) satisfy Property 9.5. Then, by Theorem 2.31, there exists a locally defined sufficiently smooth function $\bar{\mathbf{x}}_c(v,w)$ with $\bar{\mathbf{x}}_c(0,0) = 0$ such that, for $v \in V$, $w \in W$,

$$\bar{\mathbf{x}}_c(a(v),w) = f_c(\bar{\mathbf{x}}_c(v,w),v,w). \tag{9.78}$$

By partitioning $\bar{\mathbf{x}}_c(v,w) = \mathrm{col}(\bar{\mathbf{x}}(v,w), \bar{\mathbf{z}}(v,w))$, (9.78) becomes

$$\begin{aligned}
\bar{\mathbf{x}}(a(v),w) &= f(\bar{\mathbf{x}}(v,w), k(\bar{\mathbf{x}}(v,w),v,\bar{\mathbf{z}}(v,w)),v,w),\\
\bar{\mathbf{z}}(a(v),w) &= \mathcal{G}_1\bar{\mathbf{x}}(v,w) + \mathcal{G}_2\mathbf{e}(v,w),
\end{aligned} \tag{9.79}$$

where

$$\mathbf{e}(v,w) = h(\bar{\mathbf{x}}(v,w), k(\bar{\mathbf{x}}(v,w),v,\bar{\mathbf{z}}(v,w)),v,w). \tag{9.80}$$

Express $\bar{\mathbf{x}}(v,w)$, $\bar{\mathbf{z}}(v,w)$, and $\mathbf{e}(v,w)$ uniquely as

$$\bar{\mathbf{x}}(v,w) = \sum_{l=1}^{k} X_{lw} v^{[l]} + o^k(v),$$

$$\bar{\mathbf{z}}(v,w) = \sum_{l=1}^{k} Z_{lw} v^{[l]} + o^k(v),$$

$$\mathbf{e}(v,w) = \sum_{l=1}^{k} Y_{lw} v^{[l]} + o^k(v), \tag{9.81}$$

where (X_{lw}, Z_{lw}) are constant matrices of appropriate dimensions depending perhaps on w. In analogy of the derivation of equation (9.36), substituting (9.81) into (9.79) and (9.80), expanding (9.79) and (9.80) into power series in $v^{[l]}$, and identifying the coefficients of $v^{[l]}$ yield, for $l = 1, 2, \ldots, k$,

$$\begin{aligned} X_{lw}A^{[l]} &= A_c(w)X_{lw} + B_c(w)Z_{lw} + E_{lw}, \\ Z_{lw}A^{[l]} &= \mathcal{G}_1 Z_{lw} + \mathcal{G}_2(C_c(w)X_{lw} + D_c(w)Z_{lw} + F_{lw}), \end{aligned} \tag{9.82}$$

and

$$Y_{lw} = C_c(w)X_{lw} + D_c(w)Z_{lw} + F_{lw}, \tag{9.83}$$

where, for $l = 1, \ldots, k$, $A^{[l]} = M_l A_1^{(l)} N_l$, $(E_{1w}, F_{1w}) = (E(w), F(w))$, and, for $l = 2, 3, \ldots, k$, (E_{lw}, F_{lw}) depend only on $X_{1w}, \cdots, X_{(l-1)w}$ and $Z_{1w}, \cdots, Z_{(l-1)w}$.

Since, for the given l, equations (9.82) and (9.83) take the same form as (1.118) and (1.119), the fact that the closed-loop system has Property 9.5 means that the matrix

$$\begin{bmatrix} A_c & B_c \\ \mathcal{G}_2 C_c & \mathcal{G}_1 + \mathcal{G}_2 D_c \end{bmatrix}$$

is Schur. Thus, by Lemma 1.38, $Y_{lw} = 0$ for all $w \in W$ if the pair $(\mathcal{G}_1, \mathcal{G}_2)$ incorporates a p-copy internal model of the matrix $A^{[l]}$. Moreover, let

$$\mathcal{A}_{kf} = \begin{bmatrix} A^{[1]} & 0 & \cdots & 0 \\ 0 & A^{[2]} & \cdots & 0 \\ \vdots & \vdots & \vdots & \vdots \\ 0 & 0 & \cdots & A^{[k]} \end{bmatrix}. \tag{9.84}$$

If the pair $(\mathcal{G}_1, \mathcal{G}_2)$ incorporates a p-copy internal model of the matrix $\mathcal{A}_{kf}$, it also incorporates a p-copy internal model of all the matrices $A^{[l]}$ for $l = 1, \ldots, k$. Therefore, the control law renders $Y_{lw} = 0$ for all $l = 1, \ldots, k$, thereby solving the discrete-time kth-order robust output regulation problem. As a result, we have the following result.

Lemma 9.15. *Under Assumption 9.1, assume that a control law of the form (9.63) with $g(z, e) = \mathcal{G}_1 z + \mathcal{G}_2 e$ renders the closed-loop system (9.67) into Property 9.5. Then,*

(i) *for any $l \geq 1$, $Y_{lw} = 0$ for all $w \in W$ if the pair $(\mathcal{G}_1, \mathcal{G}_2)$ incorporates a p-copy internal model of the matrix $A^{[l]}$;*

(ii) *the kth-order robust output regulation problem is solved if the pair $(\mathcal{G}_1, \mathcal{G}_2)$ incorporates a p-copy internal model of the matrix $\mathcal{A}_{kf}$.*

Now consider the linear approximation of discrete-time nonlinear system (9.62) as follows:

$$\begin{aligned} x(t+1) &= A(w)x(t) + B(w)u(t) + E(w)v(t), \\ e(t) &= C(w)x(t) + D(w)u(t) + F(w)v(t). \end{aligned} \tag{9.85}$$

Let the pair (G_1, G_2) be a minimal p-copy internal model of the matrix $\mathcal{A}_{kf}$. Since the eigenvalues of the matrix $A^{[l]}$ are given by

$$\{\lambda \mid \lambda = \lambda_1^{l_1} \times \cdots \times \lambda_q^{l_q},\ l_1 + \cdots + l_q = l,\ l_1, \ldots, l_q = 0, 1, 2, \ldots\},$$

where $\lambda_1, \ldots, \lambda_q$ are eigenvalues of A_1, under Assumption 9.8, G_1 satisfies the following transmission zeros condition: for all $\lambda \in \sigma(G_1)$,

$$\text{rank}\begin{bmatrix} A - \lambda I & B \\ C & D \end{bmatrix} = n + p. \tag{9.86}$$

By Lemma 1.37, the pair

$$\left(\begin{bmatrix} A & 0 \\ G_2 C & G_1 \end{bmatrix}, \begin{bmatrix} B \\ G_2 D \end{bmatrix}\right) \tag{9.87}$$

is stabilizable. Thus, there exist feedback gains K_1 and K_2 such that the matrix

$$\begin{bmatrix} A + BK_1 & BK_2 \\ G_2(C + DK_1) & G_1 + G_2 D K_2 \end{bmatrix} \tag{9.88}$$

is Schur; that is, there exists a static state feedback control law $u(t) = K_1 x(t) + K_2 z(t)$ that exponentially stabilizes the following system:

$$\begin{aligned} x(t+1) &= A(w)x(t) + B(w)u(t) + E(w)v(t), \\ z(t+1) &= G_1 z(t) + G_2 e(t), \\ e(t) &= C(w)x(t) + D(w)u(t) + F(w)v(t). \end{aligned} \tag{9.89}$$

That is, the following dynamic state feedback control law:

$$\begin{aligned} u(t) &= K_1 x(t) + K_2 z(t), \\ z(t+1) &= G_1 z(t) + G_2 e(t), \end{aligned} \tag{9.90}$$

solves the kth-order robust output regulation problem of the discrete-time nonlinear system (9.62).

Next, assume that (9.90) solves the kth-order robust output regulation problem of the original plant. Under Assumption 9.7, there exists an L such that $A - LC$ is Schur. Let $K = [K_1,\ K_2]$,

$$\mathcal{G}_1 = \begin{bmatrix} A + BK_1 - L(C + DK_1) & (B - LD)K_2 \\ 0 & G_1 \end{bmatrix}, \quad \mathcal{G}_2 = \begin{bmatrix} L \\ G_2 \end{bmatrix}. \tag{9.91}$$

Then, by exactly the same argument as in the proof of part (ii) of Theorem 5.7 for the continuous-time case, the dynamic output feedback control law of the form

$$\begin{aligned} u(t) &= Kz(t), \\ z(t+1) &= \mathcal{G}_1 z(t) + \mathcal{G}_2 e(t). \end{aligned} \tag{9.92}$$

solves the kth-order robust output regulation problem for the discrete-time nonlinear system (9.62).

In summary, we have the following discrete-time counterpart of Theorem 5.7.

Theorem 9.16.

(i) *Under Assumptions 9.1, 9.6, and 9.8, for any positive integer k, the discrete-time kth-order robust output regulation problem is solvable by a linear state feedback controller of the form (9.90), where* (G_1, G_2) *is a minimal p-copy internal model of the matrix* $\mathcal{A}_{kf}$.

(ii) *Under Assumptions 9.1 and 9.6 to 9.8, for any positive integer k, the discrete-time kth-order robust output regulation problem is solvable by a linear output feedback controller of the form (9.92), where* $(\mathcal{G}_1, \mathcal{G}_2)$ *is given by (9.91).*

Remark 9.17. Similar to the continuous-time case, if $v(t)$ satisfies $v(t+1) = A_1 v(t)$, then we have $v^{[l]}(t+1) = A^{[l]} v^{[l]}(t)$. Let

$$v_{kf} = \begin{bmatrix} v^{[1]} \\ v^{[2]} \\ \vdots \\ v^{[k]} \end{bmatrix}. \tag{9.93}$$

Then the matrix $\mathcal{A}_{kf}$ is such that

$$v_{kf}(t+1) = \mathcal{A}_{kf} v_{kf}(t). \tag{9.94}$$

System (9.94) can be considered as a generalized exosystem which generates not only the exogenous signal v (when $a(v) = A_1 v$), but also the higher order terms of the exogenous signal v up to order k. We call system (9.94) a discrete-time k-fold exosystem. Now consider the following linear system:

$$\begin{aligned} x(t+1) &= A(w)x(t) + B(w)u(t) + E(w)v(t), \\ v_{kf}(t+1) &= \mathcal{A}_{kf} v_{kf}(t), \\ e(t) &= C(w)x(t) + D(w)u(t) + F(w)v(t). \end{aligned} \tag{9.95}$$

Lemma 9.15 effectively asserts that designing a discrete-time kth-order robust servoregulator for a discrete-time nonlinear system (9.62) is equivalent to designing a linear discrete-time robust servoregulator for the linear system (9.95). Theorem 9.16 further gives the conditions under which the above linear discrete-time robust output regulation problem is solvable. ∎

Next, we will further show that, under some additional assumptions on the solution of the discrete regulator equations, a control law solving the discrete-time kth-order robust output regulation problem for the given plant (9.62) with the exosystem (9.2) also solves the discrete-time robust output regulation problem for the same plant and the exosystem.

Lemma 9.18. *Under Assumption 9.1, suppose a control law of the form (9.63) is such that the closed-loop system satisfies Property 9.5. Then the control law solves the robust output regulation problem if there exist sufficiently smooth functions* $(\mathbf{x}(v, w), \mathbf{u}(v, w), \mathbf{z}(v, w))$

locally defined in $v \in V$, $w \in W$ *with* $(\mathbf{x}(0,0), \mathbf{u}(0,0), \mathbf{z}(0,0)) = (0,0,0)$ *such that* $\mathbf{x}(v,w)$ *and* $\mathbf{u}(v,w)$ *are the solution of the discrete-time nonlinear regulator equations (9.75) and* $\mathbf{z}(v,w)$ *satisfies*

$$\mathbf{u}(v,w) = k(\mathbf{x}(v,w), v, \mathbf{z}(v,w)), \tag{9.96}$$

$$\mathbf{z}(a(v),w) = g(\mathbf{z}(v,w), 0). \tag{9.97}$$

Proof. By Lemma 9.14, we only need to show that there exists a sufficiently smooth function $\mathbf{x}_c(v,w)$ with $\mathbf{x}_c(0,0) = 0$ that satisfies (9.73) and (9.74). To this end, define $\mathbf{x}_c(v,w) = \text{col}(\mathbf{x}(v,w), \mathbf{z}(v,w))$. Using (9.68) yields

$$h_c(\mathbf{x}_c(v,w), v, w) = h(\mathbf{x}(v,w), k(\mathbf{x}(v,w), v, \mathbf{z}(v,w)), v, w), \tag{9.98}$$

$$f_c(\mathbf{x}_c(v,w), v, w) = \begin{bmatrix} f(\mathbf{x}(v,w), k(\mathbf{x}(v,w), v, \mathbf{z}(v,w)), v, w) \\ g(\mathbf{z}(v,w), h_c(\mathbf{x}_c(v,w), v, w)) \end{bmatrix}. \tag{9.99}$$

Substituting (9.96) into (9.98) and (9.99) gives

$$h_c(\mathbf{x}_c(v,w), v, w) = h(\mathbf{x}(v,w), \mathbf{u}(v,w), v, w), \tag{9.100}$$

$$f_c(\mathbf{x}_c(v,w), v, w) = \begin{bmatrix} f(\mathbf{x}(v,w), \mathbf{u}(v,w), v, w) \\ g(\mathbf{z}(v,w), h_c(\mathbf{x}_c(v,w), v, w)) \end{bmatrix}. \tag{9.101}$$

Using the regulator equations (9.75) and equation (9.97) in (9.100) and (9.101) gives

$$\begin{aligned} h_c(\mathbf{x}_c(v,w), v, w) &= h(\mathbf{x}(v,w), \mathbf{u}(v,w), v, w) = 0, \\ f_c(\mathbf{x}_c(v,w), v, w) &= \begin{bmatrix} f(\mathbf{x}(v,w), \mathbf{u}(v,w), v, w) \\ g(\mathbf{z}(v,w), h_c(\mathbf{x}_c(v,w), v, w)) \end{bmatrix} \\ &= \begin{bmatrix} \mathbf{x}(a(v),w) \\ g(\mathbf{z}(v,w), 0) \end{bmatrix} = \begin{bmatrix} \mathbf{x}(a(v),w) \\ \mathbf{z}(a(v),w) \end{bmatrix} = \mathbf{x}_c(a(v),w). \quad \square \end{aligned}$$

To solve the discrete-time robust output regulation problem, we need to impose an additional assumption on the exosystem (9.2).

Assumption 9.9. $a(v) = A_1 v$ for some matrix A_1, and all the eigenvalues of A_1 are simple and lie on the unit circle.

Theorem 9.19.

(i) *Under Assumptions 9.5, 9.6, 9.8, and 9.9, assume the solution* $\mathbf{x}(v,w)$ *and* $\mathbf{u}(v,w)$ *of the discrete-time regulator equations (9.75) are degree k polynomials in* v. *Then if the state feedback controller (9.90) solves the discrete-time kth-order robust output regulation problem, it also solves the discrete-time robust output regulation problem.*

(ii) *Under Assumptions 9.5 to 9.9, assume the solution* $\mathbf{u}(v,w)$ *of the discrete-time regulator equations (9.75) is a degree k polynomial in* v. *Then if the output feedback controller (9.92) solves the discrete-time kth-order robust output regulation problem, it also solves the discrete-time robust output regulation problem.*

Proof. Part (i). Assume that the controller (9.90) solves the discrete-time kth-order robust output regulation problem. By Lemma 9.18, it suffices to show that there exists a sufficiently smooth function $\mathbf{z}(v, w)$ such that

$$\mathbf{u}(v, w) = K_1\mathbf{x}(v, w) + K_2\mathbf{z}(v, w), \tag{9.102}$$

$$\mathbf{z}(A_1 v, w) = G_1\mathbf{z}(v, w). \tag{9.103}$$

To this end, let $\bar{\mathbf{x}}(v, w)$ and $\bar{\mathbf{z}}(v, w)$ be sufficiently smooth functions satisfying (9.79) with $a(v) = A_1 v$, and let $\mathbf{e}(v, w)$ be as defined in (9.80). Again, express $\bar{\mathbf{x}}(v, w)$, $\bar{\mathbf{z}}(v, w)$, and $\mathbf{e}(v, w)$ as in (9.81). Since the controller (9.90) solves the discrete-time kth-order robust output regulation problem, for $l = 1, \ldots, k$, X_{lw} and Z_{lw} satisfy (9.82) and (9.83) with $Y_{lw} = 0$, where

$$A_c(w) = A(w) + B(w)K_1, \quad B_c(w) = B(w)K_2,$$
$$C_c(w) = C(w) + D(w)K_1, \quad D_c(w) = D(w)K_2.$$

Let $U_{lw} = K_1 X_{lw} + K_2 Z_{lw}$. Then (9.82) and (9.83) imply, for $l = 1, \ldots, k$,

$$X_{lw}A^{[l]} = A(w)X_{lw} + B(w)U_{lw} + E_{lw},$$
$$0 = C(w)X_{lw} + D(w)U_{lw} + F_{lw}.$$

By Lemma 9.10, there exist sufficiently smooth functions $\mathbf{x}_k(v, w) = o^k(v)$ and $\mathbf{u}_k(v, w) = o^k(v)$ such that

$$\mathbf{x}(v, w) = \sum_{l=1}^{k} X_{lw}v^{[l]} + \mathbf{x}_k(v, w),$$

$$\mathbf{u}(v, w) = \sum_{l=1}^{k} U_{lw}v^{[l]} + \mathbf{u}_k(v, w).$$

However, by the assumption of this theorem, $\mathbf{x}(v, w)$ and $\mathbf{u}(v, w)$ are degree k polynomials in v, and thus

$$\mathbf{x}(v, w) = \sum_{l=1}^{k} X_{lw}v^{[l]},$$

$$\mathbf{u}(v, w) = \sum_{l=1}^{k} U_{lw}v^{[l]}.$$

Let

$$\mathbf{z}(v, w) = \sum_{l=1}^{k} Z_{lw}v^{[l]}.$$

Clearly, (9.102) is satisfied. Now using (9.82) and (9.83) yields

$$Z_{lw}A^{[l]} = G_1 Z_{lw}, \quad l = 1, 2, \ldots, k. \tag{9.104}$$

Multiplying (9.104) from the right by $v^{[l]}$ and then summarizing from $l = 1$ to k gives

$$\sum_{l=1}^{k} Z_{lw} A^{[l]} v^{[l]} = \sum_{l=1}^{k} G_1 Z_{lw} v^{[l]}. \tag{9.105}$$

Thus,

$$\begin{aligned}
\mathbf{z}(A_1 v, w) &= \sum_{l=1}^{k} Z_{lw} (A_1 v)^{[l]} = \sum_{l=1}^{k} Z_{lw} M_l (A_1 v)^{(l)} \\
&= \sum_{l=1}^{k} Z_{lw} M_l A_1^{(l)} N_l v^{[l]} = \sum_{l=1}^{k} Z_{lw} A^{[l]} v^{[l]} \\
&= \sum_{l=1}^{k} G_1 Z_{lw} v^{[l]} = G_1 \mathbf{z}(v, w).
\end{aligned}$$

Part (ii). The proof of part (ii) is almost the same as that of part (i). Assume that a controller of the form (9.92) solves the discrete-time kth-order robust output regulation problem. By Lemma 9.18, we need to show the existence of a sufficiently smooth function $\mathbf{z}(v, w)$ with $\mathbf{z}(0, 0) = 0$, which satisfies

$$\mathbf{u}(v, w) = K\mathbf{z}(v, w), \tag{9.106}$$

$$\mathbf{z}(A_1 v, w) = \mathcal{G}_1 \mathbf{z}(v, w). \tag{9.107}$$

Let $\bar{\mathbf{x}}(v, w)$ and $\bar{\mathbf{z}}(v, w)$ be sufficiently smooth functions satisfying (9.79), and $\mathbf{e}(v, w)$ be as defined in (9.80). Again, express $\bar{\mathbf{x}}(v, w)$, $\bar{\mathbf{z}}(v, w)$, and $\mathbf{e}(v, w)$ as in (9.81). Since the controller (9.92) solves the discrete-time kth-order robust output regulation problem, for $l = 1, \ldots, k$, X_{lw} and Z_{lw} satisfy (9.82) and (9.83) with $Y_{lw} = 0$, where

$$\begin{aligned}
A_c(w) &= A(w), & B_c(w) &= B(w)K, \\
C_c(w) &= C(w), & D_c(w) &= D(w)K.
\end{aligned}$$

Let $U_{lw} = KZ_{lw}$. Then (9.82) and (9.83) imply, for $l = 1, \ldots, k$,

$$\begin{aligned}
X_{lw} A^{[l]} &= A(w) X_{lw} + B(w) U_{lw} + E_{lw}, \\
0 &= C(w) X_{lw} + D(w) U_{lw} + F_{lw}.
\end{aligned}$$

Again, by Lemma 9.10, there exist sufficiently smooth functions $\mathbf{x}_k(v, w) = o^k(v)$ and $\mathbf{u}_k(v, w) = o^k(v)$ such that

$$\mathbf{x}(v, w) = \sum_{l=1}^{k} X_{lw} v^{[l]} + \mathbf{x}_k(v, w),$$

$$\mathbf{u}(v, w) = \sum_{l=1}^{k} U_{lw} v^{[l]} + \mathbf{u}_k(v, w).$$

However, by the assumption of this theorem, $\mathbf{u}(v, w)$ is a degree k polynomial in v, and thus

$$\mathbf{u}(v, w) = \sum_{l=1}^{k} U_{lw} v^{[l]}.$$

Let

$$\mathbf{z}(v, w) = \sum_{l=1}^{k} Z_{lw} v^{[l]}.$$

Clearly, (9.106) is satisfied. The proof of satisfaction of (9.107) is the same as that of (9.103) in part (i), and thus is omitted. □

If the exogenous signal v is available for control, it is possible to somehow relax the restriction on $\mathbf{x}(v, w)$ and $\mathbf{u}(v, w)$ as shown by the following theorem.

Theorem 9.20.

(i) *Under Assumptions 9.5, 9.6, 9.8, and 9.9, suppose that there exists some integer $k > 0$ such that $\mathbf{x}(v, w)$ and $\mathbf{u}(v, w)$ take the following form:*

$$\mathbf{x}(v, w) = \mathbf{x}^{[k]}(v, w) + \mathbf{x}_{hk}(v),$$
$$\mathbf{u}(v, w) = \mathbf{u}^{[k]}(v, w) + \mathbf{u}_{hk}(v),$$

where $\mathbf{x}^{[k]}(v, w)$ and $\mathbf{u}^{[k]}(v, w)$ are degree k polynomials of v with coefficients depending on w, and $\mathbf{x}_{hk}(v)$ and $\mathbf{u}_{hk}(v)$ are some sufficiently smooth functions of v, independent of w, vanishing at the origin together with their derivatives up to order k. If the state feedback controller (9.90) solves the discrete-time kth-order robust output regulation problem, then the following controller:

$$\begin{aligned} u(t) &= K_1(x(t) - \mathbf{x}_{hk}(v(t))) + K_2 z(t) + \mathbf{u}_{hk}(v(t)), \\ z(t+1) &= G_1 z(t) + G_2 e(t) \end{aligned} \tag{9.108}$$

solves the discrete-time robust output regulation problem.

(ii) *Under Assumptions 9.5 to 9.9, suppose that there exists some integer $k > 0$ such that $\mathbf{u}(v, w)$ takes the form of*

$$\mathbf{u}(v, w) = \mathbf{u}^{[k]}(v, w) + \mathbf{u}_{hk}(v).$$

If the output feedback controller (9.92) solves the discrete-time kth-order robust output regulation problem, then the following controller:

$$\begin{aligned} u(t) &= \mathcal{K} z(t) + \mathbf{u}_{hk}(v(t)), \\ z(t+1) &= \mathcal{G}_1 z(t) + \mathcal{G}_2 e(t) \end{aligned} \tag{9.109}$$

solves the discrete-time robust output regulation problem.

Proof. The proof of this theorem is almost the same as that of Theorem 5.14 and is thus omitted. □

9.4 The Inverted Pendulum on a Cart Example

In this section, we will consider the asymptotic tracking problem for the discretized model of the inverted pendulum on a cart system. The continuous-time model is given in equation (2.110). Discretizing the continuous-time model (2.110) via Euler's method with T as the sampling period gives the discrete-time model as follows:

$$\begin{aligned}
x_1(t+1) &= x_1(t) + Tx_2(t),\\
x_2(t+1) &= x_2(t) + \frac{T}{M+m(\sin x_3(t))^2}(u(t) + mlx_4^2(t)\sin x_3(t)\\
&\qquad - bx_2(t) - mg\cos x_3(t)\sin x_3(t)),\\
x_3(t+1) &= x_3(t) + Tx_4(t),\\
x_4(t+1) &= x_4(t) + \frac{T}{l(M+m(\sin x_3(t))^2)}((M+m)g\sin x_3(t) - u(t)\cos x_3(t)\\
&\qquad + bx_2(t)\cos x_3(t) - mlx_4^2(t)\sin x_3(t)\cos x_3(t)),\\
y(t) &= x_1(t).
\end{aligned} \tag{9.110}$$

Again, consider the asymptotic tracking of the output $y(t)$ to a sinusoidal function $y_d(t) = A_m\sin(\omega t)$. Thus the exosystem is given by

$$v(t+1) = a(v(t)) = A_1 v(t) \tag{9.111}$$

with

$$A_1 = \begin{bmatrix} \cos\omega & \sin\omega \\ -\sin\omega & \cos\omega \end{bmatrix}, \quad v(t) = \begin{bmatrix} v_1(t) \\ v_2(t) \end{bmatrix}, \quad v(0) = \begin{bmatrix} 0 \\ A_m \end{bmatrix}.$$

It is clear that $v_1(t) = A_m\sin(\omega t)$. Thus, we can define the error equation as follows:

$$e(t) = y(t) - v_1(t) = x_1(t) - v_1(t).$$

It can be verified that the matrix A_1 has two distinct eigenvalues, $\cos\omega \pm j\sin\omega$, which are clearly located on the unit circle. Thus, the exosystem satisfies Assumption 9.1.

If we consider the coefficient of viscous friction b as an uncertain parameter and assume that $b = b_0 + \Delta b$ with $b_0 = 12.98$ kg/sec, then the Jacobian linearization of the discrete-time inverted pendulum on a cart system (9.110) can be calculated as follows:

$$A = \frac{\partial f(0,0,0,0)}{\partial x} = \begin{bmatrix} 1 & T & 0 & 0 \\ 0 & 1-\frac{b_0T}{M} & -\frac{mgT}{M} & 0 \\ 0 & 0 & 1 & T \\ 0 & \frac{b_0T}{lM} & \frac{(M+m)gT}{lM} & 1 \end{bmatrix},$$

$$B = \frac{\partial f(0,0,0,0)}{\partial u} = \begin{bmatrix} 0 \\ \frac{T}{M} \\ 0 \\ -\frac{T}{lM} \end{bmatrix},$$

$$C = \frac{\partial h(0,0,0,0)}{\partial x} = [\,1 \quad 0 \quad 0 \quad 0\,],$$

$$D = \frac{\partial h(0,0,0,0)}{\partial u} = 0.$$

It is now possible to verify that the pair (A, B) is controllable, and none of the transmission zeros of the linearized plant are on the unit circle. Thus the plant also satisfies Assumptions 9.6 and 9.8. By Theorem 9.16, for any $k > 0$, the discrete-time kth-order robust output regulation problem is solvable by dynamic state feedback control. Of course, the nominal plant also satisfies Assumptions 9.1, 9.2, and 9.4, and thus the kth-order output regulation problem for this system is also solvable for any integer k assuming $b = b_0$. In what follows, we will design both a third-order state feedback servoregulator and a third-order state feedback robust servoregulator for this system.

Third-Order State Feedback Servoregulator: The discrete regulator equations associated with the inverted pendulum on a cart system are

$$
\begin{aligned}
\mathbf{x}_1(A_1 v) &= \mathbf{x}_1(v) + T\mathbf{x}_2(v),\\
\mathbf{x}_2(A_1 v) &= \mathbf{x}_2(v) + \frac{T}{M + m(\sin \mathbf{x}_3(v))^2}(\mathbf{u}(v) + ml\mathbf{x}_4^2(v)\sin \mathbf{x}_3(v)\\
&\qquad - b\mathbf{x}_2(v) - mg\cos \mathbf{x}_3(v)\sin \mathbf{x}_3(v)),\\
\mathbf{x}_3(A_1 v) &= \mathbf{x}_3(v) + T\mathbf{x}_4(v),\\
\mathbf{x}_4(A_1 v) &= \mathbf{x}_4(v) + \frac{T}{l(M + m(\sin \mathbf{x}_3(v))^2)}((M+m)g\sin \mathbf{x}_3(v) - \mathbf{u}(v)\cos \mathbf{x}_3(v)\\
&\qquad + b\mathbf{x}_2(v)\cos \mathbf{x}_3(v) - ml\mathbf{x}_4^2(v)\sin \mathbf{x}_3(v)\cos \mathbf{x}_3(v)),\\
0 &= \mathbf{x}_1(v) - v_1.
\end{aligned}
\tag{9.112}
$$

By an inspection, equations (9.112) can be partially solved as follows:

$$
\mathbf{x}_1(v) = v_1, \tag{9.113}
$$

$$
\mathbf{x}_2(v) = [\cos\omega - 1, \quad \sin\omega]v/T, \tag{9.114}
$$

$$
\begin{aligned}
\mathbf{u}(v) = {} & \frac{M + m(\sin \mathbf{x}_3(v))^2}{T^2}[\cos\omega - 1 \quad \sin\omega](A_1 - I_q)v - ml\mathbf{x}_4^2(v)\sin \mathbf{x}_3(v)\\
&+ \frac{b}{T}[\cos\omega - 1 \quad \sin\omega]v + mg\cos \mathbf{x}_3(v)\sin \mathbf{x}_3(v)
\end{aligned}
\tag{9.115}
$$

with $\mathbf{x}_3(v)$ and $\mathbf{x}_4(v)$ satisfying the following equations:

$$
\begin{aligned}
\mathbf{x}_3(A_1 v) &= \mathbf{x}_3(v) + T\mathbf{x}_4(v),\\
\mathbf{x}_4(A_1 v) &= \mathbf{x}_4(v) + \frac{gT}{l}\sin \mathbf{x}_3(v) - \cos \mathbf{x}_3(v)\frac{[\cos\omega - 1 \quad \sin\omega](A_1 - I_q)v}{Tl}.
\end{aligned}
\tag{9.116}
$$

The above two equations can be viewed as center manifold equations associated with the following nonlinear difference equations:

$$
\begin{aligned}
x_3(t+1) &= x_3(t) + Tx_4(t),\\
x_4(t+1) &= \frac{gT}{l}\sin x_3(t) + x_4(t) - \cos x_3(t)\frac{[\cos\omega - 1 \quad \sin\omega](A_1 - I_q)v(t)}{Tl}.
\end{aligned}
\tag{9.117}
$$

When v is set to zero, the Jacobian linearization of (9.117) at the origin is

$$\begin{aligned} x_3(t+1) &= x_3(t) + Tx_4(t), \\ x_4(t+1) &= \frac{gT}{l}x_3(t) + x_4(t). \end{aligned} \tag{9.118}$$

It can be easily verified that the two eigenvalues of (9.118) are not on the unit circle for all $\frac{gT}{l} > 0$, and the two eigenvalues of the matrix A_1 are on the unit circle. Therefore, by Theorem 2.31 (the Center Manifold Theorem for Maps), equations (9.116) admit a solution. However, the complex nonlinearity of (9.116) precludes an attempt to obtain an analytic solution. Therefore, let us find an approximate solution of (9.116) as follows.

Eliminating $\mathbf{x}_4(v)$ from (9.116) gives

$$\begin{aligned} \mathbf{x}_3(A_1^2 v) = {} & 2\mathbf{x}_3(A_1 v) - \mathbf{x}_3(v) + \frac{gT^2}{l}\sin \mathbf{x}_3(v) \\ & + \frac{\cos \mathbf{x}_3(v)}{l}[\cos\omega - 1 \quad \sin\omega](I_q - A_1)v. \end{aligned} \tag{9.119}$$

Therefore, as long as we can obtain the function $\mathbf{x}_3(v)$ by solving (9.119), we can then obtain $\mathbf{x}_1(v)$, $\mathbf{x}_2(v)$, and $\mathbf{u}(v)$ through (9.113) to (9.115) and $\mathbf{x}_4(v)$ through

$$\mathbf{x}_4(v) = (\mathbf{x}_3(A_1 v) - \mathbf{x}_3(v))/T. \tag{9.120}$$

A third-order polynomial approximation for $\mathbf{x}_3(v)$ denoted by $\mathbf{x}_3^{(3)}(v)$ can be obtained by solving (9.119) and is given as follows:

$$\mathbf{x}_3^{(3)}(v) = a_{10}v_1 + a_{01}v_2 + a_{30}v_1^3 + a_{21}v_1^2 v_2 + a_{12}v_1 v_2^2 + a_{03}v_2^3, \tag{9.121}$$

where

$$a_{10} = \frac{b_1 b_3 - b_2 b_4}{b_3^2 + b_4^2}, \qquad a_{01} = \frac{b_1 b_4 + b_2 b_3}{b_3^2 + b_4^2},$$

$$\begin{aligned} b_1 &= (\sin^2\omega - (1-\cos\omega)^2)/l, & b_2 &= 2(1-\cos\omega)\sin\omega, \\ b_3 &= \cos(2\omega) - 2(\cos\omega) + 1 - gT^2/l, & b_4 &= 2\sin\omega - \sin(2\omega), \end{aligned}$$

and

$$\begin{bmatrix} a_{30} & a_{21} & a_{12} & a_{03} \end{bmatrix}^T = \begin{bmatrix} x_{11} & x_{12} & x_{13} & x_{14} \\ x_{21} & x_{22} & x_{23} & x_{24} \\ x_{24} & -x_{23} & x_{22} & -x_{21} \\ -x_{14} & x_{13} & -x_{12} & x_{11} \end{bmatrix}^{-1} \begin{bmatrix} -\frac{gT^2}{6l}a_{10}^3 - \frac{1}{2}a_{10}^2 b_1 \\ -\frac{gT^2}{2l}a_{10}^2 a_{01} - \frac{1}{2}(2a_{10}a_{01}b_1 + a_{10}^2 b_2) \\ -\frac{gT^2}{2l}a_{10}a_{01}^2 - \frac{1}{2}(2a_{10}a_{01}b_2 + a_{01}^2 b_1) \\ -\frac{gT^2}{6l}a_{01}^3 - \frac{1}{2}a_{01}^2 b_2 \end{bmatrix},$$

where

$$
\begin{aligned}
x_{11} &= \cos^3(2\omega) - 2\cos^3(\omega) + 1 - gT^2/l, \\
x_{12} &= -\cos^2(2\omega)\sin(2\omega) + 2\cos^2(\omega)\sin(\omega), \\
x_{13} &= \cos(2\omega)\sin^2(2\omega) - 2\cos(\omega)\sin^2(\omega), \\
x_{14} &= -\sin^3(2\omega) + 2\sin^3(\omega), \\
x_{21} &= 3\cos^2(2\omega)\sin(2\omega) - 6\cos^2(\omega)\sin(\omega), \\
x_{22} &= (\cos^3(2\omega) - 2\cos(2\omega)\sin^2(2\omega)) - 2(\cos^3(\omega) - 2\cos(\omega)\sin^2(\omega)) + 1 - gT^2/l, \\
x_{23} &= (\sin^3(2\omega) - 2\cos^2(2\omega)\sin(2\omega)) - 2(\sin^3(\omega) - 2\cos^2(\omega)\sin(\omega)), \\
x_{24} &= 3\sin^2(2\omega)\cos(2\omega) - 6\sin^2(\omega)\cos(\omega).
\end{aligned}
$$

For example, when $\omega = 0.05\pi$ rad/sec, $g = 9.8$ m/sec^2, $l = 0.325$ m, and $T = 0.1$ sec,

$$\mathbf{x}_3^{(3)}(v) = -0.2300v_1 - 0.0337v_2 + 0.0039v_1^3 + 0.001v_1^2v_2 + 0.0012v_1v_2^2 + 0.0003v_2^3,$$

and when $\omega = 0.1\pi$ rad/sec, $g = 9.8$ m/sec^2, $l = 0.325$ m, and $T = 0.1$ sec,

$$\mathbf{x}_3^{(3)}(v) = -0.7396v_1 - 0.1792v_2 + 0.1362v_1^3 + 0.0292v_1^2v_2 + 0.0734v_1v_2^2 + 0.0221v_2^3.$$

With $\mathbf{x}_3^{(3)}(v)$ at hand, we can obtain the third-order approximations of $\mathbf{x}(v)$ and $\mathbf{u}(v)$, denoted by $\mathbf{x}^{(3)}(v)$ and $\mathbf{u}^{(3)}(v)$, by using (9.113), (9.114), (9.120), and (9.115). Thus a third-order state feedback controller is given as follows:

$$u(t) = \mathbf{u}^{(3)}(v(t)) + K_x(x(t) - \mathbf{x}^{(3)}(v(t))), \tag{9.122}$$

where the feedback gain K_x is selected such that the eigenvalues of the matrix $A + BK_x$ are

$$0.7488 \pm 0.4072j, \quad 0.7679 \pm 0.1301j,$$

which are obtained by bilinear transformation from the ITAE prototype design for the continuous-time systems with the cutoff frequency equal to 4.0 rad/sec.

Third-Order Robust State Feedback Servoregulator: To design a third-order robust state feedback controller, we need to find a pair of matrices (G_1, G_2) that incorporates a one-copy internal model of $\mathcal{A}_{3f}$. Since the solution of the discrete-time regulator equations does not contain the second-order term, the output equation of the closed-loop system under any state feedback control law of the form (9.90) will not contain the second-order term either. Thus, it suffices to find a pair of matrices (G_1, G_2) that incorporates a one-copy internal model of $A^{[1]}$ and $A^{[3]}$. The minimal polynomials of $A^{[1]}$ and $A^{[3]}$ are computed as follows:

$$
\begin{aligned}
\alpha_1(\lambda) &= (\lambda - e^{j\omega})(\lambda - e^{-j\omega}), \\
\alpha_3(\lambda) &= (\lambda - e^{j\omega})(\lambda - e^{-j\omega})(\lambda - e^{j3\omega})(\lambda - e^{-j3\omega}).
\end{aligned}
$$

Thus, the minimal polynomial of the matrix block diag $\left(A^{[1]}, A^{[3]}\right)$ is

$$(\lambda - e^{j\omega})(\lambda - e^{-j\omega})(\lambda - e^{j3\omega})(\lambda - e^{-j3\omega}).$$

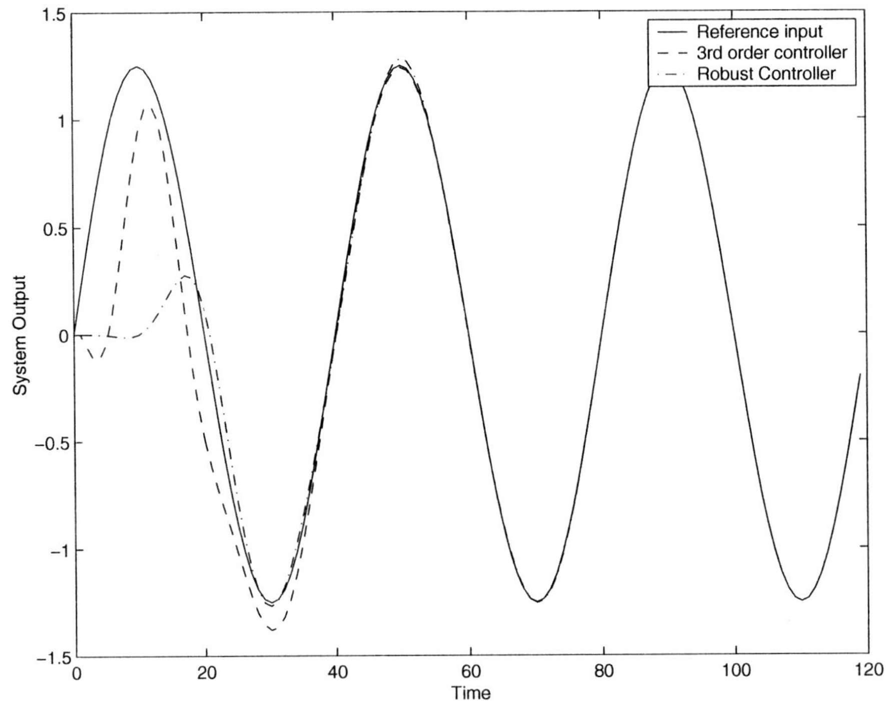

Figure 9.1. *Tracking performance: Nominal case* $A_m = 1.25$ *and* $\omega = 0.05\pi$.

Therefore, following the discussion in Section 5.5, we can specify G_1 and G_2 as follows:

$$G_1 = \begin{bmatrix} \cos\omega & \sin\omega & 0 & 0 \\ -\sin\omega & \cos\omega & 0 & 0 \\ 0 & 0 & \cos 3\omega & \sin 3\omega \\ 0 & 0 & -\sin 3\omega & \cos 3\omega \end{bmatrix}, \quad G_2 = \begin{bmatrix} 0 \\ 1 \\ 0 \\ 1 \end{bmatrix}.$$

The compensator, together with the plant, forms an eight-dimensional system. The feedback gain (K_1, K_2) is chosen such that the eigenvalues of the linearized closed-loop system are

$$0.4128, \quad 0.8283 \pm 0.4137j, \quad 0.8188 \pm 0.2521j, \quad 0.7591 \pm 0.1740j, \quad 0.7644,$$

which, again, are obtained by bilinear transformation from the ITAE prototype design for the continuous-time systems with the cutoff frequency equal to 4.0 rad/sec.

Both controllers are designed based on the nominal values of the system parameters, which are given as follows: $b_0 = 12.98$ kg/sec, $M = 1.378$ kg, $l = 0.325$ m, $g = 9.8$ m/sec^2, $m = 0.051$ kg.

Let us first compare the performance of the linear controller, the third-order controller, and the third-order robust controller for the nominal case, that is, $\Delta b = 0$. The frequency of the reference input is fixed at $\omega = 0.05\pi$ rad/sec while the amplitude A_m of the reference input takes $A_m = 0.75, 1.0, 1.25, 1.5$, respectively. Table 9.1 shows the maximal steady-state

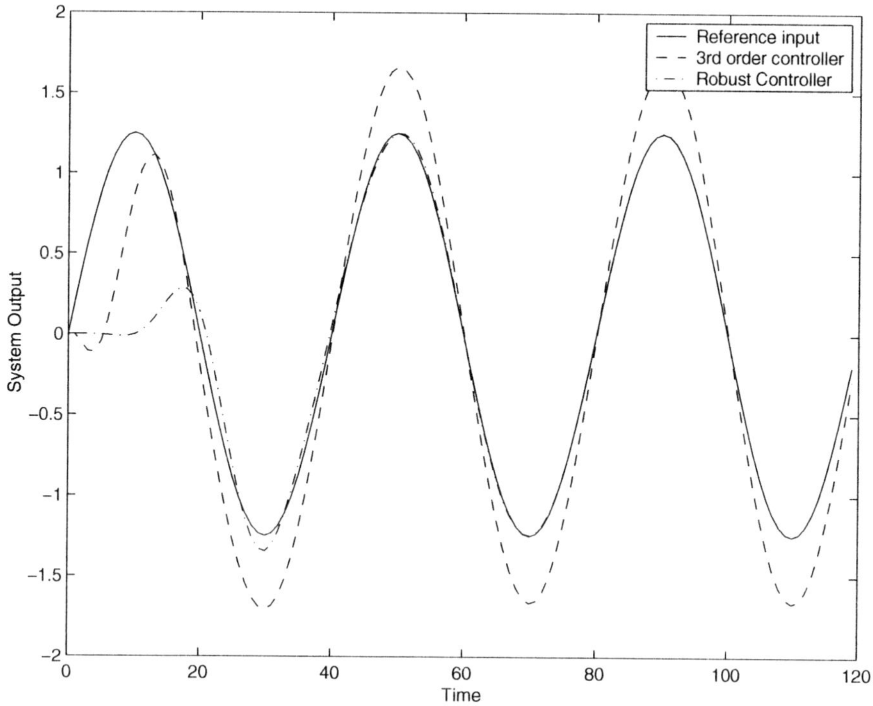

Figure 9.2. *Tracking performance: Perturbed system with* $A_m = 1.25$, $\omega = 0.05\pi$, *and* $\Delta b = 1.0$.

tracking errors of the closed-loop systems under various control laws for $\omega = 0.05\pi$ rad/sec and $A_m = 0.75, 1.0, 1.25, 1.5$. It is seen that the tracking performance of all controllers is quite good. The steady-state tracking error of the third-order robust controller is much smaller than that of the other two controllers, while the third-order controller is better than the linear controller. Figure 9.1 shows the tracking performance of the nominal closed-loop system resulting from the third-order controller and the third-order robust controller with $A_m = 1.25$ and $\omega = 0.05\pi$ rad/sec.

Next, we compare the performance of the various controllers in the presence of the parameter uncertainty with $A_m = 1.25$ and $\omega = 0.05\pi$. Assume that the parameter b is perturbed to $b = 12.98 + \Delta b$ with $\Delta b = -1.0, -0.5, 0.5, 1.0, 1.5$. Table 9.2 shows the steady-state tracking error of the perturbed closed-loop systems. As shown in Table 9.2, the third-order robust controller maintains small maximal steady-state tracking errors when the value of b varies. In contrast, the tracking performance of both the linear and the third-order controller greatly deteriorates when the parametric uncertainties are present. It is interesting to note that, while the third-order controller performs much better than the linear controller in the nominal case, it shows no advantage over the linear controller when the parameter uncertainties are present. Figure 9.2 shows the tracking performance of the perturbed closed-loop system resulting from the third-order controller and the third-order robust controller with $A_m = 1.25$, $\omega = 0.05\pi$, and $\Delta b = 1.0$.

Amplitude	ω	Linear	Third order	Third-order robust
0.75	0.05π	0.0095	0.0002	0.0000
1.00	0.05π	0.0226	0.0007	0.0000
1.25	0.05π	0.0446	0.0021	0.0002
1.50	0.05π	0.0788	0.0053	0.0008

Table 9.1. *The maximal steady-state tracking errors of the nominal system.*

Δb	Linear	Third order	Third-order robust
0.00	0.0446	0.0021	0.0002
−1.00	Unstable	Unstable	0.0014
−0.50	0.1502	0.1408	0.0006
0.50	0.1787	0.1792	0.0001
1.00	0.4125	0.4150	0.0001
1.50	0.7367	0.7399	0.0009

Table 9.2. *The maximal steady-state tracking errors of the perturbed system with* $A_m = 1.25$ *and* $\omega = 0.05\pi$.

Appendix A
Kronecker Product and Sylvester Equation

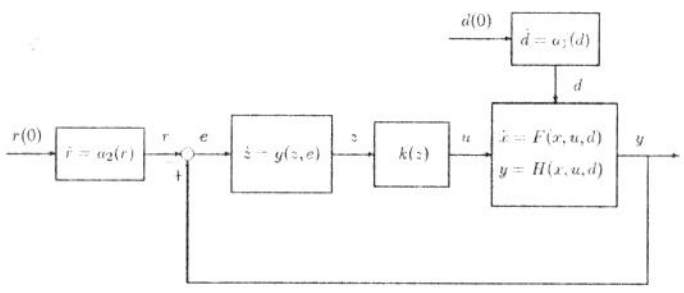

Let $A = [a_{ij}] \in \mathcal{R}^{m\times q}$ and $B = [b_{ij}] \in \mathcal{R}^{p\times n}$. Then the *Kronecker product* of A and B, denoted by $A \otimes B$, is defined by

$$A \otimes B = \begin{bmatrix} a_{11}B & \cdots & a_{1q}B \\ \vdots & \vdots & \vdots \\ a_{m1}B & \cdots & a_{mq}B \end{bmatrix}.$$

Let vec : $\mathcal{R}^{n\times m} \to \mathcal{R}^{nm\times 1}$ be a vector-valued function of a matrix such that, for any $X \in \mathcal{R}^{n\times m}$,

$$\text{vec}\,(X) = \begin{bmatrix} X_1 \\ \vdots \\ X_m \end{bmatrix},$$

where, for $i = 1, \ldots, m$, X_i is the ith column of X.

Proposition A.1.

(i) *For any matrices A, B, C, D of conformable dimensions,*

$$(A \otimes B)(C \otimes D) = (AC) \otimes (BD), \tag{A.1}$$

$$(A + B) \otimes (C + D) = (A \otimes C) + (A \otimes D) + (B \otimes C) + (B \otimes D). \tag{A.2}$$

(ii) *Let $A \in \mathcal{R}^{m\times q}$, $B \in \mathcal{R}^{p\times n}$, and $X \in \mathcal{R}^{n\times m}$. Then*

$$\text{vec}\,(BXA) = (A^T \otimes B)\text{vec}\,(X).$$

Proof. (i) follows directly from the definition of Kronecker product.

(ii)

$$\text{vec}\,(BX) = \begin{bmatrix} BX_1 \\ \vdots \\ BX_m \end{bmatrix} = \begin{bmatrix} B & 0 & \cdots & 0 \\ 0 & B & \cdots & 0 \\ \vdots & \vdots & \vdots & \vdots \\ 0 & 0 & \cdots & B \end{bmatrix} \begin{bmatrix} X_1 \\ \vdots \\ X_m \end{bmatrix} = (I_m \otimes B)\text{vec}\,(X)$$

and

$$\mathrm{vec}\,(XA) = \begin{bmatrix} \sum_{k=1}^{m} x_{1k}a_{k1} \\ \sum_{k=1}^{m} x_{2k}a_{k1} \\ \vdots \\ \sum_{k=1}^{m} x_{nk}a_{k1} \\ \sum_{k=1}^{m} x_{1k}a_{k2} \\ \sum_{k=1}^{m} x_{2k}a_{k2} \\ \vdots \\ \sum_{k=1}^{m} x_{nk}a_{k2} \\ \vdots \\ \sum_{k=1}^{m} x_{1k}a_{kq} \\ \sum_{k=1}^{m} x_{2k}a_{kq} \\ \vdots \\ \sum_{k=1}^{m} x_{nk}a_{kq} \end{bmatrix} = \begin{bmatrix} a_{11}I_n & a_{21}I_n & \cdots & a_{m1}I_n \\ a_{12}I_n & a_{22}I_n & \cdots & a_{m2}I_n \\ \vdots & \vdots & \vdots & \vdots \\ a_{1q}I_n & a_{2q}I_n & \cdots & a_{mq}I_n \end{bmatrix} \begin{bmatrix} X_1 \\ \vdots \\ X_m \end{bmatrix}$$

$$= (A^T \otimes I_n)\mathrm{vec}\,(X).$$

Thus

$$\begin{aligned} \mathrm{vec}\,(BXA) &= (A^T \otimes I_p)\mathrm{vec}\,(BX) \\ &= (A^T \otimes I_p)(I_m \otimes B)\mathrm{vec}\,(X) \\ &= (A^T \otimes B)\mathrm{vec}\,(X). \quad \square \end{aligned}$$

More detailed discussion on the properties of the Kronecker product can be found in [1]. Consider the linear matrix equation of the following form:

$$MXA - BXN = Q, \tag{A.3}$$

where $M, B \in \mathcal{R}^{p\times n}$, $A, N \in \mathcal{R}^{m\times q}$, and $Q \in \mathcal{R}^{p\times q}$ are known matrices, and $X \in \mathcal{R}^{n\times m}$ is an unknown matrix. Using property (ii) of Proposition A.1, (A.3) can be converted into the following standard form:

$$(A^T \otimes M - N^T \otimes B)\mathrm{vec}\,(X) = \mathrm{vec}\,(Q). \tag{A.4}$$

When $m = q, n = p$, and M and N are identity matrices, (A.3) becomes

$$XA - BX = Q \tag{A.5}$$

and is called the Sylvester equation. Correspondingly, (A.4) becomes the following:

$$(A^T \otimes I_n - I_m \otimes B)\mathrm{vec}\,(X) = \mathrm{vec}\,(Q). \tag{A.6}$$

The Sylvester equation has the following properties.

Proposition A.2.

(i) *The Sylvester equation (A.5), where $A \in \mathcal{R}^{m\times m}$ and $B \in \mathcal{R}^{n\times n}$, has a unique solution if and only if A and B have no eigenvalues in common.*

(ii) *Let $A \in \mathcal{R}^{m\times m}$ and $B \in \mathcal{R}^{n\times n}$. A linear mapping $\mathcal{S} : \mathcal{R}^{n\times m} \rightarrow \mathcal{R}^{n\times m}$ such that*

$$\mathcal{S}(X) = XA - BX \tag{A.7}$$

is called a Sylvester map. Let $\mathcal{K}$ be the kernel of $\mathcal{S}$, that is,

$$\mathcal{K} = \{X \in \mathcal{R}^{n\times m} \mid \mathcal{S}(X) = 0\}. \tag{A.8}$$

Let $\{\delta_i, i = 1, \dots, n_1\}$ and $\{\epsilon_j, j = 1, \dots, n_2\}$ be the lists of invariant factors of B and A, respectively. Let γ_{ij}, $i = 1, \dots, n_1$, $j = 1, \dots, n_2$, be the greatest common divisor of δ_i and ϵ_j. Then

$$\dim(\mathcal{K}) = \sum_{i=1}^{n_1}\sum_{j=1}^{n_2} \deg(\gamma_{ij}). \tag{A.9}$$

(iii) *Consider the Sylvester equation (A.5) with $m = n$. Assume A and B have no common eigenvalues and there exist $N \in \mathcal{R}^{n\times 1}$ and $\Psi \in \mathcal{R}^{1\times n}$ such that $Q = N\Psi$ with (B, N) controllable and (Ψ, A) observable. Then the Sylvester equation (A.5) has a unique solution $X \in \mathcal{R}^{n\times n}$ which is nonsingular.*

Proof. For simplicity, assume that A and B have distinct eigenvalues denoted by $\{\lambda_1, \dots, \lambda_m\}$ and $\{\mu_1, \dots, \mu_n\}$, respectively. Suppose α_i and β_j are eigenvectors of A^T and B corresponding to the eigenvalues λ_i and μ_j, respectively. By property (i) of Proposition A.1, $\alpha_i \otimes \beta_j$ is the eigenvector of $(A^T \otimes I - I \otimes B)$ corresponding to the eigenvalue $\lambda_i - \mu_j$. Thus, the eigenvalues of $(A^T \otimes I - I \otimes B)$ are given by $\{\lambda_i - \mu_j, i = 1, \dots, m, j = 1, \dots, n\}$. That is, the matrix $(A^T \otimes I - I \otimes B)$ is nonsingular if and only if the matrices A and B have no common eigenvalues.

Proof of property (ii) is suggested on page 25 of [112] and is outlined here. First show that (A.9) holds when A and B are in Jordan form. Then letting $A = T_A^{-1} J_A T_A$ and $B = T_B^{-1} J_B T_B$, where J_A and J_B are the Jordan form of A and B, respectively, gives

$$XA - BX = T_B^{-1}(YJ_A - J_B Y)T_A,$$

where $Y = T_B X T_A^{-1}$. Let $\bar{\mathcal{K}} = \{X \in \mathcal{R}^{n\times m} \mid YJ_A - J_B Y = 0\}$. Clearly $\dim(\bar{\mathcal{K}}) = \dim(\mathcal{K})$. Thus (A.9) holds for any A and B.

Property (iii) is a special case of Theorem 7-10 of [10]. The proof is outlined below. Let the characteristic polynomial of A be

$$\Delta(s) = \det(sI - A) = s^n + \alpha_1 s^{n-1} + \cdots + \alpha_n.$$

Then it can be shown that

$$X\Delta(A) - \Delta(B)X = [N \; BN \; \cdots \; B^{n-1}N] \begin{bmatrix} \alpha_{n-1} & \alpha_{n-2} & \cdots & \alpha_1 & 1 \\ \alpha_{n-2} & \alpha_{n-3} & \cdots & 1 & 0 \\ \vdots & \vdots & & \vdots & \vdots \\ \alpha_1 & 0 & \cdots & 0 & 0 \\ 1 & 0 & \cdots & 0 & 0 \end{bmatrix} \begin{bmatrix} \Psi \\ \Psi A \\ \vdots \\ \Psi^{n-2}A \\ \Psi^{n-1}A \end{bmatrix}. \tag{A.10}$$

Clearly, the right-hand side of (A.10) is invertible since (B, N) is controllable and (Ψ, A) is observable. Moreover, $\Delta(A) = 0$ by the Cayley–Hamilton theorem, and $\Delta(B)$ is invertible since the eigenvalues of $\Delta(B)$ are $\{\Delta(\mu_1), \dots, \Delta(\mu_n)\}$ and A and B have no common eigenvalues. Thus X is invertible. $\square$

Appendix B

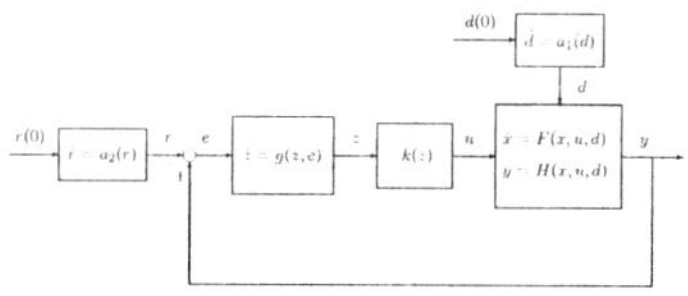

ITAE Prototype Design

A convenient way to select the desirable pole locations for a closed-loop system is to make a member of a set of the so-called prototype polynomials as the characteristic polynomial of the closed-loop system. There are several sets of prototype polynomials, one of which is shown in Table B.1.

k	Pole locations for $\omega_0 = 1$ rad/sec
1	$s + 1$
2	$s + 0.7071 \pm 0.7071j$
3	$(s + 0.7081)(s + 0.5210 \pm 1.068j)$
4	$(s + 0.4240 \pm 1.2630j)(s + 0.6260 \pm 0.4141j)$
5	$(s + 0.8955)(s + 0.3764 \pm 1.2920j)(s + 0.5758 \pm 0.5339j)$
6	$(s + 0.3099 \pm 1.2634i)(s + 0.5805 \pm 0.7828j)(s + 0.7346 \pm 0.2873j$
7	$(s + 0.6816)(s + 1.2123 \pm 1.0070j)(s + 0.2492 \pm 1.0707j)(s + 0.4214 \pm 0.5579j)$
8	$(s + 2.0782)(s + 0.6675)(s + 0.2031 \pm 1.1774j)(s + 0.3945 \pm 0.7479j)(s + 0.6296 \pm 0.5567j)$

Table B.1. *Pole locations of ITAE prototype design.*

This table was worked out by Graham and Lathrop [30] based on the criterion of minimizing the integral of the time multiplied by the absolute value of the error (ITAE), that is,

$$P = \int_0^\infty t|e|dt.$$

In Table B.1, the nominal cutoff frequency is $\omega_0 = 1$ rad/sec. Pole locations for other values of ω_0 can be obtained by substituting s/ω_0 for s everywhere [27].

Notes and References

Chapter 1. Various versions of the linear output regulation problem have been thoroughly studied since the early 1970s. The problem was first treated for the special case where both the reference input and disturbance are step functions by Johnson [71] and Smith and Davison [99]. Extension to the general case with various versions can be found in Cheng and Pearson [18], Davison [21], [22], [23], Francis [28], Francis and Wonham [29], and Wonham and Pearson [113], to name just a few. A self-contained treatment on this topic was given by Desoer and Wang [26]. Extensive exposition on this topic can be found in several textbooks, such as Chen [10], Knobloch, Isidori, and Flockerzi [77], Saberi, Stoorvogel, and Sannuti [94], and Wonham [112]. The main references for this chapter are Davison [23], Desoer and Wang [26], and Wonham [112]. Most results in Section 1.3 can be found in [28], in which the solvability of the regulator equations is tied to the solvability of the regulation problem. Most results in Section 1.4 can be found in [23] and [26], but the exposition is more close, in spirit, to Huang [41]. The exposition in Section 5 is based on the work of [28] and [29]. Frequency domain synthesis of linear regulators can be found in [18]. Output regulation of linear systems with input saturation was studied by Lin and Saberi [85].

Chapter 2. The materials in Section 2.2 are quite standard. The notion of input-to-state stability summarized in Section 2.3 was first proposed by Sontag [100] and [101]. Properties of input-to-state stability were further elaborated by Krichman, Sontag, and Wang [79], Sontag [102], Sontag and Wang [103], [104], and [105]. A discrete-time version of the concept of input-to-state stability was treated by Jiang and Wang [70]. A concise yet quite self-contained introduction to input-to-state stability concepts was given by Isidori [64] for autonomous systems and by Khalil for nonautonomous systems [74]. The exposition and notation of Section 2.3 is quite close to Section 10.4 of the book [64] with a major difference that nonautonomous systems are treated here. Various versions of the Small Gain Theorem can be found in [36], [37], [64], [66], and [69]. Theorem 2.18 as well as Corollaries 2.19 and 2.20 are taken from [16], and it can be viewed as a special case of Theorem 1 of [66]. Sections 2.4 and 2.5 are mainly based on Carr [7]. Normal form and zero dynamics as summarized in Sections 2.6 and 2.7 have now become a standard topic in nonlinear control textbooks after the trendsetting book of Isidori [63]. Other main references for these two sections are Khalil [74], Nijmeijer and van der Schaft [88], and Slotine and Li [98]. The models of the three typical nonlinear systems, that is, the RTAC system, the inverted pendulum on a cart, and the ball and beam system of Section 2.8 are taken from [2], [31], and [32], respectively.

Chapter 3. The output regulation problem for nonlinear systems was first treated for the special case in which the exogenous signals are constant by Francis and Wonham [29]. Further elaboration of this case was given by Hepburn and Wonham [33] to [35], Desoer and Lin [25], and Huang and Rugh [58]. In particular, Huang and Rugh tied the solvability of the nonlinear robust output regulation with constant exogenous signals to the solvability of a set of nonlinear algebraic equations, which are a special case of the nonlinear regulator equations. The nonlinear output regulation problem with time-varying exogenous signals was first studied in 1990 by Isidori and Byrnes without considering parameter uncertainty [65]. They fundamentally established the solvability of the nonlinear output regulation problem in terms of the solvability of the nonlinear regulator equations. The formulation of the nonlinear output regulation given in Section 3.2 is slightly more general than what was given in [65]. Results in Section 3.3 were basically covered in [65]. Solvability of the nonlinear regulator equations was investigated in several papers by Cheng, Tarn, and Spurgeon [17], Huang [47], Huang and Lin [56], and Isidori and Byrnes [65]. Section 3.4 is essentially taken from the work of Huang [47]. The output regulation of nonlinear systems with nonhyperbolic zero dynamics was studied by Huang in [38] and [45]. Section 3.5 is based on the work of [38]. The output regulation of nonlinear systems was studied in Wang and Huang [111], and Section 3.6 is a refinement of the work in [111]. The estimate of the convergence region of output regulation, an important issue but not touched on in this book, was addressed by very recent work of Pavlov, van de Wouw, and H. Nijmeijer in [90].

Chapter 4. This chapter is mainly based on two papers by Huang and Rugh [59], [60]. Similar work on the formal Taylor series solution of the regulator equations can be found in Krener [78]. The proof of Lemma 4.8 in Section 4.2 is from [72]. Section 4.3 is an expansion of Theorem 1 of [45]. Section 4.4 on the approximation solution of the asymptotic tracking of the inverted pendulum on a cart system is based on the work of Huang [44]. Approximation approaches based on neural networks were studied in [109] and [110].

Chapter 5. Francis and Wonham discovered as early as 1976 that, for the special case where the exogenous signals are constant, the linear internal model that works for linear systems also works for nonlinear systems. However, this technique does not work for the general case where the exogenous signals are time-varying, as shown by a counterexample by Byrnes and Isidori [6]. Huang and Lin first revealed in 1991 that the linear internal model principle fails because, unlike for linear systems, the steady-state tracking error of a nonlinear system is a nonlinear function of the exogenous signals [51]. They also introduced the notion of kth-order robust output regulation in [51], [54]. Huang and Lin further showed that when the solution of the regulator equations is polynomial, the kth-order robust regulator also solves the robust output regulation problem [39], [53]. Other aspects of robust output regulation were studied in Byrnes et al. [4], [5], Delli Prescoli [24], Huang [40], [43], and Khalil [75]. Sections 5.1 to 5.3 are essentially based on the work of Huang [39], [43]. Section 5.4 is taken from the work of [40]. kth-order robust control of the ball and beam system was studied in Huang and Lin [57]. A frequency approach can be found in [42].

Chapter 6. This chapter is mainly based on the papers by Huang [46], Huang and Chen [49], and Chen and Huang [14]. The new design framework presented in Section 6.1 was first proposed in Huang and Chen [48]. The notion of the steady-state generator is closely related to the concept of system immersion suggested by Byrnes et al. [5]. Using the system immersion concept, Byrnes et al. gave an alternative sufficient condition for solvability of the robust output regulation problem, which requires that the solution of the regulator equations satisfy some partial differential equation [5]. This result leads directly to Proposition 6.12. Proposition 6.14 is based on the work of Huang [46]. Lemma 6.17 of Section 6.2 and most parts of Section 6.3 are based on the work of Chen and Huang [14]. The example on the RTAC system is based on the work of Huang and Hu [50].

Chapter 7. The formulation of the global robust output regulation problem for general nonlinear systems given in Section 7.1 is taken from Huang and Chen [49]. The main references for Section 7.2 are [19], [64], [66], [68], and [84]. In particular, the paper by Jiang and Mareels [66] studied the robust stabilization of lower triangular continuous systems with dynamic uncertainties. Theorem 7.6 can be viewed as a refinement of the results given in [68]. The robust stabilization problem of lower triangular continuous systems without dynamic uncertainties was also treated in Section 11.4 of the book by Isidori [64]. Further extensions of the results in Section 11.4 of the book [64] can be found in [12] and [84]. Use of the inequality given in Lemma 7.8 and its variations has been made in several papers, such as [84] and [91]. The proof of Lemma 7.8 was also suggested in [84] and [91]. The robust stabilization of the systems in output feedback form was studied by Marino and Tomei in [86]. A somewhat alternative treatment is also given in Section 11.3 of the book by Isidori [64]. The global robust regulation of systems in output feedback form for the special case where the system admits a linear internal model was studied by Serrani and Isidori [95], and the more general case was studied by Chen and Huang [15]. The result in Section 7.4 is mainly taken from [49]. Examples 7.26 and 7.32 are worked out by my Ph.D. student Zhiyong Chen. The semiglobal robust output regulation problem for various nonlinear systems was studied by Isidori in [62], Serrani, Isidori, and Marconi in [96], and Khalil in [75] and [76]. The adaptive output regulation for systems with uncertain exosystems was studied in Chen and Huang [13], Nikiforov [89], Serrani, Isidori, and Marconi [97], and Ye and Huang [107]. A more extensive exposition of global robust stabilization of nonlinear systems can be found in books by Kristic, Kanellakopoulos, and Kokotovic [80], Marino and Tomei [87], and Qu [92], and in the papers [73], [67], and [106].

Chapter 8. Output regulation of nonlinear singular systems was first studied in Huang and Zhang in [61]. A comprehensive treatment for singular linear systems was given by Dai [20], which is also the main reference for Section 8.2. Section 8.3 is based on the work of Huang and Zhang in [61]. A major portion of Section 8.4 is taken from [11]. Output regulation of linear singular systems with input saturation is studied by Lan and Huang in [81].

Chapter 9. The output regulation for discrete-time nonlinear systems was studied by Castillo et al. [8], [9] and Huang and Lin [52], [55]. The approximate output regulation problem for discrete-time nonlinear systems is treated by Wang and Huang [108]. Robust output regulation for discrete-time nonlinear systems was given by Lan and Huang [82]. The output regulation and the robust output regulation of the inverted pendulum on a cart example were studied in [82] and [108], respectively.

Bibliography

[1] J. W. Brewer, "Kronecker products and matrix calculus in systems theory," *IEEE Transactions on Automatic Circuits and Systems*, Vol. 25, pp. 772–781, 1979.

[2] R. T. Bupp, D. S. Bernstein, and V. T. Coppola, "A benchmark problem for nonlinear control design," *International Journal of Robust and Nonlinear Control*, Vol. 8, pp. 307–310, 1998.

[3] R. T. Bupp, D. S. Bernstein, and V. T. Coppola, "Experimental implementation of integrator backstepping and passive nonlinear controllers on the RTAC testbed," *International Journal of Robust and Nonlinear Control*, Vol. 8, pp. 435–457, 1998.

[4] C. I. Byrnes, F. Delli Priscoli, and A. Isidori, *Output Regulation of Uncertain Nonlinear Systems*, Birkhäuser, Boston, 1997.

[5] C. I. Byrnes, F. Delli Priscoli, A. Isidori, and W. Kang, "Structurally stable output regulation of nonlinear systems," *Automatica*, Vol. 33, pp. 369–385, 1997.

[6] C. I. Byrnes and Isidori, "Nonlinear output regulation: Remarks on robustness," *Proceedings of 27th Allerton Conference on Communication, Control, and Computing*, Allerton, IL, pp. 150–158, 1989.

[7] J. Carr, *Applications of the Center Manifold Theory*, Springer-Verlag, New York, 1981.

[8] B. Castillo and S. D. Gennaro, "Asymptotic output tracking for SISO nonlinear discrete time systems," *Proceedings of the 30th IEEE Conference on Decision and Control*, pp. 1802–1806, 1991.

[9] B. Castillo, S. D. Gannaro, S. Monaco, and D. Normand-Cyrot, "Nonlinear regulation for a class of discrete-time systems," *Systems and Control Letters*, Vol. 20, pp. 57–66, 1993.

[10] C.-T. Chen, *Linear System Theory and Design*, 3rd ed., Oxford University Press, Oxford, UK, 1998.

[11] Z. Chen and J. Huang, "Solution of output regulation of singular nonlinear systems by normal output feedback," *IEEE Transactions on Automatic Control*, Vol. 47, pp. 803–813, 2002.

[12] Z. Chen and J. Huang, "Global robust stabilization of cascaded polynomial systems," *Systems and Control Letters*, Vol. 47, pp. 445–453, 2002.

[13] Z. Chen and J. Huang, "Global tracking of uncertain nonlinear cascaded systems with adaptive internal model," *Proceedings of the 41st IEEE Conference on Decision and Control*, pp. 3855–3862, 2002.

[14] Z. Chen and J. Huang, "Nonlinear internal model and the robust output regulation problem," *Proceedings of 2003 American Control Conference*, pp. 1560–1565, 2003.

[15] Z. Chen and J. Huang, "Global robust output regulation for output feedback systems," *Proceedings of IEEE International Conference on Robotics, Intelligent Systems and Signal Processing*, pp. 25–30, 2003.

[16] Z. Chen and J. Huang, "A variation of the small gain theorem," *Proceedings of the Eighth International Conference on Control, Automation, Robotics and Vision*, Dec. 2004, to appear.

[17] D. Cheng, T. J. Tarn, and S. K. Spurgeon, "On the design of output regulators for nonlinear systems," *Systems and Control Letters*, Vol. 43, pp. 167–179, 2001.

[18] L. Cheng and J. B. Pearson, "Frequency-domain synthesis of multivariable linear regulators," *IEEE Transactions on Automatic Control*, Vol. 23, pp. 3–15, 1978.

[19] J. M. Coron, L. Praly, and A. R. Teel, "Feedback stabilization of nonlinear systems: Sufficient conditions and Lyapunov and input-output techniques," *Trends in Control*, A. Isidori, ed., Springer-Verlag, Berlin, pp. 293–348, 1995.

[20] L. Dai, *Singular Control Systems*, Lecture Notes in Control and Information Sciences, Vol. 118, Springer-Verlag, Berlin, 1989.

[21] E. J. Davison, "The output control of linear time-invariant multivariable systems with unmeasurable arbitrary disturbances," *IEEE Transactions on Automatic Control*, Vol. 17, pp. 621–630, 1972.

[22] E. J. Davison, "A generalization of the output control of linear time-invariant multivariable systems with unmeasurable arbitrary disturbances," *IEEE Transactions on Automatic Control*, Vol. 20, pp. 788–792, 1975.

[23] E. J. Davison, "The robust control of a servomechanism problem for linear time-invariant multivariable systems," *IEEE Transactions on Automatic Control*, Vol. 21, pp. 25–34, 1976.

[24] F. Delli Priscoli, "Robust tracking for polynomial plants," *Proceedings of the European Control Conference*, pp. 369–373, 1993.

[25] C. A. Desoer and C. A. Lin, "Tracking and disturbance rejection of MIMO nonlinear systems with PI controller," *IEEE Transactions on Automatic Control*, Vol. 30, pp. 861–867, 1985.

[26] C. A. Desoer and Y. T. Wang, "Linear time-invariant robust servomechanism problem: A self-contained exposition," *Control and Dynamic Systems*, Vol. 16, pp. 81–129, 1980.

[27] G. F. Franklin, J. D. Powell, and A. Emami-Nieini, *Feedback Control of Dynamic Systems*, 3rd ed., Addison Wesley, Reading, MA, 1994.

[28] B. A. Francis, "The linear multivariable regulator problem," *SIAM Journal on Control and Optimization*, Vol. 15, pp. 486–505, 1977.

[29] B. A. Francis and W. M. Wonham, "The internal model principle of control theory," *Automatica*, Vol. 12, pp. 457–465, 1976.

[30] D. Graham and R. C. Lathrop, "The synthesis of optimum response: Criteria and control," *Transactions of the American Institute for Electrical Engineering*, Vol. 72, pt. 1, pp. 273–288, 1953.

[31] R. Gurumoorthy and S. R. Sanders, "Controlling nonminimum phase nonlinear systems—the inverted pendulum on a cart example," *Proceedings of American Control Conference*, pp. 680–685, June 1993.

[32] J. Hauser, S. Sastry, and P. Kokotovic, "Nonlinear control via approximate input-output linearization: The ball and beam example," *IEEE Transactions on Automatic Control*, Vol. 37, pp. 392–398, March 1992.

[33] J. S. A. Hepburn and W. M. Wonham, "Error feedback and internal models on differential manifolds," *IEEE Transactions on Automatic Control*, Vol. 29, pp. 397–403, 1984.

[34] J. S. A. Hepburn and W. M. Wonham, "Structural stable nonlinear regulation with step inputs," *Mathematical Systems Theory*, Vol. 17, pp. 319–333, 1984.

[35] J. S. A. Hepburn and W. M. Wonham, "The semistable-center-unstable manifold near a critical element," *Journal of Mathematical Analysis and Applications*, Vol. 103, pp. 321–331, 1984.

[36] D. Hill and P. J. Moylan, "The stability of nonlinear dissipative systems," *IEEE Transactions on Automatic Control*, Vol. 21, pp. 708–711, 1976.

[37] D. Hill and P. J. Moylan, "Stability results for nonlinear feedback systems," *Automatica*, Vol. 13, pp. 377–382, 1977.

[38] J. Huang, "Output regulation of nonlinear systems with nonhyperbolic zero dynamics," *IEEE Transactions on Automatic Control*, Vol. 40, pp. 1497–1500, 1995.

[39] J. Huang, "Asymptotic tracking and disturbance rejection in uncertain nonlinear systems," *IEEE Transactions on Automatic Control*, Vol. 40, pp. 1118–1122, 1995.

[40] J. Huang, "Calculation of the minimal order nonlinear robust servo-regulator," *Systems and Control Letters*, Vol. 26, pp. 313–320, 1995.

[41] J. Huang, "A simple proof of the output feedback linear robust regulator," *Control-Theory and Advanced Technology*, Vol. 10, pp. 1499–1504, 1995.

[42] J. Huang, "Asymptotic tracking in uncertain Volterra systems," *Systems and Control Letters* Vol. 31, pp. 215–223, 1997.

[43] J. Huang, "K-fold exosystem and the robust nonlinear servomechanism problem," *ASME Journal of Dynamics, Control, and Measurement*, pp. 149–153, March 1998.

[44] J. Huang, "On a control problem of the inverted-pendulum on a cart system," *Proceedings of the 4th International Conference on Information Systems, Analysis and Synthesis*, pp. 51–56, 1998.

[45] J. Huang, "Asymptotic tracking of a nonminumum phase nonlinear system with nonhyperbolic zero dynamics," *IEEE Transactions on Automatic Control*, Vol. 45, pp. 542–546, 2000.

[46] J. Huang, "Remarks on robust output regulation problem for nonlinear systems," *IEEE Transactions on Automatic Control*, Vol. 46, pp. 2028–2031, 2001.

[47] J. Huang, "On the solvability of the regulator equations for a class of nonlinear systems," *IEEE Transactions on Automatic Control*, Vol. 48, pp. 880–885, 2003.

[48] J. Huang and Z. Chen, "A general framework for output regulation problem," *Proceedings of 2002 American Control Conference*, pp. 102–109, 2002.

[49] J. Huang and Z. Chen, "A general framework for tackling output regulation problem," *IEEE Transactions on Automatic Control*, to appear.

[50] J. Huang and G. Hu, "A control design for the nonlinear benchmark problem via the output regulation method," *Journal of Control Theory and Applications*, Vol. 2, pp. 11–19, 2004.

[51] J. Huang and C.-F. Lin, "On a robust nonlinear servomechanism problem," *Proceedings of the 30th IEEE Conference on Decision and Control*, pp. 2529–2530, December 1991.

[52] J. Huang and C.-F. Lin, "On the discrete-time nonlinear servomechanism problem," *Proceedings of American Control Conference*, pp. 844–848, June 1993.

[53] J. Huang and C.-F. Lin, "Internal model principle and robust control of nonlinear systems," *Proceedings of the 32nd IEEE Conference on Decision and Control*, pp. 1501–1513, December 1993.

[54] J. Huang and C.-F. Lin, "On a robust nonlinear servomechanism problem," *IEEE Transactions on Automatic Control*, Vol. 39, pp. 1510–1513, 1994.

[55] J. Huang and C.-F. Lin, "A stability property and its application to discrete-time nonlinear system control," *IEEE Transactions on Automatic Control*, Vol. 39, pp. 2307–2311, 1994.

[56] J. Huang and C.-F. Lin, "On the solvability of the general nonlinear servomechanism problem," *Control-Theory and Advanced Technology*, Vol. 10, pp. 1253–1262, 1995.

[57] J. Huang and C.-F. Lin, "Robust control of the ball and beam system," *Proceedings of American Control Conference*, pp. 833–837, 1995.

[58] J. Huang and W. J. Rugh, "On a nonlinear multivariable servomechanism problem," *Automatica*, Vol. 26, pp. 963–972, 1990.

[59] J. Huang and W. J. Rugh, "Stabilization on zero-error manifolds and the nonlinear servomechanism problem," *IEEE Transactions on Automatic Control*, Vol. 37, pp. 1009–1013, 1992.

[60] J. Huang and W. J. Rugh, "An approximation method for the nonlinear servomechanism problem," *IEEE Transactions on Automatic Control*, Vol. 37, pp. 1395–1398, 1992.

[61] J. Huang and J.-F. Zhang, "Impulse-free output regulation of singular nonlinear systems," *International Journal of Control*, Vol. 71, pp. 789–806, 1998.

[62] A. Isidori, "A remark on the problem of semiglobal nonlinear output regulation," *IEEE Transactions on Automatic Control*, Vol. 42, pp. 1734–1738, 1997.

[63] A. Isidori, *Nonlinear Control Systems*, 3rd ed., Springer-Verlag, New York, 1995.

[64] A. Isidori, *Nonlinear Control Systems, Volume* II, Springer-Verlag, New York, 1999.

[65] A. Isidori and C. I. Byrnes, "Output regulation of nonlinear systems," *IEEE Transactions on Automatic Control*, Vol. 35, pp. 131–140, 1990.

[66] Z. P. Jiang and I. Mareels, "A small-gain control method for nonlinear cascaded systems with dynamic uncertainties," *IEEE Transactions on Automatic Control*, Vol. 42, pp. 292–308, 1997.

[67] Z. P. Jiang and I. Mareels, "Robust nonlinear integral control," *IEEE Transactions on Automatic Control*, Vol. 46, No. 8, pp. 1336–1342, 2001.

[68] Z. P. Jiang and L. Praly, "Design of robust adaptive controllers for nonlinear systems with dynamic uncertainties," *Automatica*, Vol. 34, pp. 825–840, 1998.

[69] Z. P. Jiang, A. R. Teel, and L. Praly, "Small-gain theorem for ISS systems and applications," *Mathematics of Control, Signals and Systems*, Vol. 7, pp. 95–120, 1994.

[70] Z. P. Jiang and Y. Wang, "Input-to-state stability for discrete-time nonlinear systems," *Automatica*, Vol. 37, pp. 857–869, 2001.

[71] C. D. Johnson, "Accommodation of external disturbances in linear regulator and servomechanism problems," *IEEE Transactions on Automatic Control*, Vol. 16, pp. 535–644, 1971.

[72] W. Kang and J. Huang, "Calculation of the minimal dimension kth-order robust servoregulator," *IEEE Transactions on Automatic Control*, Vol. 47, pp. 382–386, 1997.

[73] I. Kanellakopoulos, P. Kokotović, and A. S. Morse, "A toolkit for nonlinear feedback design," *Systems and Control Letters*, Vol. 18, pp. 83–92, 1992.

[74] H. Khalil, *Nonlinear Systems*, Prentice Hall, Englewood Cliffs, NJ, 1996.

[75] H. Khalil, "Robust servomechanism output feedback controllers for feedback linearizable systems," *Automatica*, Vol. 30, pp. 1587–1589, 1994.

[76] H. Khalil, "On the design of robust servomechanisms for minimum phase nonlinear systems," *International Journal of Robust Nonlinear Control*, Vol. 10, pp. 339–361, 2000.

[77] H. W. Knobloch, A. Isidori, and D. Flockerzi, *Topics in Control Theory*, Birkhäuser, Boston, 1993.

[78] A. J. Krener, "The construction of optimal linear and nonlinear regulators," In *Systems, Models, and Feedback* (A. Isidori and T. J. Tarn, eds.), pp. 301–322, Birkhäuser, Boston, 1992.

[79] M. Krichman, E. D. Sontag, and Y. Wang, "Input-output-to-state stability," *SIAM Journal on Control and Optimization*, Vol. 39, pp. 1874–1928, 2001.

[80] M. Kristic, I. Kanellakopoulos, and P. Kokotovic, *Nonlinear and Adaptive Control Design*, John Wiley & Sons, New York, 1995.

[81] W. Y. Lan and J. Huang, "Semi-global stabilization and output regulation of singular linear systems with input saturation," *IEEE Transactions on Automatic Control*, Vol. 48, pp. 1274–1279, 2003.

[82] W. Y. Lan and J. Huang, "On the discrete-time robust nonlinear servomechanism problem," *Communications in Information and Systems*, Vol. 3, pp. 1–26, 2003.

[83] W. Lin and L. Dai, "Solutions to the output regulation problem of linear singular systems," *Automatica*, Vol. 32, pp. 1713–1718, 1996.

[84] W. Lin and Q. Gong, "A remark on partial-state feedback stabilization of cascade systems using small gain theorem," *IEEE Transactions on Automatic Control*, Vol. 48, pp. 497–500, 2003.

[85] Z. Lin and A. Saberi, "Output regulation for linear systems subject to input saturation," *Automatica*, Vol. 32, pp. 29–47, 1996.

[86] R. Marino and P. Tomei, "Global adaptive output feedback control of nonlinear systems. Part I: Linear parameterization," *IEEE Transactions on Automatic Control*, Vol. 38, pp. 17–32, 1993.

[87] R. Marino and P. Tomei, *Nonlinear Control Design: Geometric, Adaptive and Robust*, Prentice-Hall, Englewood Cliffs, NJ, 1995.

[88] H. Nijmeijer and A. J. van der Schaft, *Nonlinear Dynamical Control Systems*, Springer-Verlag, New York, 1990.

[89] V. O. Nikiforov, "Adaptive non-linear tracking with complete compensation of unknown disturbances," *European Journal of Control*, Vol. 4, pp. 132–139, 1998.

[90] A. Pavlov, N. van de Wouw, and H. Nijmeijer, "The local out regulation problem: convergence region estimates," *IEEE Transactions on Automatic Control*, Vol. 49, pp. 814–819, 2004.

[91] L. Praly and Z. P. Jiang, "Stabilization by output feedback for systems with ISS inverse dynamics," *Systems and Control Letters*, Vol. 21, pp. 19–33, 1993.

[92] Z. Qu, *Robust Control of Nonlinear Uncertain Systems*, John Wiley & Sons, Interscience Division, New York, 1998.

[93] W. Rudin, *Principles of Mathematical Analysis*, 3rd ed., McGraw-Hill, New York, 1976.

[94] A. Saberi, A. A. Stoorvogel, and P. Sannuti, *Control of Linear Systems with Regulation and Input Constraints*, Springer-Verlag, New York, 2000.

[95] A. Serrani and A. Isidori, "Global robust output regulation for a class of nonlinear systems," *Systems and Control Letters*, Vol. 39, pp. 133–139, 2000.

[96] A. Serrani, A. Isidori, and L. Marconi, "Semiglobal robust output regulation of minimum-phase nonlinear systems," *International Journal of Robust and Nonlinear Control*, Vol. 10, pp. 379–396, 2000.

[97] A. Serrani, A. Isidori, and L. Marconi, "Semiglobal nonlinear output regulation with adaptive internal model," *IEEE Transactions on Automatic Control*, Vol. 46, pp. 1178–1194, 2001.

[98] J. J. Slotine and W. Li, *Applied Nonlinear Control*, Prentice-Hall, Englewood Cliffs, NJ, 1991.

[99] H. W. Smith and E. J. Davison, "Design of industrial regulators: Integral feedback and feedforward control," *Proceedings of the IEE*, Vol. 199, pp. 1210–1216, 1972.

[100] E. D. Sontag, "Smooth stabilization implies coprime factorization," *IEEE Transactions on Automatic Control*, Vol. 34, pp. 435–443, 1989.

[101] E. D. Sontag, "Further facts about input to state stabilization," *IEEE Transactions on Automatic Control*, Vol. 34, pp. 473–476, 1990.

[102] E. D. Sontag, "On the input-to-state stability property," *European Journal of Control*, Vol. 1, pp. 24–36, 1995.

[103] E. D. Sontag and Y. Wang, "New characterizations of input-to-state stability," *IEEE Transactions on Automatic Control*, Vol. 41, pp. 1283–1294, 1996.

[104] E. Sontag and Y. Wang, "Notions of input to output stability," *Systems and Control Letters*, Vol. 38, pp. 351–359, 1999.

[105] E. Sontag and Y. Wang, "Lyapunov characterizations of input to output stability," *SIAM Journal on Control and Optimization*, Vol. 39, pp. 226–249, 2000.

[106] J. Tsinias, "Sufficient Lyapunov-like conditions for stabilization," *Mathematics of Control, Signals, and Systems*, Vol. 2, pp. 343–357, 1989.

[107] X. D. Ye and J. Huang, "Decentralized adaptive output regulation for large-scale nonlinear systems," *IEEE Transactions on Automatic Control*, Vol. 48, pp. 276–281, 2003.

[108] D. Wang and J. Huang, "Solving the discrete-time output regulation problem with Taylor series Method," *Proceedings of the Chinese Control Conference*, pp. 700–704, 2000.

[109] J. Wang and J. Huang, "Neural network enhanced output regulation in nonlinear systems," *Automatica*, Vol. 37, No. 8, pp. 1189–1200, 2001.

[110] J. Wang, J. Huang, and S. T. T. Yau, "Approximate output regulation based on universal approximation theorem," *International Journal of Robust and Nonlinear Control*, Vol. 10, pp. 439–456, 2000.

[111] J. Wang and J. Huang, "A neural-aided controller for asymptotic disturbance rejection of the RTAC system," *Proceedings of the International Forum cum Conference on Information Technology and Communication at the Dawn of the New Millennium*, pp. 571–580, 2000.

[112] W. M. Wonham, *Linear Multivariable Control: A Geometric Approach*, 3rd ed., Springer-Verlag, New York, 1985.

[113] W. M. Wonham and J. B. Pearson, "Regulation and internal stabilization in linear multivariable systems," *SIAM Journal on Control and Optimization*, Vol. 12, pp. 5–18, 1974.

Index